Chemistry

for the IB Diploma

Exam Preparation Guide

First edition

Steve Owen

Chris Martin

Cambridge University Press's mission is to advance learning, knowledge and research worldwide.

Our IB Diploma resources aim to:

- encourage learners to explore concepts, ideas and topics that have local and global significance
- help students develop a positive attitude to learning in preparation for higher education
- assist students in approaching complex questions, applying critical-thinking skills and forming reasoned answers.

CAMBRIDGE
UNIVERSITY PRESS

CAMBRIDGE
UNIVERSITY PRESS

University Printing House, Cambridge CB2 8BS, United Kingdom

Cambridge University Press is part of the University of Cambridge.

It furthers the University's mission by disseminating knowledge in the pursuit of education, learning and research at the highest international levels of excellence.

Information on this title: education.cambridge.org

First published 2015

Printed in the United Kingdom by Latimer Trend

A catalogue record for this publication is available from the British Library

ISBN 978-1-107-49580-7 Paperback

CONTENTS

Contents

INTRODUCTION

If you are reading this you are probably in the final stages of preparing for your IB examinations in Chemistry and are looking for some help with revision and exam technique. The first piece of advice we can offer you is that there is no substitute for thorough preparation and in order to maximise your chance of success you should use this book in conjunction with past examination papers, mark schemes, IB textbooks, the IB subject guide (syllabus) and your notes from school. This examination preparation guide provides a summary of the material you should need to get a good grade and we have tried to concentrate especially on the topics that come up most often in examinations.

Revision should be an active process so don't just read through worked examples and model answers – cover up the answer and try to do the question. If you can't do it or get stuck, then have a look at the solution. Once you have learnt the technique for doing a question, cover up the answer again and make sure that you can *really* do the question. There is no substitute for repetition so if you have learnt a particular topic on one day go back the next day and have a go at the Test Yourself questions again just to make sure that everything is fresh in your memory. There is very little in this guide that is not required specifically for the examination and you must be prepared to make sure you learn the material thoroughly.

General tips for revision

- Make sure you take lots of breaks when revising and be aware of yourself – stop if you are not taking anything in – go and do something else for a little while then come back to revision.
- Practise questions as well as learning material. There are some types of questions that come up regularly – learn the methods for doing these.
- Try to understand the topics – the more you can understand, the less you will have to learn by rote.
- You will not be allowed a calculator for Paper 1 (the multiple choice paper) so when you do past papers do them under timed conditions without a calculator.
- Do not learn mark schemes – use them as a guide to the type of answers required – you are unlikely to get an identical question to one in a past paper.

General tips for the examinations

- Get a good night's sleep so that you are fresh for the exams!
- Be aware of time – in the IB Chemistry examinations you get, on average, about 1.5 minutes for each mark so plan out how long you should spend on each question.
- Be aware of the number of marks for each question – if there are 3 marks available for a particular question then you should make sure that you make at least three points in your answer.
- Think carefully about your answers and try to make them as concise and clear as possible – you do not have to write in complete sentences.

Finally – Good Luck!

HOW TO USE THIS BOOK: A GUIDED TOUR

Introduction – sets the scene of each chapter, helps with navigation through the book and gives a reminder of what's important about each topic.

2 ATOMIC STRUCTURE

All matter is made up of atoms and Chemistry is basically a study of how these atoms combine to make the world around us. In this chapter we will delve into the internal structure of atoms, which is fundamental to the study of many aspects of Chemistry.

This chapter covers the following topics:

- ☐ The structure of atoms
- ☐ Atomic emission spectra
- ☐ Electron configurations
- ☐ Atomic spectra calculations (HL only)
- ☐ Ionisation energy (HL only)

Definitions – clear and straightforward explanations of the most important words in each topic.

DEFINITIONS

RELATIVE ATOMIC MASS (A_r) of an element is the average mass of the naturally occurring isotopes of the element relative to the mass of 1/12 of an atom of carbon-12.

RELATIVE MOLECULAR MASS (M_r) of a compound is the mass of a molecule of the compound relative to the mass of 1/12 of an atom of carbon-12.

Model answer – an example of an answer that would score full marks to show you exactly what an examiner wants to see.

☆ Model answer 2.1

Calculate the relative atomic mass of chlorine given that it has two naturally occurring isotopes:
^{35}Cl...natural abundance 75.77%　^{37}Cl...natural abundance 24.23%

Assuming we have 100 Cl atoms; 75.77 will have mass 35 and 24.23 will have mass 37.
The average mass of these atoms is: $\frac{75.77 \times 35 + 24.23 \times 37}{100}$ i.e. 35.48, therefore, the A_r of Cl is **35.48**

Annotated exemplar answer – a question with a sample answer plus examiners' comments about what was good and what could be improved. An excellent way to see how to snap up extra marks.

Annotated exemplar answer 2.1

Describe the atomic emission spectrum of hydrogen in the visible region and explain how a line in the spectrum arises. **[3]**

The spectrum is a series of coloured lines that get closer together. Each line occurs when an electron falls from a higher to a lower shell.

1/3

No mention of a photon being given out to produce the line.

State that this occurs at higher frequency.

'Energy level' would be better than 'shell' but this still gets the mark.

The question mentions the visible region of the spectrum – state that the electrons fall to the second energy level.

Worked examples – a step by step approach to answering exam-style questions, guiding you through from start to finish.

Worked example 1.1

What is the total number of atoms present in 0.0100 mol of propane (C_3H_8)?

Avogadro's constant is 6.02×10^{23} mol^{-1}.

A 6.02×10^{21} **B** 5.47×10^{20} **C** 6.62×10^{22} **D** 1.02×10^{23}

Before starting the question: **read the question carefully** – do you need atoms, molecules, ions...?

It is handy to underline/highlight key words in the question.

There are 0.0100 mol propane molecules, therefore the number of **molecules** is $0.0100 \times 6.02 \times 10^{23}$. There are 11 atoms per molecule (3+8) and so the total number of atoms is going to be $0.0100 \times 6.02 \times 10^{23} \times 11$. This calculation looks quite daunting without a calculator but it is made simpler by the fact that answers are given in the question – if we realise that 11×6 is 66, then answer C must be the correct one.

Test yourself questions – check your own knowledge and see how well you're getting on by answering questions.

TEST YOURSELF 1.6

1 Calculate the volume of O_2 produced (measured at STP) when 5.00 g of $KClO_3$ decomposes according to the following equation:
$2KClO_3(s) \rightarrow 2KCl(s) + 3O_2(g)$

2 Calculate the volume (in cm^3) of 0.250 mol dm^{-3} hydrochloric acid required to react exactly with 1.50 g $CaCO_3$ and the volume (in cm^3) of $CO_2(g)$ produced at STP.
$CaCO_3(s) + 2HCl(aq) \rightarrow CaCl_2(aq) + CO_2(g) + H_2O(l)$

Hints – quick suggestions to remind you about key facts and highlight important points.

hint

If the initial data is given in terms of percentage composition the calculation is done in exactly the same way – just start with the percentage composition of each element (instead of the mass as here) and divide by the A_r.

Nature of Science – these discuss particular concepts or discoveries from the point of view of one or more aspects of 'Nature of Science'.

Nature of Science. Scientific discoveries in one field can lead to advances in other areas. For instance, the discovery of radioactivity allowed the design of the Geiger–Marsden experiment from which Rutherford developed a new model of the atom.

Checklist

At the end of this chapter you should be able to:

- ☐ Describe the differences between compounds and mixtures (heterogeneous and homogeneous), and recognise examples of each.
- ☐ Describe the three states of matter and explain changes of state.
- ☐ Balance a range of chemical equations.
- ☐ Carry out calculations involving numbers of particles (atoms/ions/molecules).
- ☐ Work out empirical and molecular formulas.
- ☐ Carry out calculations involving masses of substances, volumes of ideal gases and concentrations of solutions.
- ☐ Describe the relationships between pressure, volume and temperature for an ideal gas and carry out calculations based on these.
- ☐ Carry out calculations involving the ideal gas equation, including working out the relative molecular mass of a gas.

11

Checklist – at the end of each chapter so you can check off the topics as you revise them.

ACKNOWLEDGEMENTS

The authors and publisher acknowledge the following sources of copyright material and are grateful for the permissions granted. While every effort has been made, it has not always been possible to identify the sources of all the material used, or to trace all copyright holders. If any omissions are brought to our notice, we will be happy to include the appropriate acknowledgements on reprinting.

Artwork illustrations throughout © Cambridge University Press

Cover: Leigh Prather/Shutterstock

1

STOICHIOMETRIC RELATIONSHIPS

Most chemical reactions involve two or more substances reacting with each other. Substances react with each other in certain ratios, and stoichiometry is the study of the ratios in which chemical substances combine.

This chapter covers the following topics:

- ☐ The particulate nature of matter and chemical change
- ☐ The mole concept
- ☐ Empirical and molecular formulas
- ☐ Reacting masses and volumes

1.1 The particulate nature of matter and chemical change

DEFINITIONS

COMPOUND a pure substance formed when two or more elements combine chemically.

The chemical and **physical properties** of a compound are very different to those of the elements from which it is formed. When elements combine to form compounds, they always combine in fixed ratios.

MIXTURE two or more substances mixed together.

The components of a mixture are not chemically bonded together and so retain their individual properties. The components of a mixture can be mixed together in any proportion.

HOMOGENEOUS MIXTURE a mixture that has the same (uniform) composition throughout. It consists of only one phase, such as a solution (e.g. NaCl(aq)) or a mixture of gases (e.g. the air).

These can be separated by physical means, e.g. distillation/crystallisation for NaCl(aq) and fractional distillation of liquid air.

HETEROGENEOUS MIXTURE a mixture that does not have uniform composition and consists of separate phases (regions of uniform composition). Examples include sand in water or a mixture of two or more solids (e.g. a mixture of iron and sulfur).

Heterogeneous mixtures can be separated by mechanical means, e.g. filtration or using a magnet (for a mixture of iron and sulfur).

1.2 States of matter (solids, liquids and gases)

Particle diagrams for the three states of **matter** and their interconversions are shown in Figure **1.1a**.

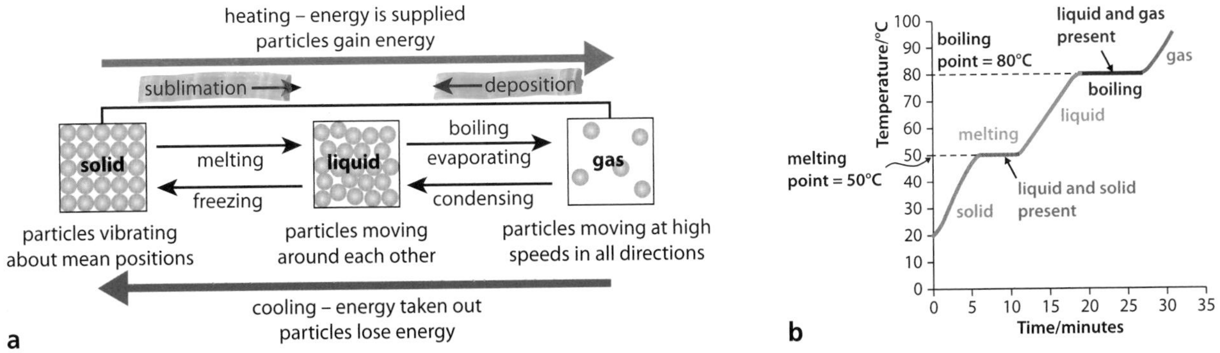

Figure 1.1

Figure **1.1b** shows how the temperature changes as a solid is heated until it becomes a gas. As a substance is heated the particles gain energy and move more quickly (vibrate more quickly for a solid). At the melting and boiling points all the energy being supplied is used to overcome forces between particles and the temperature does not rise again until the change of state is complete.

1.3 Balancing equations

When a chemical equation is required in answer to a question it must always be balanced - the number of atoms of each type (and total charge) must be the same on both sides.

How to write and balance equations:

- Never write a 'word equation' unless it is specifically requested – always write a symbol equation.
- Double check that all formulas are correct, then do not change them.
- Only large numbers (coefficients) can be inserted/changed – never change subscripts or superscripts.
- Balance compounds first and then elements last (they do not depend on anything else).
- Think about whether a full equation is required or an ionic equation (if writing an ionic equation you must also make sure that all charges balance).
- Think about the type of arrow required – a reversible arrow can sometimes be marked wrong – only use reversible arrows for questions on equilibrium and weak acids/bases unless you are absolutely sure that the reaction is reversible.
- Does the question request state symbols? If not, it is probably better to leave them out (better to leave them out than get them wrong). State symbols are essential in some equations, e.g. energetics questions (Chapter 5).

hint

Water is a liquid and not an aqueous solution – $H_2O(l)$.

State symbols:	(s) = solid	(l) = liquid	(g) = gas	(aq) = aqueous (dissolved in water)

hint

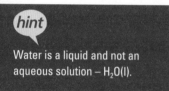

If asked to balance an equation in a multiple choice question and then work out the sum of the coefficients, do not forget the '1's: e.g. the sum of the coefficients in the equation $4Na + O_2 \rightarrow 2Na_2O$ is 7 because the coefficient of O_2 is 1.

TEST YOURSELF 1.1

State the sum of the coefficients when each of the following is balanced using the smallest possible integers.

1 $SF_4 + H_2O \rightarrow SO_2 + HF$ 8

2 $Fe_2O_3 + CO \rightarrow Fe + CO_2$ 9

3 $NH_3 + O_2 \rightarrow N_2 + H_2O$ 15

1.4 Moles

DEFINITIONS

RELATIVE ATOMIC MASS (A_r) of an element is the average mass of the naturally occurring isotopes of the element relative to the mass of $\frac{1}{12}$ of an atom of carbon-12.

RELATIVE MOLECULAR MASS (M_r) of a compound is the mass of a molecule of the compound relative to the mass of $\frac{1}{12}$ of an atom of carbon-12.

The M_r is the sum of the relative atomic masses for the individual atoms making up a molecule. Therefore, the relative molecular mass of methane (CH_4) is 16.05.

DEFINITIONS

A MOLE is the amount of substance that contains 6.02×10^{23} particles.
AVOGADRO'S CONSTANT has the value 6.02×10^{23} mol^{-1} and is sometimes given the symbol L or N_A.

For example, since the A_r of Si is 28.09 the molar mass of Si is 28.09 g mol^{-1}. This means that 28.09 g of Si contains 6.02×10^{23} Si atoms.

CO_2 – M_r 44.01 – molar mass 44.01 g mol^{-1} – 44.01 g of CO_2 contains 6.02×10^{23} **molecules**.

The number of moles present in a certain mass of substance can be worked out using the equation:

$$\text{Number of moles } (n) = \frac{\text{mass of substance}}{\text{molar mass}}$$

hint

Anything **relative** does not have units but molar mass does have units (g mol^{-1}) – the units of molar mass are not g.

TEST YOURSELF 1.2

1 Work out the number of moles in 2.00 g of methane (CH_4). 0.125 mol
2 Work out the mass of 0.0100 mol calcium sulfate. 1.36g

The mass of a molecule

The mass of 1 mole of water is 18.02 g. This contains 6.02×10^{23} molecules of water.

$$\text{mass of 1 molecule} = \frac{18.02}{6.02 \times 10^{23}} \quad \text{i.e. } 2.99 \times 10^{-23} \text{ g}$$

hint

Remember that the mass of a molecule is going to be a very **small** number!

TEST YOURSELF 1.3

1 Work out the mass of 1 molecule of propan-1-ol (C_3H_7OH). 9.99×10^{-23}
2 Work out the mass of 1 molecule of CO_2. 7.31×10^{-3}

The number of particles

If we multiply the number of moles of molecules by the number of a particular type of atom in a molecule (i.e. by the subscript of the atom), we get the number of moles of that type of atom. Thus, in 0.25 mol H_2SO_4 there are $4 \times 0.25 = 1.0$ mol oxygen atoms. If we now multiply the number of moles of atoms by 6.02×10^{23} we get the total number of atoms of that element present in the molecule - there are 6.02×10^{23} O atoms in 0.25 mol H_2SO_4.

Worked example 1.1

What is the total number of atoms present in 0.0100 mol of propane (C_3H_8)?
Avogadro's constant is 6.02×10^{23} mol^{-1}.

A 6.02×10^{21} **B** 5.47×10^{20} **C** 6.62×10^{22} **D** 1.02×10^{23}

Before starting the question: **read the question carefully** – do you need atoms, molecules, ions...?

It is handy to underline/highlight key words in the question.

There are 0.0100 mol propane molecules, therefore the number of **molecules** is $0.0100 \times 6.02 \times 10^{23}$. There are 11 atoms per molecule (3+8) and so the total number of atoms is going to be $0.0100 \times 6.02 \times 10^{23} \times 11$. This calculation looks quite daunting without a calculator but it is made simpler by the fact that answers are given in the question – if we realise that 11×6 is 66, then answer C must be the correct one.

TEST YOURSELF 1.4

1 Work out the number of oxygen atoms in 0.200 mol HNO_3. 3.61×10^{23}
2 Work out the number of hydrogen atoms in 0.0461 g of ethanol (C_2H_5OH). 3.61×10^{21}

1.5 Empirical and molecular formulas

hint

The empirical formula must involve whole numbers, therefore $CH_{2.5}$ is not an empirical formula – double everything to get C_2H_5.

DEFINITION

EMPIRICAL FORMULA the simplest whole number ratio of the elements present in a compound.

Therefore, CH_2 is an empirical formula but C_2H_4 is not.

DEFINITION

MOLECULAR FORMULA the total number of atoms of each element present in a molecule of the compound.

The molecular formula is a whole number multiple of the empirical formula. For example, if the empirical formula of a compound is CH_2, the molecular formula is $(CH_2)_n$; i.e. C_2H_4 or C_3H_6 or C_4H_8 etc.

Worked example 1.2

hint

If the initial data is given in terms of percentage composition the calculation is done in exactly the same way – just start with the percentage composition of each element (instead of the mass as here) and divide by the A_r.

A compound has the following composition by mass: C 1.665 g; H 0.280 g; O 0.555 g.

a Calculate the empirical formula of the compound.

b If the relative molecular mass of the compound is 144.24, calculate the molecular formula.

a

	C	H	O
mass/g	1.665	0.280	0.555
divide by relative atomic mass to give number of moles	$\frac{1.665}{12.01}$	$\frac{0.280}{1.01}$	$\frac{0.555}{16.00}$
number of moles/mol	0.139	0.277	0.0347
divide by smallest to get ratio	$\frac{0.139}{0.0347}$	$\frac{0.277}{0.0347}$	$\frac{0.0347}{0.0347}$
ratio	4	8	1

Therefore, the empirical formula is C_4H_8O.

b The empirical formula mass is 72.12.

Divide the relative molecular mass by the empirical formula mass: $\frac{144.24}{72.12} = 2$

The empirical formula must therefore be multiplied by 2 to get the molecular formula – $C_8H_{16}O_2$.

hint

Take care when rounding numbers – 1.02 can usually be rounded to 1 but 1.33 cannot, e.g. if the final ratio comes out as Pb 1 : O 1.33 this cannot be rounded down to PbO but rather both numbers must be multiplied by 3 to give Pb_3O_4.

TEST YOURSELF 1.5

1 Work out which of the following are empirical formulas and which are molecular formulas: Circled = Molecular

C_2H_2 CH_2 N_2H_3 C_3H_5 C_6H_6 NHO

2 A compound contains 93.88% P and 6.12% H. Calculate the empirical formula of the compound. PH_2

3 A compound, X, contains 24.7% K, 34.8% Mn and 40.5% O by mass. Work out the empirical formula of X. $KMnO_4$

1.6 Using moles in calculations

There are three main steps to doing a moles calculation:

1. Work out the number of moles of anything you can.
2. Use the chemical (stoichiometric) equation to work out the number of moles of the quantity you require.
3. Convert moles to the required quantity – volume, mass etc.

There are three ways of working out the number of moles of a substance:

Masses are given

$$\text{Number of moles (mol)} = \frac{\text{mass of substance (g)}}{\text{molar mass (g mol}^{-1})}$$

Volumes of gases are given

$$\text{Number of moles (mol)} = \frac{\text{volume of gas (dm}^3)}{\text{molar volume of the gas (dm}^3\text{ mol}^{-1})}$$

The molar volume of a gas at STP (273 K, 100 kPa) is 22.7 $dm^3\ mol^{-1}$.
STP is **standard temperature and pressure**.

Concentrations and volumes of solutions are given

Number of moles (mol) = concentration ($mol\ dm^{-3}$) × volume (dm^3)

> **hint**
>
> When working out the number of moles of a gas having been given a volume you must be careful with the units – check whether the volume is given in dm^3 or cm^3. If the volume is given in cm^3 it must be divided by 1000 before working out the number of moles using the molar volume.

☆Model answer 1.1

Consider the reaction: $CS_2(l) + 3Cl_2(g) \rightarrow CCl_4(l) + S_2Cl_2(l)$

What volume of Cl_2 (measured at STP) reacts exactly with 0.7615 g of CS_2?

$$\text{Number of moles (mol)} = \frac{\text{mass of substance (g)}}{\text{molar mass (g mol}^{-1})}$$

Number of moles of CS_2:

$$\frac{0.7615}{76.15} = 0.01000 \text{ mol}$$

From the chemical equation 1 mol CS_2 reacts with 3 mol Cl_2.

Therefore, 0.01000 mol CS_2 reacts with 3×0.01000 mol Cl_2, i.e. 0.03000 mol Cl_2

Volume of $Cl_2 = 0.03000 \times 22.7 = 0.681\ dm^3$.

> **hint**
>
> Mass of substance must be in g but, if you are doing a multiple choice question where masses are given, for instance, in tonnes or kg, there is no need to convert so long as the answers are in the same units as the quantities in the question.

TEST YOURSELF 1.6

1. Calculate the volume of O_2 produced (measured at STP) when 5.00 g of $KClO_3$ decomposes according to the following equation:
   $2KClO_3(s) \rightarrow 2KCl(s) + 3O_2(g)$ 1.39
2. Calculate the volume (in cm^3) of 0.250 $mol\ dm^{-3}$ hydrochloric acid required to react exactly with 1.50 g $CaCO_3$ and the volume (in cm^3) of $CO_2(g)$ produced at STP. 120 & 340
   $CaCO_3(s) + 2HCl(aq) \rightarrow CaCl_2(aq) + CO_2(g) + H_2O(l)$

Calculating the yield of a chemical reaction

DEFINITIONS

THEORETICAL YIELD the maximum possible amount of product formed.

EXPERIMENTAL YIELD the actual amount of product formed in the reaction.

PERCENTAGE YIELD $\frac{\text{experimental yield}}{\text{theoretical yield}} \times 100$

TEST YOURSELF 1.7

1 Calculate the percentage yield of ethyl ethanoate given that 10.0 g of ethanol reacts with excess ethanoic acid to produce 15.0 g of ethyl ethanoate.

$C_2H_5OH + CH_3COOH \rightarrow CH_3COOC_2H_5 + H_2O$

Ethanol + ethanoic acid → ethyl ethanoate + water 78.4%

2 The yield of iron in the following reaction is 60.0%. Calculate how much iron is produced from 1.00 tonne of Fe_2O_3.

$Fe_2O_3 + 3CO \rightarrow 2Fe + 3CO_2$ 0.420 tonne

Limiting reactant

To do a moles question you only need to know the number of moles of one of the reactants. If you are given enough information to work out the number of moles of more than one reactant you must consider that one of these reactants will be the **limiting reactant**.

The amount of product formed is determined by the amount of the limiting reactant. The other reactants (not the limiting reactant) are present in excess – there is more than enough to react.

To find the limiting reactant, divide the number of moles of each reactant by its coefficient in the chemical equation and the smallest number indicates the limiting reactant.

Worked example 1.3

Consider the reaction between magnesium and hydrochloric acid:

$$Mg(s) + 2HCl(aq) \rightarrow MgCl_2(aq) + H_2(g)$$

Calculate the volume of hydrogen gas produced (at STP) when 0.100 g Mg reacts with 50.0 cm³ of 0.100 mol dm⁻³ hydrochloric acid.

We have been given enough information to work out the number of moles of both magnesium and hydrochloric acid and must therefore consider that one of the reactants is limiting:

$$\text{number of moles of Mg} = \frac{0.100}{24.31} = 4.11 \times 10^{-3}\,\text{mol}$$

$$\text{number of moles of HCl} = \frac{50.0}{1000} \times 0.100 = 5.00 \times 10^{-3}\,\text{mol}$$

The coefficient of magnesium in the equation is 1 and that of HCl is 2.

$\frac{4.11 \times 10^{-3}}{1} = 4.11 \times 10^{-3}$ and $\frac{5.00 \times 10^{-3}}{2} = 2.50 \times 10^{-3}$, therefore Mg is in excess (larger number) and HCl is the limiting reactant (smaller number).

For the rest of the question we must work with the limiting reactant.

From the chemical equation, 5.00×10^{-3} mol HCl produces $\frac{5.00 \times 10^{-3}}{2}$ mol H_2, i.e. 2.50×10^{-3} mol

Volume of $H_2 = 2.50 \times 10^{-3} \times 22.7 = 0.0568\,\text{dm}^3$.

TEST YOURSELF 1.8

1. Work out the limiting reactant when 0.1 mol Sb_4O_6 reacts with 0.5 mol H_2SO_4.
   $Sb_4O_6 + 6H_2SO_4 \rightarrow 2Sb_2(SO_4)_3 + 6H_2O$
2. What is the limiting reactant when 2.7 mol O_2 reacts with 2.7 mol SO_2?
   $2SO_2 + O_2 \rightarrow 2SO_3$

Ideal gases

An **ideal gas** is used to model the behaviour of real gases. Two assumptions made when defining an ideal gas are that the molecules themselves have no volume and that no forces exist between them (except when they collide).

Gases deviate most from ideal behaviour at high pressure and low temperature (when a gas is most like a liquid). Under these conditions the particles will be close together – the forces between molecules and the volumes occupied by the molecules will be significant.

Nature of Science. Scientists use simplified models to describe more complex systems and make predictions – ideal gases can be described fairly simply using mathematical equations, which allows predictions to be made about the properties of real gases. Agreement/disagreement between models and observations/measurements can lead to refinement of models. Quite complex models are commonly used in describing climate change.

Calculations involving gases

Avogadro's Law: equal volumes of ideal gases measured at the same temperature and pressure contain the same number of particles. Another way of saying this is that the number of moles of an ideal gas is proportional to its volume (as long as temperature and pressure remain constant).

If you are given a volume of gas and the answer requires a volume of gas you do not have to convert to moles. So, for instance, from the equation

$H_2(g) + Cl_2(g) \rightarrow 2HCl(g)$

1 mole of H_2 reacts with 1 mole of Cl_2 to give 2 moles of HCl, or

1 volume of H_2 reacts with 1 volume of Cl_2 to give 2 volumes of HCl, e.g.

50 cm^3 of H_2 reacts with 50 cm^3 of Cl_2 to give 100 cm^3 of HCl.

TEST YOURSELF 1.9

1. Work out the volume of oxygen required to react exactly with 10 cm^3 of methane in the following reaction:
   $CH_4(g) + 2O_2(g) \rightarrow CO_2(g) + 2H_2O(l)$ 20 cm^3
2. What volume of SO_3 is produced when 20 cm^3 of SO_2 reacts with 5 cm^3 of O_2?
   $2SO_2(g) + O_2(g) \rightarrow 2SO_3(g)$ 10 cm^3

Nature of Science. Scientific laws develop from observations and careful measurements. A scientific law is a description of a phenomenon – a theory is a proposed explanation of that phenomenon. Avogadro's Law and the other gas laws can be explained in terms of the kinetic theory.

Relationship between pressure (*P*), volume (*V*) and temperature (*T*) for an ideal gas

The gas laws can be summarised as:

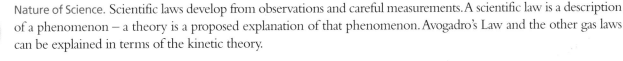

$P \propto \frac{1}{V}$	if the pressure of a gas is doubled at constant temperature then the volume will be halved and vice versa	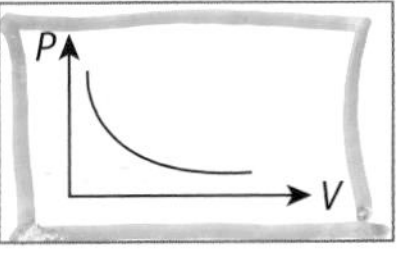

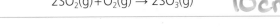

hint

A graph of *PV* (*y*-axis) against *P* is a horizontal straight line.

hint

Remember that V and P are only proportional to temperature in K – the relationship does **not** work for temperatures in °C. For instance, if the volume of an ideal gas at 25 °C is 500 cm^3 the volume it will occupy at 50 °C will be about 560 cm^3 – i.e. not double.

hint

Any units may be used for P and V as long as they are consistent between both sides of the equation, but temperature must be in K.

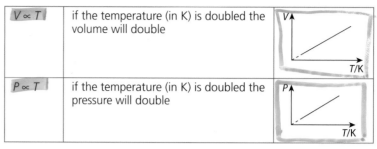

$V \propto T$	if the temperature (in K) is doubled the volume will double	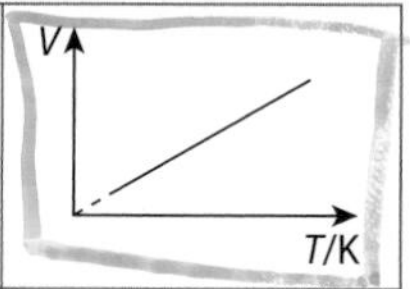
$P \propto T$	if the temperature (in K) is doubled the pressure will double	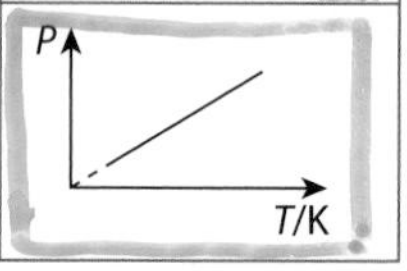

The overall gas law equation

If you are given pressure, volume and temperature for an ideal gas you can use this equation to work out how changing some of these affects the other quantities:

$$\frac{P_1V_1}{T_1} = \frac{P_2V_2}{T_2}$$

TEST YOURSELF 1.10

Work out the final volume when 200 cm^3 of hydrogen gas is heated from 27 °C to 327 °C at constant pressure. 400

The ideal gas equation

$PV = nRT$

R (8.31 J K^{-1} mol^{-1}) is the **gas constant** (ideal gas constant), n = number of moles

A consistent set of units must be used with this equation:

Pressure in Pa	or	Pressure in kPa
Volume in m^3		Volume in dm^3
Temperature in K		

Worked example 1.4

0.120 g of an ideal gas was introduced into a gas syringe. The volume occupied by the gas at a pressure of 1.02×10^5 Pa and temperature 20 °C was 49.3 cm^3. Calculate the molar mass of the gas.

Since we know the mass of the gas we can work out the molar mass if we can calculate how many moles are present. $PV = nRT$ can be used to work out the number of moles.

It is important when using $PV = nRT$, that the units are correct, so the first step should always be to write out all quantities with a consistent set of units:

$P = 1.02 \times 10^5$ Pa ✓ units fine

$V = 49.3$ cm^3 ✗ units not consistent – must be changed to m^3

1 m^3 is $100 \times 100 \times 100$ cm^3 i.e. 1×10^6 cm^3, 1 $m^3 \Leftrightarrow 1 \times 10^6$ cm^3

To convert cm^3 to m^3 we go from 1×10^6 to 1, therefore we must divide by 1×10^6

$V = 49.3/(1 \times 10^6) = 4.93 \times 10^{-5}$ m^3 $n = ?$ $R = 8.31$ J $K^{-1}mol^{-1}$

$T = 20$ °C ✗ units not consistent – must be changed to K: i.e. 293 K

$PV = nRT$

$1.02 \times 10^5 \times 4.93 \times 10^{-5} = n \times 8.31 \times 293$

Re-arranging the equation gives $n = 2.07 \times 10^{-3}$ mol

This number of moles has a mass of 0.120 g

$$\text{Molar mass} = \frac{\text{mass}}{\text{number of moles}}$$

$$\text{i.e. molar mass} = \frac{0.120}{2.07 \times 10^{-3}}$$

i.e. 58.1 g mol^{-1}

TEST YOURSELF 1.11

1 Calculate the volume occupied by 0.100 mol of an ideal gas at 25 °C and 1.2×10^5 Pa. 2·06 dm³

2 Calculate the molar mass of an ideal gas given that 0.586 g of the gas occupies a volume of 282 cm³ at a pressure of 1.02×10^5 Pa and a temperature of −18 °C. 43·2g mol⁻¹

1.7 Solutions

DEFINITION

A STANDARD SOLUTION a solution of known concentration.

The equation for working out concentrations in mol dm^{-3} is:

$$\text{concentration (mol dm}^{-3}) = \frac{\text{number of moles (mol)}}{\text{volume (dm}^3)}$$

If the concentration is expressed in g dm^{-3}, the relationship is:

$$\text{concentration (g dm}^{-3}) = \frac{\text{mass (g)}}{\text{volume (dm}^3)}$$

Square brackets indicate the concentration of a substance e.g. [HCl] = 2.0 mol dm^{-3}

Worked example 1.5

A solution of sodium thiosulfate was made up by dissolving 2.48 g of $Na_2S_2O_3\cdot5H_2O$ in water and making up to a total volume of 100.0 cm³. What is the concentration, in mol dm^{-3}, of sodium ions in the solution?

A 0.100 B 0.0100 C 0.200 D 0.314

The key thing to notice here is that the concentration of **sodium ions** is required and not that of sodium thiosulfate.

$Na_2S_2O_3\cdot5H_2O$ has a molar mass of 248 g mol^{-1} (do not worry about the decimal places in the A_r values when doing multiple choice questions).

$\frac{2.48}{248} = 0.0100$ mol

This number is the same as Answer B – do not just pounce on the first number you get that looks like one of the answers – you must realise that the concentration is required, not the number of moles – therefore dismiss answer B.

To work out a concentration in mol dm^{-3} the number of moles must be divided by the volume in dm³ (0.1).

0.0100/0.1 = 0.100 mol dm^{-3}

This looks like answer A but, again, you must not be tempted – this is the concentration of sodium thiosulfate and not the concentration of sodium ions. There are two sodium ions per formula unit and so this concentration must be multiplied by 2 to give answer C as the correct answer.

Answer D is incorrect – it would be obtained if the water of crystallisation is not included in the molar mass of sodium thiosulfate.

hint

If you rip the front page off Paper 1 you can keep the periodic table in front of you during the whole examination.

Titrations

Titration is a technique for finding out the volumes of solutions that react exactly with each other.

Worked example 1.6

A student carried out an experiment to determine the concentration of a solution of sodium hydroxide using a standard solution of ethanedioic acid. She made up a standard solution of ethanedioc acid by weighing out 1.600 g of $(COOH)_2{\cdot}2H_2O$, dissolving it in distilled water then making it up to a total volume of 250.0 cm^3 with distilled water in a volumetric flask.

The student then carried out a series of titrations to determine the concentration of the unknown solution of sodium hydroxide. She measured out 25.00 cm^3 of the ethanedioic acid solution using a pipette and transferred it to a conical flask. She then added two drops of phenolphthalein indicator. A burette was filled with the sodium hydroxide solution and this was titrated against the ethanedioic acid solution. The end point was indicated by the indicator changing colour to pink. The procedure was repeated three times and the results are shown in the table.

	Trial 1	Trial 2	Trial 3
initial burette reading /cm^3 +/− 0.03	0.00	1.50	2.20
final burette reading /cm^3 +/− 0.03	29.40	30.10	30.80

The equation for the reaction is: $2NaOH + (COOH)_2 \rightarrow Na_2C_2O_4 + 2H_2O$

a Calculate the concentration of the solution of ethanedioic acid. [2]

b Determine the concentration of the sodium hydroxide solution. [4]

a In moles questions such as this the first stage is usually to determine the number of moles of anything you can. In this case we have enough information to determine the number of moles of ethanedioic acid.

M_r of $(COOH)_2{\cdot}2H_2O = 126.08$

$$\text{Number of moles} = \frac{\text{mass}}{\text{molar mass}} = \frac{1.600}{126.08} = 0.01269 \text{ mol}$$

$$\text{Concentration (mol dm}^{-3}) = \frac{\text{number of moles (mol)}}{\text{volume (dm}^3)} = \frac{0.01269}{0.250} = 0.05076 \text{ mol dm}^{-3}$$

b We need to know the volume of sodium hydroxide used in the titrations (titre) so we need to subtract the initial burette readings from the final ones (ignore uncertainties unless asked specifically about them).

	Trial 1	Trial 2	Trial 3
titre/cm^3	29.40	28.60	28.60

When experimental data is given for titrations you should always check whether the first titration was just a rough one – sometimes it is labelled as such, but not always – look to see if one titration involves a volume significantly different to the others. Here, the first titration appears to be a rough one and should be ignored in subsequent calculations.

Again, the first stage is to work out the number of moles of whatever you can – you know the volume and concentration of the ethanedioic acid solution used in each experiment so the moles of this can be worked out:

Number of moles = concentration × volume in dm^3

$$\text{Number of moles} = 0.05076 \times \frac{25}{1000} \quad \text{i.e. } 1.269 \times 10^{-3} \text{ mol}$$

The chemical equation must be used to work out the number of moles of NaOH.

$2NaOH + (COOH)_2 \rightarrow Na_2C_2O_4 + 2H_2O$

Moles of NaOH = $2 \times 1.269 \times 10^{-3} = 2.538 \times 10^{-3}$ mol

$$\text{Concentration of NaOH} = \frac{\text{number of moles (mol)}}{\text{volume (dm}^3)} = \frac{2.538 \times 10^{-3}}{0.02860} = 0.08874 \text{ mol dm}^{-3}$$

TEST YOURSELF 1.12

1 Calculate the concentration of a sodium hydroxide solution given that 28.0 cm^3 of the solution reacts with 25.0 cm^3 of 0.100 M sulfuric acid according to the equation:
$2NaOH(aq) + H_2SO_4(aq) \rightarrow Na_2SO_4(aq) + 2H_2O(l)$ 0.179

2 Calculate the concentration of sodium ions in mol dm^{-3} when 1.42 g Na_2SO_4 is dissolved in water and made up to a total volume of 50.0 cm^3. 0.400

Concentrations of very dilute solutions

When dealing with very small concentrations, the unit parts per million, ppm, may be used.

$$\text{concentration in ppm} = \frac{\text{mass of solute (g)}}{\text{mass of solution (g)}} \times 10^6$$

hint

A percentage is 'parts per hundred' – this equation is basically the same as working out the percentage of a solute in a solution except we multiply by a million instead of 100.

Model answer 1.2

If a sample of 200.00 g of water is found to contain 5.78 mg of copper ions, what is the concentration of copper ions in ppm?

mass of copper ions in grams = 5.78×10^{-3} g

$$\text{concentration of copper ions in ppm} = \frac{5.78 \times 10^{-3}}{200.00} \times 10^6 = 28.9 \text{ ppm}$$

hint

Take care with units – the mass of solute and solution must have the same units.

The ppm notation is also used when discussing the concentrations of pollutant gases in air:

$$\text{concentration in ppm} = \frac{\text{volume of gas}}{\text{volume of air}} \times 10^6$$

TEST YOURSELF 1.13

Calculate the concentration of hydrogen sulfide (in ppm) if 0.120 g is present in 15 000 g water. 8

Checklist

At the end of this chapter you should be able to:

- ☐ Describe the differences between compounds and mixtures (heterogeneous and homogeneous), and recognise examples of each.
- ☐ Describe the three states of matter and explain changes of state.
- ☐ Balance a range of chemical equations.
- ☐ Carry out calculations involving numbers of particles (atoms/ions/molecules).
- ☐ Work out empirical and molecular formulas.
- ☐ Carry out calculations involving masses of substances, volumes of ideal gases and concentrations of solutions.
- ☐ Describe the relationships between pressure, volume and temperature for an ideal gas and carry out calculations based on these.
- ☐ Carry out calculations involving the ideal gas equation, including working out the relative molecular mass of a gas.

2

ATOMIC STRUCTURE

All matter is made up of atoms and Chemistry is basically a study of how these atoms combine to make the world around us. In this chapter we will delve into the internal structure of atoms, which is fundamental to the study of many aspects of Chemistry.

This chapter covers the following topics:

- ☐ The structure of atoms
- ☐ Atomic emission spectra
- ☐ Electron configurations
- ☐ Atomic spectra calculations (HL only)
- ☐ Ionisation energy (HL only)

2.1 The structure of atoms

Virtually all the mass of the **atom** is concentrated in the nucleus (protons and neutrons). Most of the volume of the atom is due to the electrons – the nucleus is very small compared to the total size of the atom.

The properties of sub-atomic particles:

hint

The relative masses and charges should be learnt.

Particle	Relative mass	Relative charge
proton	1	+1
neutron	1	0
electron	5×10^{-4} (negligible)	–1

Nature of Science. Scientific discoveries in one field can lead to advances in other areas. For instance, the discovery of radioactivity allowed the design of the Geiger–Marsden experiment from which Rutherford developed a new model of the atom.

DEFINITIONS

NUCLEONS protons and neutrons – the particles in the nucleus of an atom.

ATOMIC NUMBER the number of protons in the nucleus of an atom.

MASS NUMBER the number of protons+neutrons in the nucleus of an atom.

hint

You need to remember that *Z* stands for the atomic number and *A* the mass number.

The number of protons in an atom = number of electrons (because the atom is overall neutral)

The number of neutrons = mass number − atomic number

$${}^{A}_{Z}X$$

Figure 2.1

Ions

DEFINITIONS

POSITIVE ION charged particle formed when atoms lose electron(s) – has fewer electrons than protons.

NEGATIVE ION charged particle formed when atoms gain electron(s) – has more electrons than protons.

Worked example 2.1

What is the number of protons, electrons and neutrons in $^{45}Sc^{3+}$?

A	**45 protons**	**42 electrons**	**45 neutrons**
B	**24 protons**	**21 electrons**	**21 neutrons**
C	**21 protons**	**18 electrons**	**24 neutrons**
D	**21 protons**	**24 electrons**	**45 neutrons**

The first thing you need to do is look at the periodic table to get the atomic number (21) for scandium.

- Answer A would be obtained if you used the mass number instead of the atomic number to give the number of protons and is therefore incorrect.
- Answer B would be obtained if an atom forms a positive ion by gaining protons instead of losing electrons and is therefore incorrect.
- Answer C is correct – the number of electrons is given by the atomic number, three electrons have been lost to form a positive ion.

  The number of neutrons = the mass number – the atomic number.
- Answer D would be obtained if the student thought that a positive ion is formed when an atom gains electrons and that the number of neutrons is given by the mass number – this is incorrect.

TEST YOURSELF 2.1

State the number of protons, neutrons and electrons in each of the following:

	^{65}Cu	$^{15}N^{3-}$	$^{137}Ba^{2+}$
P	29	7	56
N	36	8	81
E	29	10	54

2.2 Isotopes and relative atomic masses

Isotopes

DEFINITION

ISOTOPES different atoms of the same element with different mass numbers, i.e. different numbers of neutrons in the nucleus.

Isotope	Protons	Neutrons	Electrons
$^{12}_{6}C$	6	**6**	6
$^{14}_{6}C$	6	**8**	6

This means that isotopes have the same atomic number but different mass numbers.

A naturally occurring sample of an element will contain different isotopes – the proportion of each isotope present in the sample of an element can be measured using a **mass spectrometer**.

In the mass spectrum of an element, we get one peak for each individual isotope. The height of each peak is proportional to the number of atoms of this isotope in the sample tested. For example, the mass spectrum of magnesium would look like Figure **2.2**.

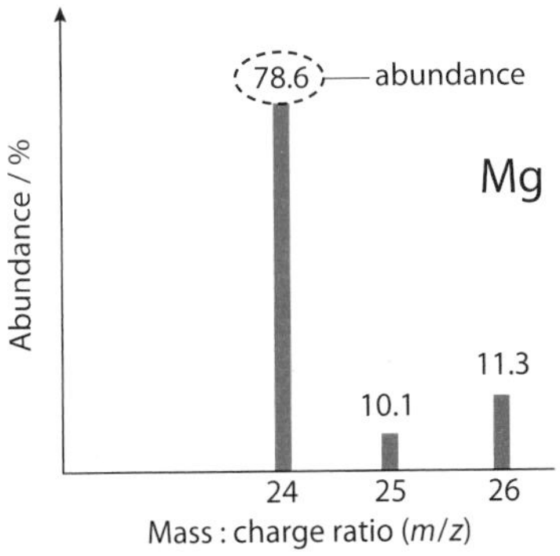

Figure 2.2

Relative atomic masses

hint

Relative atomic masses are given in the periodic table – check your answer against this – is your answer roughly what it should be?

Model answer 2.1

Calculate the relative atomic mass of chlorine given that it has two naturally occurring isotopes:
^{35}Cl...natural abundance 75.77% **^{37}Cl...natural abundance 24.23%**

Assuming we have 100 Cl atoms; 75.77 will have mass 35 and 24.23 will have mass 37.

The average mass of these atoms is: $\frac{75.77 \times 35 + 24.23 \times 37}{100}$ i.e. 35.48, therefore, the A_r of Cl is **35.48**

Worked example 2.2

Gallium has a relative atomic mass of 69.72 and consists of Ga-69 and Ga-71 isotopes. Calculate the percentage composition of a naturally occurring sample of gallium.

Assume that we have 100 atoms and that x of these have a mass of 69. This means that there will be $(100-x)$ atoms that have a mass of 71.

The average mass of the 100 atoms will be: $\frac{69x + 71(100-x)}{100} = 69.72$

Re-arranging this gives $x = 64$

This means that the naturally occurring sample of gallium contains **64% Ga-69** and **36% Ga-71**.

Check your final answer against the relative atomic mass – 69.72 is closer to 69 than 71, therefore there must be more Ga-69 than Ga-71 present in the sample.

TEST YOURSELF 2.2

1 Determine the relative atomic mass of copper (to 2 decimal places) given the following natural abundances: ^{63}Cu 76.00% ^{65}Cu 24.00% 63.48g

2 Determine the natural abundance of ^{11}B given that boron consists of two isotopes, ^{10}B and ^{11}B, and the relative atomic mass is 10.80. ^{10}B = 20% ^{11}B = 80%

2.3 Electron configurations

At the simplest level the electrons in an atom are arranged in **energy levels (shells)** about the nucleus. The maximum number of electrons in each main energy level is given by $2n^2$, where n is the number of the main energy level (shell number).

The main energy levels (shells) in atoms are divided up into sub-levels (sub-shells).

Sub-level	s	p	d	f
Maximum number of electrons	2	6	10	14

Within any main energy level (shell) the ordering of the sub-levels is always: s < p < d < f

Remember, however, that the 4s sub-level is lower in energy than the 3d sub-level – see Figure **2.3**.

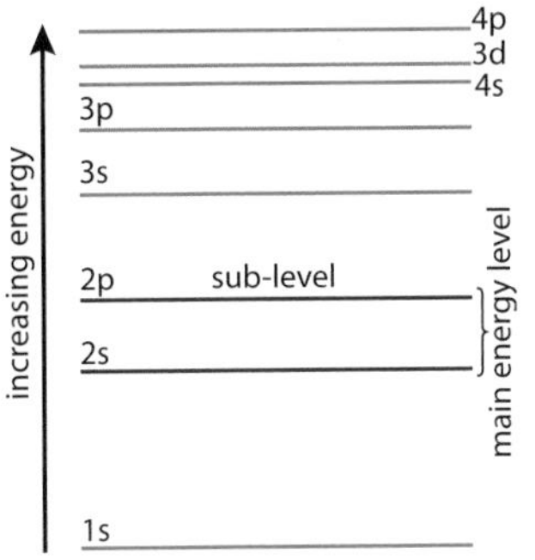

Figure 2.3

The **Aufbau** (building-up) **principle**

The full electron configuration for an atom can be worked out directly from the periodic table – starting at H follow the arrows through to the target atom.

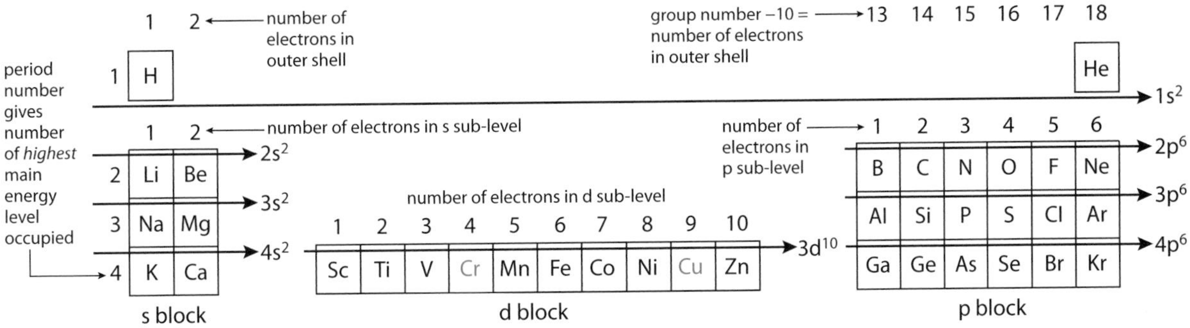

Figure 2.4

For example, the full electron configuration of selenium can be worked out as:

H→He	Li→Be	B→Ne	Na→Mg	Al→Ar	K→Ca	Sc→Zn	Ga→Se
$1s^2$	$2s^2$	$2p^6$	$3s^2$	$3p^6$	$4s^2$	**$3d^{10}$**	$4p^4$

Therefore, the full electron configuration is: $1s^2 2s^2 2p^6 3s^2 3p^6 4s^2 3d^{10} 4p^4$

Condensed electron configurations – the electron configuration of the previous noble gas is assumed and everything after that is given in full, for example, for Se: [Ar] $4s^2 3d^{10} 4p^4$.

Exceptions – **chromium** and **copper**, which, instead of having electron configurations of the form [Ar] $4s^2 3d^n$ have only one electron in the 4s sub-level.

i.e. $_{24}$Cr: [Ar] $3d^5 4s^1$ $_{29}$Cu: [Ar] $3d^{10} 4s^1$

TEST YOURSELF 2.3

1 Give the full electron configurations of:
B P Ti Cr Cu Se

B $1s^2 2s^2 2p^1$
P $1s^2 2s^2 2p^6 3s^2 3p^3$
Ti $1s^2 2s^2 2p^6 3s^2 3p^6 4s^2 3d^2$
Cr $1s^2 2s^2 2p^6 3s^2 3p^6 4s^1 3d^5$
Cu $1s^2 2s^2 2p^6 3s^2 3p^6 4s^1 3d^{10}$
Se $1s^2 2s^2 2p^6 3s^2 3p^6 4s^2 3d^{10} 4p^4$

2 Give the condensed electron configurations of:
Al As

Al: [Ne] $3s^2 3p^1$ As: [Ar] $4s^2 3d^{10} 4p^3$

Orbitals

Electrons in atoms occupy atomic orbitals.

DEFINITION

ORBITAL a region of space where there is a high probability of finding an electron.

hint

Superscripts are essential – if you write $1s_2$ or 1s2, you will lose marks!

hint

Don't forget that between Sc and Zn we are filling the remaining sub-level in the third main energy level and so it is $3d^n$ and not $4d^n$.

hint

Is the exam question asking you for a full or condensed electron configuration? If the question just states 'electron configuration' it is safer to write out the full one.

Each orbital represents a discrete energy level. An orbital can contain a maximum of two electrons with opposite spins.

You need to be able to recognise and draw the shapes of s (spherical) and p (dumbbell-shaped) orbitals.

s orbital	p orbitals		
	p_x	p_y	p_z

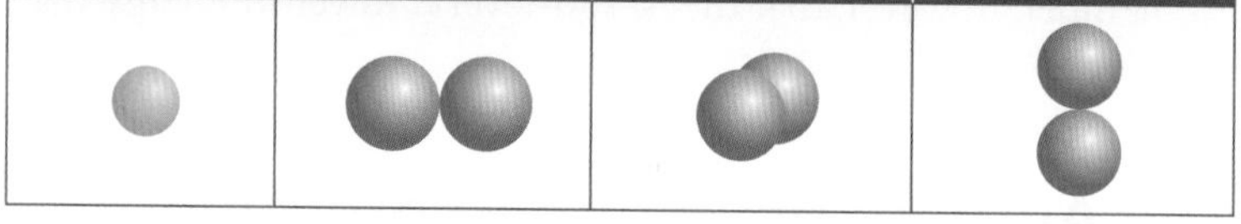

The p_x orbital points along the x-axis, the p_y along the y-axis and the p_z along the z-axis.

Each sub-level is made up of a set of degenerate (same energy) orbitals.

Sub-level	Number of orbitals
s	1
p	3
d	5
f	7

hint

Take care with technical terms – make sure that you are clear about the differences between main energy levels, sub-levels and orbitals.

Orbital diagrams

Orbital diagrams show the electrons as single-headed arrows (fish hooks) in boxes. Each box represents an orbital. Arrows pointing either up or down represent the electrons spinning in opposite directions.

Pauli Exclusion Principle: The maximum number of electrons in an orbital is two. If there are two electrons in an orbital they must have opposite spins.

Hund's Rule: Electrons fill orbitals of the same energy (degenerate orbitals) so as to give the maximum number of electrons with spins the same. The orbital diagram for O is shown in Figure **2.5**.

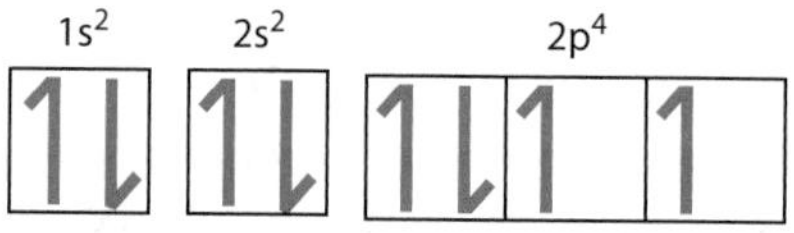

Figure 2.5

TEST YOURSELF 2.4

Draw orbital diagrams for Si, Fe and Cr.

Si 1s² 2s² 2p⁶ 3s² 3p²

Fe 1s 2s 2p 3s 3p 4s 3d

Cr 1s 2s 2p 3s 3p 4s 3d

Electron configurations of ions

The highest energy electrons are removed first when positive ions are formed – for most atoms this basically means 'last in, first out' – the last electrons to be put in to the atom are the first ones to be removed when a positive ion is formed, e.g.

Mg	$1s^2\ 2s^2\ 2p^6\ \mathbf{3s^2}$	Mg^{2+}	$1s^2\ 2s^2\ 2p^6$

For transition metals and the elements in Groups 13 and 14, the situation is slightly different and you must remember that the 4s electrons are removed before the 3d electrons.

Ga	$1s^2\ 2s^2\ 2p^6\ 3s^2\ 3p^6\ \mathbf{4s^2}\ 3d^{10}\ \mathbf{4p^1}$	Ga^{3+}	$1s^2\ 2s^2\ 2p^6\ 3s^2\ 3p^6\ 3d^{10}$
Mn	$1s^2\ 2s^2\ 2p^6\ 3s^2\ 3p^6\ \mathbf{4s^2}\ 3d^5$	Mn^{2+}	$1s^2\ 2s^2\ 2p^6\ 3s^2\ 3p^6\ 3d^5$

TEST YOURSELF 2.5

Give the full electron configurations of: Na^+ Cl^- Fe^{3+}

$Na^+ \rightarrow 1s^2\ 2s^2\ 2p^6$

$Cl^- \rightarrow 1s^2\ 2s^2\ 2p^6\ 3s^2\ 3p^6$

$Fe^{3+} \rightarrow 1s^2\ 2s^2\ 2p^6\ 3s^2\ 3p^6\ 3d^5$

2.4 Atomic emission spectra

The electromagnetic spectrum

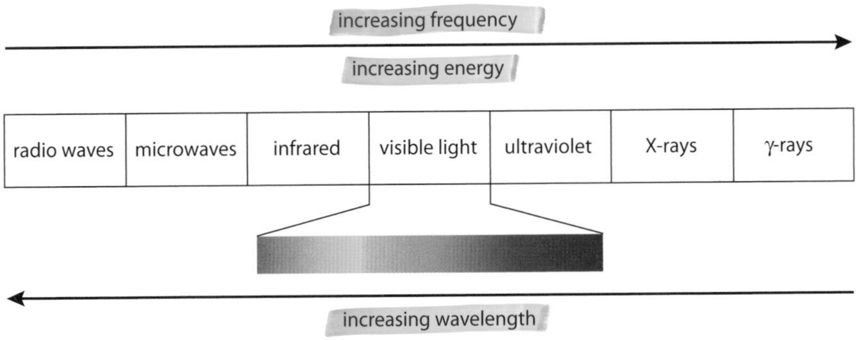

Figure 2.6 Light is a form of energy. Visible light is just one part of the electromagnetic spectrum.

> **hint**
> The electromagnetic spectrum is given in the *IB Data Booklet* but frequency is not mentioned – remember that frequency is proportional to energy and inversely proportional to wavelength – the higher the frequency the shorter the wavelength.

Evidence for energy levels in atoms – the hydrogen atom spectrum

When hydrogen gas at low pressure is subjected to a very high voltage in a discharge tube the gas glows pink – light is emitted by the gas so this is called an **emission spectrum**. The emission spectrum of an element such as hydrogen is a line spectrum – the spectrum consists of a series of sharp, bright, lines on a dark background – **the lines get closer together at higher frequency/energy**. The fact that a line spectrum is produced means that electrons in an atom are only allowed to have certain amounts of energy.

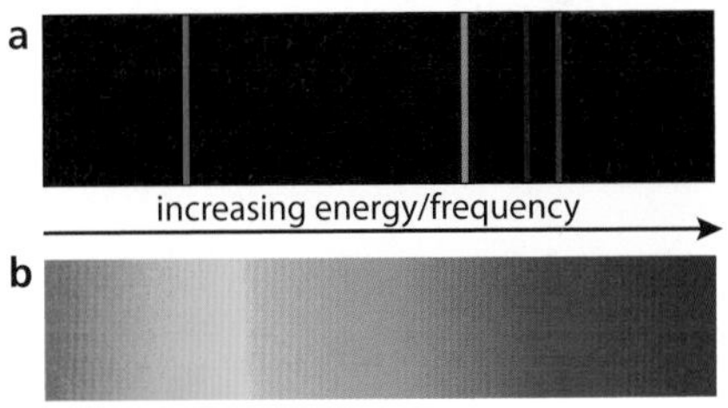

Figure 2.7 **a Line spectrum** – only certain frequencies/wavelengths of light present. **b Continuous spectrum** – all frequencies/wavelengths of light present.

How an emission spectrum arises

Passage of the electric discharge causes an electron in an atom to be promoted to a higher energy level (shell). Each line in the emission spectrum arises when an electron falls from a **higher to a lower energy level**. As it returns to the lower energy level the extra energy is given out in the form of a **photon of light** – this gives a line in the spectrum.

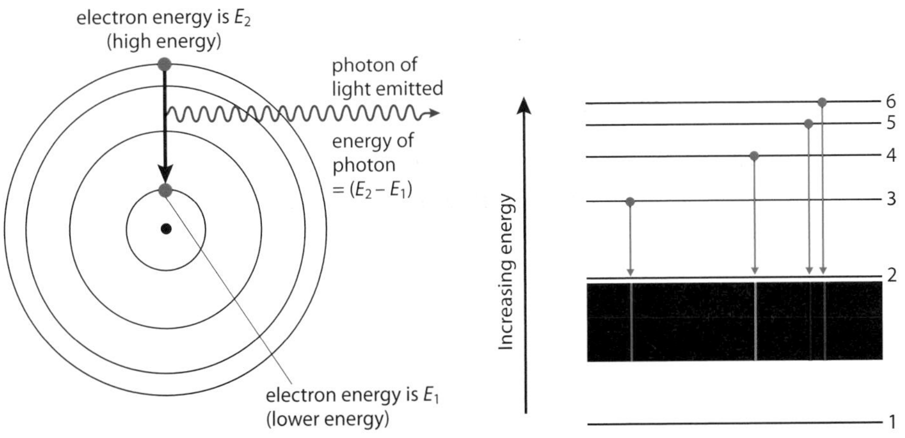

Figure 2.8

Different series of lines

Different series of lines arise when the electron falls down to different energy levels – these can be in the visible, infrared or UV part of the electromagnetic spectrum.

Area of spectrum	Level to which electron falls back
ultraviolet	1
visible	2
infrared	3

Annotated exemplar answer 2.1

Describe the atomic emission spectrum of hydrogen in the visible region and explain how a line in the spectrum arises. **[3]**

The spectrum is a series of coloured lines that get closer together. Each line occurs when an electron falls from a higher to a lower shell.

1/3

No mention of a photon being given out to produce the line.

State that this occurs at higher frequency.

'Energy level' would be better than 'shell' but this still gets the mark.

The question mentions the visible region of the spectrum – state that the electrons fall to the second energy level.

Convergence

The lines in the emission spectrum get closer together at higher energy (converge), therefore, the energy levels in the atom must get closer together in energy at higher energy. Eventually, at the **convergence limit**, the lines merge to form a continuum. Beyond this point the electron can have any energy – the electron is no longer in the atom.

TEST YOURSELF 2.6

Select the highest energy transition from the following list:

$n = 4 \rightarrow n = 2$ $n = 12 \rightarrow n = 3$ $n = 2 \rightarrow n = 1$ $n = 15 \rightarrow n = 2$.

2.5 Atomic spectra calculations (HL only)

The energy (E) of a photon is related to the frequency (ν) of the electromagnetic radiation:

$$E = h\nu$$

where ν is the frequency of the electromagnetic radiation (Hz or s^{-1})

h is Planck's constant (6.63×10^{-34} J s)

This equation can be used to work out the differences in energy between various levels in the hydrogen atom.

hint

The formula $E = h\nu$ and the value of Planck's constant are given in the *IB Data Booklet*.

☆ Model answer 2.2

The frequency of a line in the emission spectrum of hydrogen when an electron falls from level 2 to level 1 is 2.47×10^{15} Hz. Calculate the energy difference between the two levels.

$E = h\nu$

Therefore, $E = 6.63 \times 10^{-34} \times 2.47 \times 10^{15} = 1.64 \times 10^{-18}$ J

The wavelength λ of the electromagnetic radiation is related to the frequency: $\nu = \dfrac{c}{\lambda}$

c is the speed of light (3.0×10^{8} m s^{-1})

TEST YOURSELF 2.7

The wavelength of a line in the emission spectrum of hydrogen is 6.56×10^{-7} m.

1 Calculate the frequency of the light emitted.

2 Calculate the energy difference between the energy levels in the hydrogen atom.

1. c/υ = λ
(3.0×10⁸)/6.56×10⁻⁷ = 4.57×10¹⁴
2. E = hυ
= 6.63×10⁻⁵ ×
= 3.03×10⁻¹⁹

hint

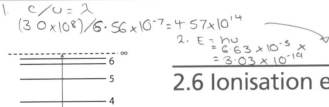

Make sure that you know conversion factors for the different units – the wavelength could be given, for example, in nm.

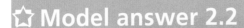

2.6 Ionisation energy (HL only)

DEFINITION

IONISATION ENERGY is the minimum amount of energy required to remove an electron from a gaseous atom:

$M(g) \rightarrow M^{+}(g) + e^{-}$

hint

The gaseous symbols, (g), are essential in this equation.

The ionisation energy for hydrogen represents the minimum energy for the removal of an electron [from level 1 to ∞ (where the energy levels come together)].

The ionisation energy can be worked out from the frequency of the light at the convergence limit using $E = h\nu$.

hint

Remember – the ionisation energy of hydrogen can only be obtained from a study of the series of lines where the electron falls back to its ground state ($n = 1$) energy level.

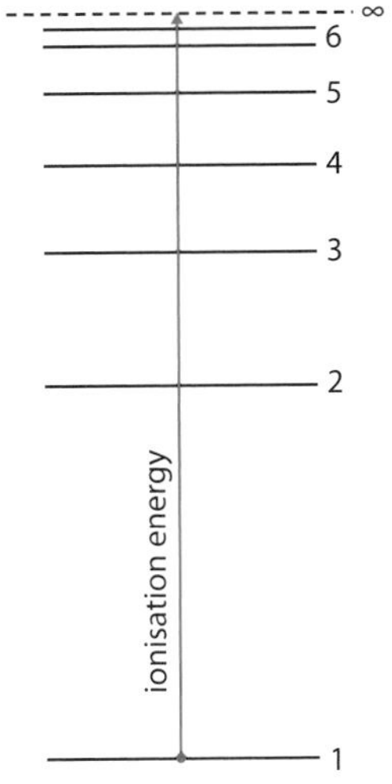

Figure 2.9

Worked example 2.3

If the frequency of the convergence limit of the series of lines in the UV region of the emission spectrum of hydrogen is 3.28×10^{15} Hz, calculate the ionisation energy of hydrogen in kJ mol^{-1}.

$E = h\nu$

Therefore: $E = 6.63 \times 10^{-34} \times 3.28 \times 10^{15} = 2.17 \times 10^{-18}$ J

This is the amount of energy to remove an electron from one atom of hydrogen – multiply by Avogadro's constant to calculate the total energy required to remove one electron from each atom in 1 mole of hydrogen atoms:

$2.17 \times 10^{-18} \times 6.02 \times 10^{23} = 1.31 \times 10^{6}$ J mol^{-1}.

Dividing by 1000 gives the answer in kJ mol^{-1}, so the ionisation energy of hydrogen is 1.31×10^{3} kJ mol^{-1}.

Successive ionisation energies

First ionisation energy	Second ionisation energy	n^{th} ionisation energy
$M(g) \rightarrow M^{+}(g)+e^{-}$	$M^{+}(g) \rightarrow M^{2+}(g)+e^{-}$	$M^{(n-1)+}(g) \rightarrow M^{n+}(g)+e^{-}$

The second ionisation energy is always higher than the first

hint

Either explanation is OK to answer an exam question – you do not need both.

- Once an electron has been removed from an atom a positive ion is created. A positive ion attracts a negatively charged electron more strongly than a neutral atom does, therefore more energy is needed to remove an electron from a positive ion than from a neutral atom.
- Once an electron has been removed from an atom there is less repulsion between the remaining electrons. They are therefore pulled in closer to the nucleus. If they are closer to the nucleus they are more strongly attracted and more difficult to remove.

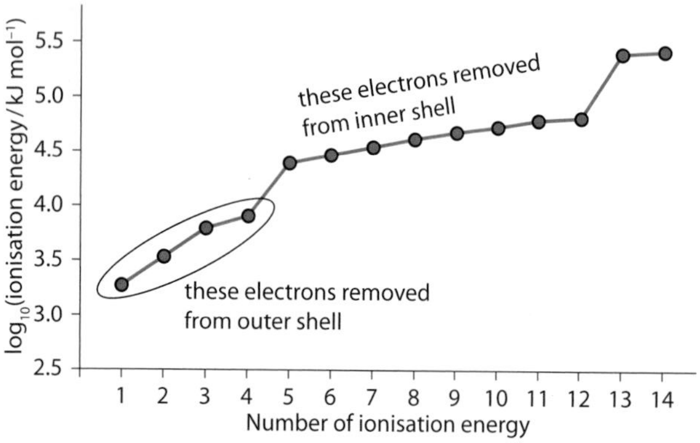

Figure 2.10 This is a graph of successive ionisation energies for Si (group 14 therefore $4e^{-}$ in the outer shell) – there is a large jump after the fourth ionisation energy – the fourth electron is removed from the third main energy level and the fifth electron is removed from the second main energy level – an electron in the second main energy level is closer to the nucleus and less shielded, therefore it is more strongly attracted.

TEST YOURSELF 2.8

1 Write equations to represent the first and second ionisation energies of potassium.

2 Which of the following could be the first four ionisation energies (in $kJ\ mol^{-1}$) of boron?

A	900	1760	14 800	21 000
B	799	2420	3660	25 000
C	1400	2860	4590	7480
D	494	4560	6940	9540

1. $k(g) \rightarrow k^{+}(g) + e^{-}$
$k^{+}(g) \rightarrow k^{2+}(g) + e^{-}$

Variation of ionisation energy across a period

General trend – ionisation energy increases from left to right across a period.

Model answer 2.3

Explain the general trend in first ionisation energy across period 2 in the periodic table.

- Across period 2, the nuclear charge increases from Li (3+) to Ne (10+) as protons are added to the nucleus.
- The electrons are removed from the same main energy level and electrons in the same energy level do not shield each other very well.
- Therefore, the force of attraction, from the nucleus, on the outer electrons increases from left to right across the period and the outer electron is more difficult to remove for neon.
- The neon atom is also smaller than the lithium atom and, therefore, the outer electron is closer to the nucleus and more strongly held.

Exceptions to the general increase in ionisation energy across a period

Model answer 2.4

The graph shows the variation in first ionisation energy across period 2 in the periodic table. Explain in terms of electron configurations why boron has a lower first ionisation energy than beryllium.

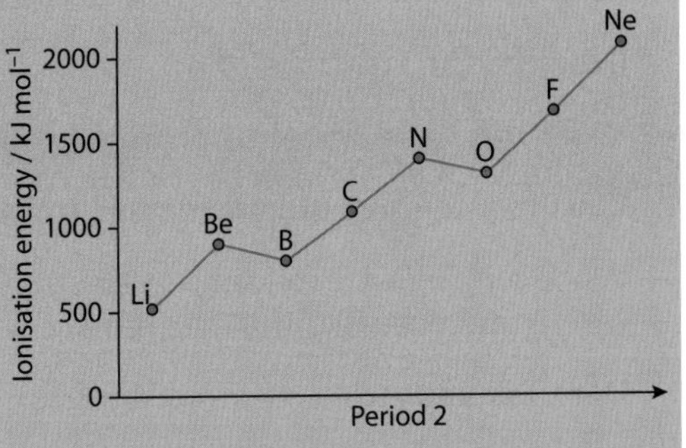

- The electron configurations of Be and B are:
  Be $1s^2 2s^2$ and B $1s^2 2s^2 2p^1$
- The electron to be removed from the boron atom is in a 2p sub-level whereas it is in a 2s sub-level in beryllium.
- The 2p sub-level in B is higher in energy than the 2s sub-level in Be and therefore less energy is required to remove an electron from B.

Annotated exemplar answer 2.2

Explain in terms of electron configurations why oxygen has a lower first ionisation energy than nitrogen. **[3]**

The question asks about electron configurations so these should be stated.

This is just repeating the question – you will not get any marks and are just wasting time.

The first ionisation energy of oxygen is lower than that of nitrogen. The nitrogen p orbital has two opposing spinning electrons which repel and make ionisation easier. The lower nuclear charge of N does not make the energy lower. The electrons of N are further apart so there is less repulsion between them.

1/3

Good – comparison here and the idea of less repulsion in N.

Yes, but compare with oxygen – draw an orbital diagram showing electrons in boxes for N and O – O has $2e^-$ in the same p orbital but N does not. $2e^-$ in the same p orbital repel more.

Checklist

At the end of this chapter you should be able to:

- ☐ Describe the structure of an atom.
- ☐ Work out the number of protons, neutrons and electrons in atoms and ions.
- ☐ Calculate relative atomic masses from isotopic composition and vice versa.
- ☐ Work out full and condensed electron configurations for atoms and ions up to $Z = 36$ and draw orbital diagrams.
- ☐ Describe and explain atomic emission spectra.

Higher Level only

- ☐ Carry out calculations using $E = h\nu$, including working out ionisation energies of atoms.
- ☐ Explain graphs of successive ionisation energies and work out which group an atom is in from these graphs.
- ☐ Explain the graph of first ionisation energy across a period.

PERIODICITY

3

The periodic table is one of the great unifying principles in Chemistry – so much chemical knowledge is encompassed in the table and an understanding of the trends, and exceptions to these trends, has been fundamental to the understanding of Chemistry over the last 150 years.

This chapter covers the following topics:

- ☐ The periodic table
- ☐ Periodic trends
- ☐ First-row transition metals (d block elements) (HL only)
- ☐ Colours of complex ions (HL only)

3.1 The periodic table

The elements in the periodic table are arranged in order of atomic number.

Groups – the vertical columns in the periodic table.

Periods – the horizontal rows in the periodic table.

The periodic table is arranged into s, p, d and f blocks according to the highest occupied sub-level.

Groups 1 and 2: the group number indicates the number of electrons in the outer shell.

Groups 13–18: group number −10 = number of electrons in the outer shell.

The period number indicates the number of shells (main energy levels) in an atom, the highest energy level that is occupied or the number of the outer shell.

The patterns in properties repeat in basically the same way across each period – this is known as periodicity.

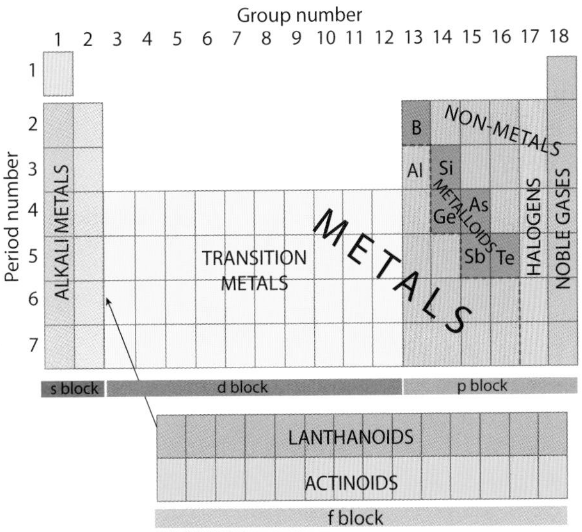

Figure 3.1

hint

Groups are numbered from 1 to 18 including the transition metals – take care here – you will see periodic tables on websites and in books that are numbered differently.

hint

Don't forget that H and He form the first period, so the elements Li to Ne are in period 2.

TEST YOURSELF 3.1

Use a periodic table to classify each of the elements below in one or more of the following categories:
a d block element, a noble gas, an s block element, a halogen, an f block element, a lanthanoid
Elements: bromine, magnesium, europium, titanium, plutonium, iridium, radon, radium

Bromine → halogen
Magnesium → s block
Europium → f block
lanthanoid
titanium → d block
plutonium → f block
iridium → d block
radon → noble gas
radium → s block

3.2 Variation of properties down a group and across a period

There are various trends in properties across a period and down a group.

	Definition	Variation down a group	Variation from left to right across a period
electronegativity	the attraction of an atom for the electron pair in the covalent bond of which it is a part	decreases	increases
atomic radius	the radius of an atom	increases	decreases
ionic radius	the radius of an ion	increases	decreases for positive ions (as charge gets higher), increases for negative ions (as charge gets higher)
ionisation energy	the energy required to remove an electron from a gaseous atom: $M(g) \rightarrow M^{+}(g) + e^{-}$	decreases	increases – but beware of exceptions (see section on 'Exceptions to the general increase in ionisation energy across a period' in Chapter 2)
electron affinity	the energy given out when an electron is added to a gaseous atom: $M(g) + e^{-} \rightarrow M^{-}(g)$	decreases (becomes less exothermic) – but beware of exceptions	increases (becomes more exothermic) – but beware of exceptions – there is not a very clear trend

The trends **down a group** can be explained as … down a group there are more shells of electrons and so the atoms get bigger. This means that the outer electrons are further from the nucleus and less strongly attracted.

The trends **across a period** can be explained as … across a period there is an increase in the number of protons in the nucleus with no significant increase in shielding, therefore the attraction of the nucleus for the outer electrons increases.

Noble gases (group 18) are not included in most of these trends because they are inert and do not generally form compounds.

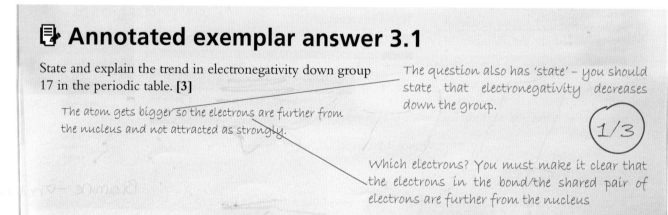

Annotated exemplar answer 3.1

State and explain the trend in electronegativity down group 17 in the periodic table. **[3]**

The atom gets bigger so the electrons are further from the nucleus and not attracted as strongly.

The question also has 'state' – you should state that electronegativity decreases down the group.

1/3

Which electrons? You must make it clear that the electrons in the bond/the shared pair of electrons are further from the nucleus

Nature of Science. Scientists look for patterns in data – this can be useful in developing theories and allowing scientists to make predictions. You could be asked to make predictions about the properties of elements that you are not familiar with from data about those with which you are familiar.

Atomic and ionic radius

The sizes of atoms and ions can be explained in terms of a balance between attraction (nucleus for electrons) and repulsion (between electrons) – electrons will be pulled in closer to the nucleus until the attraction of the nucleus for the electrons balances the repulsion between the electrons.

hint

Do not forget that atomic radius decreases across a period – as the nuclear charge increases for the same number of shells of electrons (shielding stays approximately constant), the outer electrons are pulled in more closely.

The radii of positive ions are smaller than their atomic radii.

Consider M and M^+:

- M and M^+ have the same number of protons therefore the same amount of nuclear attraction.
- M^+ has one fewer electron therefore less electron-electron repulsion so the electrons are pulled in more closely and the size is smaller.

The radii of negative ions are larger than their atomic radii.

Consider X and X^-:

- Same number of protons so the nuclear attraction is the same.
- X^- has one more electron so there is greater electron-electron repulsion.

Worked example 3.1

In which of the following are the species arranged in order of increasing radius?

A	Na^+	Mg^{2+}	Al^{3+}	**B**	Na	Cl	Cl^-
C	Cl	Br	Br^-	**D**	O^{2-}	F^-	Na^+

Answer A – this is not the correct answer – all the ions have the same number of electrons (10) but the nuclear charge increases from Na (11+) to Mg (12+) to Al (13+). Because there is the same number of electrons the repulsion between electrons is approximately the same but because the nuclear charge increases the electrons are pulled in closer to the nucleus and the ions get smaller along this series.

Answer B – this is not the correct answer – Cl is smaller than Na, they are both in the same period and atomic radius decreases from left to right across a period.

Answer C – this is the correct answer – Br is larger than Cl, atomic radius increases down a group. Br^- is larger than Br because they have the same number of protons but Br^- has an extra electron and therefore there is more electron-electron repulsion and the electron cloud expands.

Answer D – this is not the correct answer – same explanation as A.

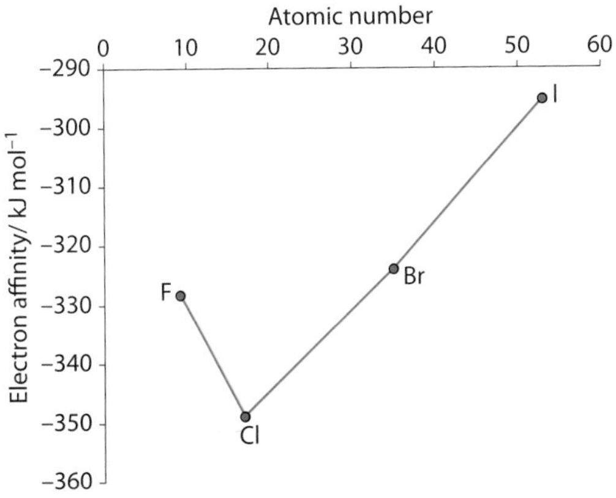

Figure 3.2

Electron affinity down a group

The electron affinity becomes less exothermic from Cl to I as the size of the atom increases – the electron is brought into the outer shell of the atom and as the atom gets bigger there is a weaker attraction between the added electron and the nucleus as it is brought to a position that is further from the nucleus.

The decrease in electron affinity between F and Cl is due to a decrease in electron-electron repulsion as the size of the atom/ion increases.

hint

Electron affinity values are all exothermic so you must be careful using the words 'larger' and 'smaller' because, e.g. −350 is actually a smaller value than −300. It is therefore safer to use the terms 'more exothermic' and 'less exothermic'.

TEST YOURSELF 3.2

1 Which of the following increases down a group?
electronegativity, ionisation energy, ionic radius

2 Which of the following increase(s) across a period?
atomic radius, electronegativity, electron affinity

Trends in metallic behaviour

A metallic structure consists of a regular lattice of positive ions in a sea of delocalised electrons. To form a metallic structure, an element must be able to lose electrons fairly readily to form positive ions; that is, have low ionisation energy.

Metallic structures are less likely to be formed by elements on the right-hand side of the periodic table – ionisation energy increases across a period.

Down a group, ionisation energy decreases, therefore elements are much more likely to exhibit metallic behaviour lower down a group.

Oxides of period 3 elements

DEFINITIONS

BASIC OXIDE (e.g. Na_2O) – one that will react with an acid to form a salt and, if soluble in water, will produce an alkaline solution.

AMPHOTERIC OXIDE (e.g. Al_2O_3) – one that reacts with both acids and bases.

ACIDIC OXIDE (e.g. SO_2) – one that reacts with bases/alkalis to form a salt and, if soluble in water, will produce an acidic solution.

Oxides change from basic, to **amphoteric**, to acidic, across a period.

hint

In general, metallic oxides are basic and non-metallic oxides are acidic but watch out for Al_2O_3 – it is amphoteric.

	Sodium	Magnesium	Aluminium	Silicon	Phosphorus	Sulfur
Formula of oxide	Na_2O	MgO	Al_2O_3	SiO_2	P_4O_{10}	SO_2 SO_3
Nature of element	metal			non-metal		
Nature of oxide	basic		amphoteric	acidic		

Some of the oxides react with water:

Na_2O	$Na_2O(s)+H_2O(l) \rightarrow 2NaOH(aq)$
MgO	$MgO(s)+H_2O(l) \rightarrow Mg(OH)_2(aq)$
P_4O_{10}	$P_4O_{10}(s)+6H_2O(l) \rightarrow 4H_3PO_4(aq)$
NO_2	$2NO_2(g)+H_2O(l) \rightarrow HNO_2(aq)+HNO_3(aq)$
SO_2	$SO_2(g)+H_2O(l) \rightarrow H_2SO_3(aq)$
SO_3	$SO_3(g)+H_2O(l) \rightarrow H_2SO_4(aq)$

hint

These equations are all mentioned specifically on the syllabus – make sure you learn them!

Nitrogen and sulfur oxides contribute to **acid deposition** – e.g. acid rain, which can cause damage to trees and kill fish in lakes.

acidic basic

TEST YOURSELF 3.3

1 Classify each of these oxides as acidic, basic or amphoteric.
Na_2O SO_3 MgO Al_2O_3 Cl_2O SiO_2 P_4O_6 SO_2 NO_2 CaO

2 Write equations for the reactions of the following with water:
Na_2O P_4O_{10} SO_3

$Na_2O + H_2O \rightarrow 2NaOH$

$P_4H_{10} + 6H_2O \rightarrow 4H_3PO_4$

$SO_3 + H_2O \rightarrow H_2SO_4$

3.3 Chemical properties of elements in the same group

The reactions of an atom are determined by the number of electrons in the outer shell (highest main energy level) and since elements in the same group in the periodic table have the same number of electrons in their outer shell they will react in basically the same way.

> **hint**
> You are only expected to know specific reactions and trends for groups 1 and 17.

Reactions of the elements in group 1 (the alkali metals)

- The atoms all have one electron in their outer shell and reactions virtually all involve the loss of this outer shell electron to form the positive ion M^+.
- Reactivity increases down the group because the ionisation energy decreases as the size of the atom increases – Cs loses its outer electron to form a positive ion much more easily than Na and will react more vigorously.

The alkali metals react rapidly with water: $2M(s) + 2H_2O(l) \rightarrow 2MOH(aq) + H_2(g)$

The alkali metals all react with halogens to form salts: $2M(s) + X_2(g) \rightarrow 2MX(s)$

How vigorous the reaction is depends on the particular halogen and alkali metal used; the most vigorous reaction occurs between fluorine and caesium and the least vigorous reaction between lithium and iodine.

TEST YOURSELF 3.4

Write equations for:

1 the reaction between lithium and water $2Li + 2H_2O \rightarrow 2LiOH + H_2$

2 the reaction between potassium and bromine. $2K + Br_2 \rightarrow 2KBr$

Reactions of the elements in group 17

The atoms of the elements in group 17 all have seven electrons in their outer shell and react, either by gaining an electron to form the X^- ion or by forming covalent compounds. Reactivity decreases down the group.

Displacement reactions of halogens

These are reactions between a solution of the halogen and a solution containing halide ions.

	KCl(aq) (colourless)	KBr(aq) (colourless)	KI(aq) (colourless)
$Cl_2(aq)$ (pale green)	no reaction	orange solution $Cl_2(aq) + 2Br^-(aq) \rightarrow 2Cl^-(aq) + Br_2(aq)$	brown solution $Cl_2(aq) + 2I^-(aq) \rightarrow 2Cl^-(aq) + I_2(aq)$
$Br_2(aq)$ (orange)	no reaction	no reaction	brown solution $Br_2(aq) + 2I^-(aq) \rightarrow 2Br^-(aq) + I_2(aq)$
$I_2(aq)$ (brown)	no reaction	no reaction	no reaction

The more reactive halogen displaces the halide ion of the less reactive halogen from solution. Thus chlorine displaces bromide ions and iodide ions from solution and bromine displaces iodide ions from solution. Chlorine is a stronger **oxidising agent** than bromine, which is a stronger oxidising agent than iodine.

> **hint**
> You should learn these reactions and colour changes – you could be asked about them.

TEST YOURSELF 3.5

1 State the colour changes that are observed when chlorine solution is added to potassium bromide solution and bromine solution is added to potassium chloride solution. colour change to orange from pale green
colour change to no colour change

2 Write an ionic equation for the reaction that occurs when chlorine solution is added to potassium bromide solution. $Cl_2 + 2KBr \rightarrow$ ~~KClBr~~ $2KCl + Br_2$

3.4 The transition metals (d block) (HL only)

The first row transition elements (transition metals) are:

> **hint**
> Remember the 4s electrons are removed before the 3d when an ion is formed.

Transition elements

Sc	Ti	V	Cr	Mn	Fe	Co	Ni	Cu	Zn
$[Ar]4s^2 3d^1$	$[Ar]4s^2 3d^2$	$[Ar]4s^2 3d^3$	$[Ar]4s^1 3d^5$	$[Ar]4s^2 3d^5$	$[Ar]4s^2 3d^6$	$[Ar]4s^2 3d^7$	$[Ar]4s^2 3d^8$	$[Ar]4s^1 3d^{10}$	$[Ar]4s^2 3d^{10}$

Figure 3.3

DEFINITION

TRANSITION METAL an element which forms at least one stable ion with a **partially-filled** d sub-shell.

Zinc is not classified as a transition metal according to this definition – the only ion formed by Zn is the 2+ ion, with electron configuration $1s^2 2s^2 2p^6 3s^2 3p^6 3d^{10}$ (full 3d sub-shell).

> **hint**
> There is no need to learn the common oxidation states of transition metals as they are given in the *IB Data Booklet* – but you should remember that they all show +2 for Paper 1.

Characteristics of transition metals

- Transition elements can exhibit more than one **oxidation state** in compounds/complexes.
- Transition elements form complex ions.
- Transition elements form coloured compounds/complexes.
- Transition elements and their compounds/complexes can act as catalysts in many reactions.
- Compounds of transition elements can exhibit magnetic properties.

Oxidation states

> **hint**
> The term 'oxidation number' may be used in questions about transition metal complex ions – this is written using Roman numerals, e.g. II for +2 and –III for –3.

All transition elements exhibit an oxidation state of +2 when the 4s electrons are removed.

The maximum oxidation state exhibited by a transition metal ion is given by the sum of the number of 4s and 3d electrons – e.g. +4 for Ti and +7 for Mn.

Why more than one oxidation state?

The 4s and 3d sub-shells are close in energy and there are no big jumps in the successive ionisation energies when the 4s and 3d electrons are removed.

Complex ions

> **hint**
> Using the Lewis acid/base definition (electron pair acceptor/donor), a ligand can also be classified as a Lewis base.

DEFINITIONS

COMPLEX ION a central transition metal ion surrounded by **ligands**.

LIGANDS negative ions or neutral molecules which use lone pairs of electrons to bond to a transition metal ion to form a complex ion.

A ligand must possess a lone pair of electrons.

Coordinate bonds are formed between the ligand and the transition metal ion – the ligand donates a pair of electrons to the transition metal ion: e.g. $[Fe(H_2O)_6]^{2+}$.

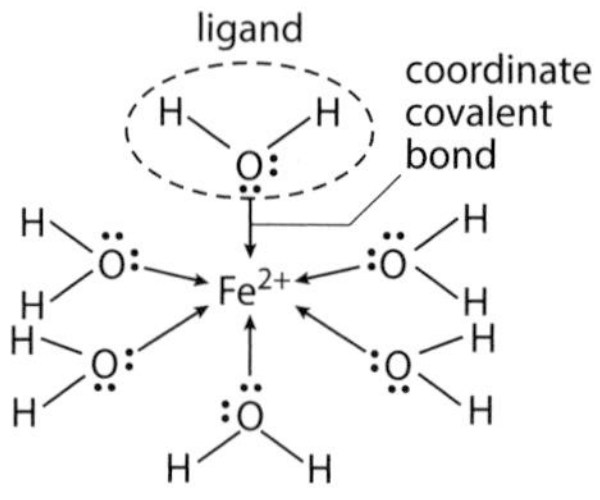

Figure 3.4

The total charge on a complex ion

The total charge on a complex ion can be worked out from the charges on the ligands and the oxidation state of the transition metal.

Ligands may be either neutral or negatively charged, see table here:

Neutral Ligands	1- ligands
H_2O	Cl^-
NH_3	CN^-
CO	OH^-

hint

The most common neutral ligands are NH_3 and H_2O – for other ligands think about whether they would be a neutral molecule or an ion if they were not part of the complex ion.

Worked example 3.2

Titanium(III) forms a complex ion $[Ti(H_2O)_5OH]^{n+}$. Deduce the value of *n*.

$[Ti(H_2O)_5OH]^{n+}$ contains five neutral ligands (H_2O) and one charged ligand (OH^-).

Overall charge on the ion = $(5\times0) \quad +(1\times1^-) \quad +3+ = 2+$

$$(5\times H_2O)+(1\times OH^-)+Ti^{3+}$$

Therefore, *n* is 2.

TEST YOURSELF 3.6

The formula of a complex ion containing iron(II) is $[Fe(CN)_5H_2O]^{x-}$. Deduce the charge on the complex ion.

Fe^{2+} 5^- charge = -3

Magnetic properties

DEFINITIONS

PARAMAGNETIC SUBSTANCES are attracted by a magnetic field. Paramagnetism is caused by the presence of unpaired electrons.

DIAMAGNETIC SUBSTANCES are repelled slightly by a magnetic field. Diamagnetism is caused by the presence of paired electrons.

All compounds have some paired electrons and so exhibit diamagnetism. The diamagnetic effect is much smaller than the paramagnetic effect and so, if there are any unpaired electrons present, the substance will be paramagnetic overall and attracted by a magnetic field.

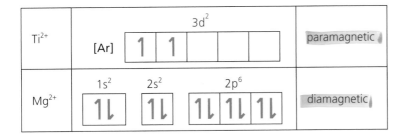

Ti^{2+}	[Ar] $3d^2$: [1][1][ ][ ][ ]	paramagnetic
Mg^{2+}	$1s^2$ [1l] $2s^2$ [1l] $2p^6$ [1l][1l][1l]	diamagnetic

hint

Remember that compounds containing Sc^{3+}, Ti^{4+}, Cu^+ and Zn^{2+} have no unpaired electrons and are diamagnetic.

TEST YOURSELF 3.7

Which of the following are paramagnetic?

$FeCl_2$ $CaCl_2$ $MnCl_2$ CuO Li_2O Cu_2O

3.5 Formation of coloured complexes (HL only)

Model answer 3.1

Explain in terms of electrons why $[Ni(H_2O)_6]^{2+}$ is green.

- The d orbitals are split into two groups at different energies.
- A partially-filled d sub-shell is required – the Ni^{2+} ion has 8 d electrons.
- Certain frequencies of **visible light** can be absorbed to promote an electron from the lower set of d orbitals to the higher set.
- The light that is transmitted/reflected is the complementary colour to the light absorbed.
- Since $[Ni(H_2O)_6]^{2+}$ is green it must absorb red light and transmit green.

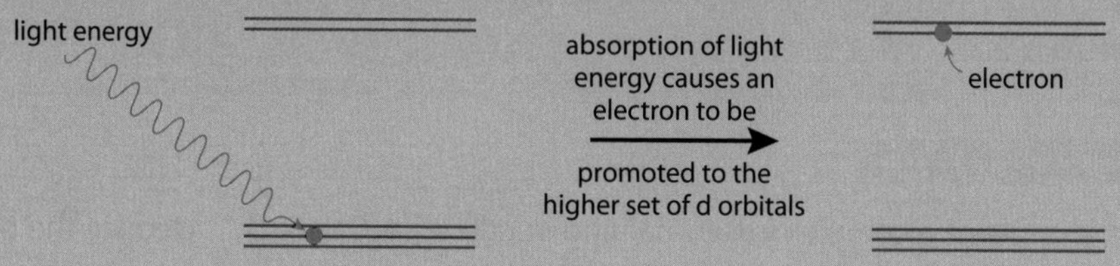

Figure 3.5

hint

Questions about the colours of transition metal ions can be awarded varying amounts of marks – to avoid dropping marks it is worth writing down all of these points – just in case – because you do not know what is on the mark scheme.

Complementary colours can be obtained from a colour wheel (given in the *IB Data Booklet*) – they are opposite each other.

Complexes/compounds containing the Sc^{3+} ion or the Ti^{4+} ion are colourless – no electrons in the 3d sub-shell.

Complexes/compounds containing the Cu^{+} ion or the Zn^{2+} ion are colourless – full 3d sub-shell.

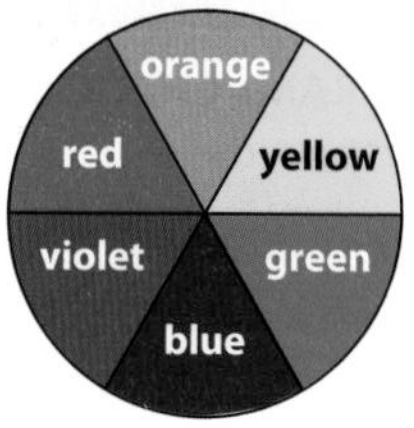

Figure 3.6

Annotated exemplar answer 3.2

Explain why $[Cu(H_2O)_6]^{2+}$ is blue but CuI is white/colourless. **[4]**

$[Cu(H_2O)_6]^{2+}$ has a partially filled d shell so can absorb blue light but CuI doesn't, so it can't absorb light.

sub-shell not shell

1/4

You must mention d orbital splitting – in $[Cu(H_2O)_6]^{2+}$ the d orbitals are split into two groups at different energies and certain frequencies of visible light are absorbed to promote an electron from the lower set of d orbitals to the higher set.

Complementary colour absorbed – Cu^{2+} absorbs certain frequencies of visible light at the red-orange end of the spectrum

Factors affecting the colours of a complex ion

The colours of transition metal complex ions can be related to the amount of splitting of the d orbitals – the bigger the difference in energy between the lower and higher set of d orbitals, the higher the frequency (shorter wavelength) of light absorbed.

Factors that affect the colour are:

- The identity of the metal (different metals have different electron configurations).
- The oxidation state of the metal (higher charge on the ion causes greater splitting of d orbitals).
- The nature of the ligand.

The nature of the ligand

Ligands can be arranged into a **spectrochemical series** according to how much they split d orbitals (this is given in the *IB Data Booklet*):

$I^- < Br^- < Cl^- < F^- < OH^- < H_2O < NH_3 < CO \approx CN^-$

For example, CN^- causes greater splitting of the d orbitals than Cl^-, therefore a higher frequency of light will be absorbed by complexes containing CN^- ions.

The position of some ligands in the spectrochemical series ($I^- < Br^- < Cl^- < F^-$) can be explained in terms of charge density (charge per unit volume) – the F^- ion is the smallest and therefore has the highest charge density – this means that it causes greater splitting of the d orbitals than I^-.

TEST YOURSELF 3.8

In each of the following pairs identify the complex ion with the greater d orbital splitting:

1 $[Ti(H_2O)Cl_5]^{2-}$ or $[Ti(H_2O)Br_5]^{2-}$
2 $[Cr(NH_3)_6]^{3+}$ or $[Cr(H_2O)_6]^{3+}$

Checklist

At the end of this chapter you should be able to:

- ☐ Identify elements in the s, p, d and f blocks in the periodic table.
- ☐ Identify alkali metals, halogens, transition metals, lanthanoids and actinoids in the periodic table.
- ☐ Describe and explain trends in atomic radius, ionic radius, ionisation energy, electron affinity and electronegativity down a group and across a period.
- ☐ Describe the changes in acid–base nature of oxides across a period and write equations for the reactions of oxides with water.
- ☐ Describe and explain the chemical properties of elements in group 1 and group 17.

Higher Level only

- ☐ Describe the properties of transition elements.
- ☐ Explain variable oxidation states, the bonding in complex ions, why transition metal complexes are coloured and whether compounds are paramagnetic or diamagnetic.
- ☐ Explain the factors that affect the colours of complex ions.

4

CHEMICAL BONDING AND STRUCTURE

The world around us has been formed when atoms join together to form compounds. Atoms are held together by chemical bonds and here we will study the nature of these bonds.

This chapter covers the following topics:

- ☐ Ionic bonding and structure
- ☐ Covalent bonding
- ☐ Lewis structures and shapes
- ☐ Intermolecular forces
- ☐ Giant structures
- ☐ Metallic bonding
- ☐ Lewis structures and shapes for molecules/ ions with five or six electron domains (HL only)
- ☐ Sigma and pi bonding (HL only)
- ☐ Hybridisation (HL only)

4.1 Types of bonding

hint

Take care, there are some exceptions to these rules for predicting the type of bonding – e.g. ammonium nitrate (NH_4NO_3) does not contain a metal but it does have ionic bonding between the NH_4^+ and NO_3^- ions.

Compounds may be either **ionic** or **covalent**. An ionic compound usually contains a metal and a non-metal, e.g. NaCl.

Covalent compounds contain two or more non-metals, e.g. CH_4.

TEST YOURSELF 4.1

Classify each of the following as having ionic or covalent bonding:

CO_2 CaS PCl_3 OF_2 MgO

covalent ionic covalent \ ionic
covalent

4.2 Ionic bonding and structure

DEFINITIONS

POSITIVE IONS (CATIONS) formed when atoms (usually of metallic elements) lose electron(s).

NEGATIVE IONS (ANIONS) formed when atoms (usually of non-metallic elements) gain electron(s).

hint

Remember:

Negative ions ending in 'ide' are just the element, e.g. nitride is N^{3-}.

Negative ions ending in 'ate' also contain oxygen, e.g. nitrate is NO_3^-.

When an ionic compound is formed **electrons are transferred** from one atom (metal) to another atom (non-metal) to form positive and negative **ions**.

The formulas of many ions can be worked out from the periodic table.

Group	1	2	13	15	16	17
Ion	1+	2+	3+	3–	2–	1–

You must remember the name, formulas and charges of these ions:

ammonium	NH_4^+	**hydroxide**	OH^-	**carbonate**	CO_3^{2-}	**phosphate**	PO_4^{3-}
		nitrate	NO_3^-	**sulfate**	SO_4^{2-}		
		hydrogencarbonate	HCO_3^-				

Working out the formula of ionic compounds

The overall charge on an ionic compound is always zero, therefore, in aluminium fluoride, the 3+ charge on the Al^{3+} ion must be cancelled out by $3 \times 1-$ charges on $3F^{-}$ ions, i.e. Al^{3+} $(F^{-})_3$ or AlF_3.

A shortcut to working out a formula is to switch over the charges on the ions to generate the formula.

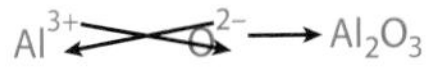

Figure 4.1

TEST YOURSELF 4.2

Work out the formulas of the following ionic compounds:
Lithium fluoride, magnesium chloride, potassium carbonate, calcium hydroxide, ammonium sulfate, sodium hydrogencarbonate, iron(II) phosphate

Ionic structures

DEFINITION

IONIC BONDING is the electrostatic attraction between oppositely-charged ions (*electrostatic attraction* means positive charges attract negative charges and vice versa).

A crystal of sodium chloride consists of a giant lattice of Na^{+} and Cl^{-} ions. There are strong electrostatic forces of attraction between the oppositely-charged ions. Sodium chloride has a high melting point because a lot of energy must be supplied to overcome the strong electrostatic forces.

Magnesium oxide has a higher melting point than sodium chloride because the electrostatic attractions between the 2+ and 2– ions in the magnesium oxide lattice are much stronger than between the 1+ and 1– ions in the sodium chloride lattice. The ions are also smaller in MgO, which also leads to stronger electrostatic attractions.

Ionic compounds do not conduct electricity in the solid state (ions not free to move) but do conduct when molten or in aqueous solution (ions free to move around).

4.3 Covalent bonding

Covalent bonding occurs when atoms **share** electrons.

DEFINITION

COVALENT BOND the electrostatic attraction between the shared pair of electrons and the nuclei of the atoms making up the bond.

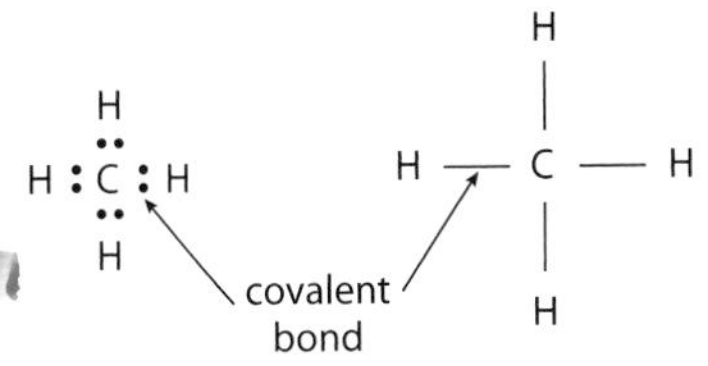

Figure 4.2

Single bond	Double bond	Triple bond
1 shared pair of electrons	2 shared pairs of electrons	3 shared pairs of electrons

Assuming the same atoms are involved (e.g. either C-C bonds or C-O bonds):

Strength	single bonds < double bonds < triple bonds
Length	single bonds > double bonds > triple bonds

The attraction of the two nuclei for two electron pairs (4 electrons) in a double bond is greater than the attraction for one electron pair (2 electrons) in a single bond.

Nature of Science. Scientists develop theories to explain natural phenomena, e.g. observation of the different properties of substances required the development of two different theories of bonding – ionic and covalent.

Coordinate (dative) covalent bond

This is a type of covalent bond where **both electrons come from the same atom**.

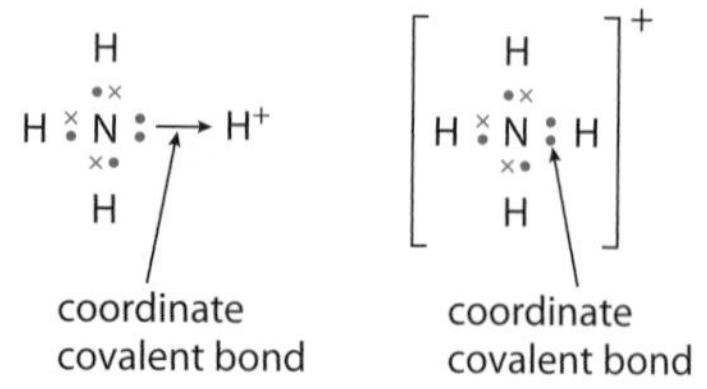

Figure 4.3

> **hint**
>
> Further molecules/ions containing coordinate covalent bonds that come up regularly in exam questions are: H_3O^+, CO and O_3.

Lewis (electron dot) structures

DEFINITIONS

LEWIS STRUCTURES diagrams showing **all the valence (outer shell) electrons** in a molecule (or ion)

THE OCTET RULE the tendency of an atom (with the exception of H) in a molecule/ion to have 8 electrons in its outer shell.

> **hint**
>
> For drawing Lewis structures:
> - don't forget to show all outer shell electrons – remember lone pairs on outer atoms
> - don't forget to show the charge on an ion (with brackets around the ion).

Rules for working out Lewis structures

- Make sure that the **outer** atoms have 8 electrons (octet) in their outer shell (except hydrogen, which should have 2). This can be done by forming single bonds, double bonds, triple bonds and/or adding/removing electrons for ions.
- If the central atom is from **period 2** it should have **no more than 8 electrons** in its outer shell. It should generally (but not always – see exceptions later in the text) have 8 electrons in its outer shell. If the central atom is from period 3 it may have up to 18 electrons in its outer shell.

> **hint**
>
> Most of the exam questions on bonding are about covalent molecular substances so you will need to spend some time practising drawing Lewis structures and working out shapes of molecules.

Worked example 4.1

Draw a Lewis structure for CO_2.

The first step is to make sure the outer atoms (O in this case) have eight electrons in their outer shell. An O atom has six electrons in its outer shell and so it needs two more to complete its octet. Since C has four outer shell electrons it can share two with each O to form a double bond with each.

As a final check we must count the electrons on the central atom – C is a period 2 atom and so cannot have more than eight outer-shell electrons – it has eight electrons in its outer shell in this structure.

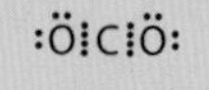

Figure 4.4

Alternative method for working out Lewis structures

This works for molecules/ions just containing period 2 atoms.

> **hint**
>
> Electron pairs may be shown using dots, crosses, lines (one line = 1 pair of electrons) – there is no need to indicate, using different symbols, where the electrons originally came from.

Worked example 4.2

Deduce the Lewis structure for NO_3^-.

Total number of valence electrons = $5+3\times6+1 = 24\ e^-$

N 3×O negative charge

There are, therefore $\frac{24}{2}$, i.e. 12 pairs of electrons

Three electron pairs (lines) must be now used to join all the atoms together. This leaves nine lines (electron pairs) that will be distributed as lone pairs of electrons on the O atoms (outer atoms) to give each an **octet** (Figure **4.5**). Now each O is 'attached to four lines' (4 pairs of electrons in its outer shell) and all the electrons have been used.

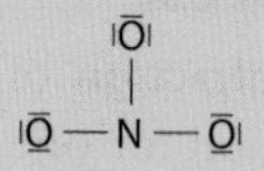

Figure 4.5

> **hint**
>
> Some of the Lewis structures involving nitrogen can be quite tricky to work out – it may be better to memorize some of the common ones (e.g. NO_2^-, NO_2^+, NO_3^-).

The nitrogen, however, only has three pairs of electrons in its outer shell therefore one of the lines needs to be moved from being a lone pair on the O to make a double bond between the N and an O. This does not change the number of electrons in the outer shell of the O but will increase the number of electrons in the outer shell of the N to 8.

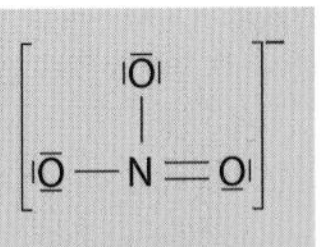

Figure 4.6

A central atom does not always have eight electrons in its outer shell, e.g. $BeCl_2(g)$ (4 electrons in the outer shell of Be) and BCl_3 (6 electrons in the outer shell of B). When the central atom is from period 3 it may have up to 18 electrons in its outer shell.

Resonance structures

For an ion such as CO_3^{2-} we can draw several structures where the only difference between them is where we put the double bond (and lone pairs if appropriate). The individual Lewis structures are called resonance structures (Figure **4.7**).

The actual structure of a molecule/ion can be described as a hybrid of these structures.

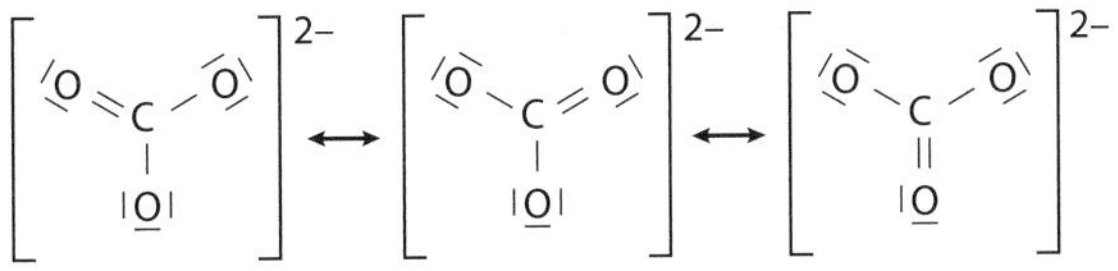

Figure 4.7

hint

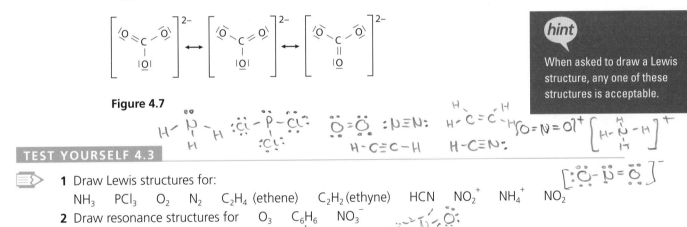

When asked to draw a Lewis structure, any one of these structures is acceptable.

TEST YOURSELF 4.3

1 Draw Lewis structures for:
NH_3 PCl_3 O_2 N_2 C_2H_4 (ethene) C_2H_2 (ethyne) HCN NO_2^+ NH_4^+ NO_2

2 Draw resonance structures for O_3 C_6H_6 NO_3^-

Shapes of molecules – Valence Shell Electron Pair Repulsion (VSEPR) theory

The shapes of molecules/ions can be predicted using **v**alence **s**hell **e**lectron **p**air **r**epulsion (VSEPR). The shape of a molecule depends on the number of electron domains in the outer shell of the **central atom**.

hint

Remember, when answering questions that ask you to describe/explain/discuss VSEPR, the words 'central atom' are essential.

An electron domain could be:

- a lone pair
- an electron pair that makes up a single bond
- the electron pairs (together) that make up a multiple bond – a double bond or a triple bond counts as one electron domain

Electron domains in the valence (outer) shell of the central atom in a molecule repel each other and will therefore take up positions in space to minimise these repulsions, i.e. to be as far apart as possible.

The repulsion between lone pairs and bonding pairs of electrons is greater than between just bonding pairs – this affects the bond angles.

> **hint**
> Remember if there are lone pairs present in the molecule you must state the **actual shape** of the **molecule** and not the electron domain geometry.

How to predict the shapes of molecules

- Draw a Lewis structure for the molecule or ion.
- Count the number of electron domains in the outer shell of the central atom.
- Work out the basic shape (electron domain geometry).
- Work out the actual shape – a lone pair is just an electron pair in the outer shell of an atom, and, as such, contributes to the overall shape of the molecule but cannot, itself, be 'seen' – it is not included in the actual shape.

You need to know the following electron domain geometries and the names of the actual shapes:

> **hint**
> If you are asked to predict a bond angle in a particular molecule, just take two or three degrees off the bond angle in the basic shape (electron domain geometry) for every lone pair. For example, a bent molecule based on a trigonal planar structure could have a bond angle of 120 – 3, i.e. 117°. There is no scientific basis for doing this but it is useful for exam questions!

Total electron pairs	Bonding pairs	Lone pairs	Basic shape (electron domain geometry)	Actual shape	Bond angle	Example
2	2	0	linear Y—X—Y	linear	180°	CO_2, NO_2^+, HCN
3	3	0	trigonal planar 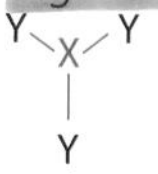	trigonal planar	120°	SO_3, NO_3^-, CO_3^{2-}
3	2	1	trigonal planar	bent, V-shaped,	117° (predicted)	SO_2, O_3, NO_2^-
4	4	0	tetrahedral 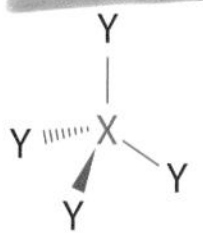	tetrahedral	109.5°	CCl_4, XeO_4, NH_4^+, SO_4^{2-}
4	3	1	tetrahedral	trigonal pyramidal	107° (predicted)	NH_3, PCl_3, XeO_3, H_3O^+
4	2	2	tetrahedral	bent, V-shaped, angular	105° (predicted)	H_2O, SCl_2, ClF_2^+

☆ Model answer 4.1

Predict and explain the shape and bond angle of NO_2^-.

The Lewis structure for NO_2^- is shown in Figure **4.8**.

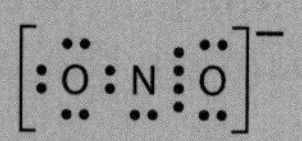

Figure 4.8

1 double bond + 1 single bond + 1 lone pair = 3 electron domains on central atom

The three electron domains repel each other and take up positions in space to be as far apart as possible to minimise repulsions.

The electron domains are distributed in a trigonal planar arrangement.

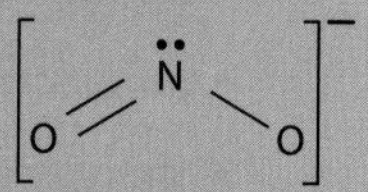

Figure 4.9

One of the electron domains is a lone pair, so the actual shape of the ion is **bent**.

This structure is based on trigonal planar (bond angle 120°) with one lone pair, so a bond angle of about 120 – 3 = 117° is predicted.

TEST YOURSELF 4.4

Predict the shapes and bond angles of: CH_4 NH_3 H_2O CO_2 NO_2^+ NH_4^+ H_3O^+ BF_3 CO_3^{2-}

CH_4 → tetrahedral NH_3 → trigonal planar H_2O → bent CO_2 → linear
NO_2^+ → linear NH_4^+ → tetrahedral H_3O^+ → trigonal pyramidal
BF_3 → trigonal planar CO_3^{2-} → trigonal planar

Electronegativity and polarity

A polar bond arises when two atoms have different electronegativities – the atom with higher electronegativity will be slightly negative (δ–), the atom with lower electronegativity will be slightly positive (δ+) – this is a **dipole**. The bigger the difference in electronegativity, the more polar the bond.

The H-F bond is polar because F is more electronegative than H.

$^{\delta+}$H — F$^{\delta-}$

Figure 4.10

For a molecule to be polar there must be a difference in electronegativity between atoms and the shape should be such that dipoles do not cancel out, e.g. HCl, NH_3 and H_2O are all **polar molecules**.

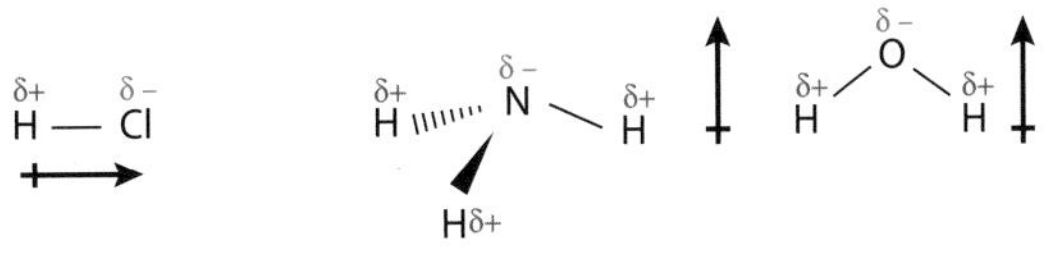

Figure 4.11

> **hint**
>
> For Papers 2 and 3 you do not need to remember electronegativity values because they are given in the *IB Data Booklet*. For Paper 1 remember that F is the most electronegative element – electronegativity increases across a period towards F and decreases down a group from F.

CO_2 is a non-polar molecule – each C-O bond is polar because oxygen is more electronegative than carbon but overall, due to the symmetry, the dipoles cancel so that there is no overall **dipole moment** and the molecule is non-polar.

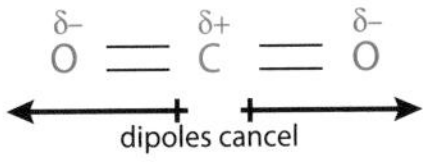

Figure 4.12

TEST YOURSELF 4.5

1 Which of the following bonds is most polar? H–C H–Cl H–F

2 Which of the following molecules is/are polar? N_2 CO NH_3 CO_2 BF_3 PCl_3

4.4 Intermolecular forces

Only intermolecular forces are broken when covalent molecular substances are melted or boiled – covalent bonds (intramolecular forces) are not broken. Generally, the stronger the intermolecular forces, the higher the boiling point – more energy must be supplied to overcome the forces between molecules.

London forces are the only intermolecular forces between non-polar molecules. London forces are stronger if more electrons are present in the molecules – this is generally related to the relative molecular mass – the higher the relative molecular mass, the stronger the London forces – e.g. Br_2 has a higher boiling point than Cl_2.

Dipole–dipole forces exist between polar molecules – a δ+ atom in one molecule attracts a δ– atom in another molecule. If molecules with **similar relative molecular** masses are compared (London forces are of similar strength), polar molecules have stronger intermolecular forces and therefore higher melting and boiling points than non-polar molecules. For example ICl (polar) has a higher boiling point than Br_2 (non-polar).

Hydrogen bonding occurs between molecules if they contain an H atom joined directly to a very electronegative atom (N, O or F) with a lone pair. It is an **intermolecular** force between the lone pair of the δ– **oxygen, nitrogen** or **fluorine** atom in one molecule and a δ+ hydrogen atom in another molecule.

Intermolecular forces increase in strength as: London forces < dipole-dipole forces < hydrogen bonding

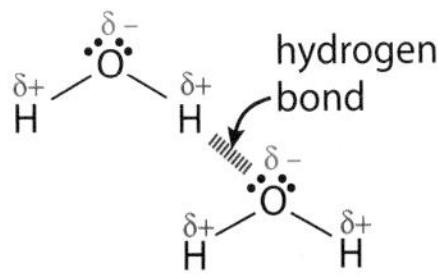

Figure 4.13

> **hint**
>
> In questions about boiling points of covalent substances always look for molecules that could exhibit hydrogen bonding.

Annotated exemplar answer 4.1

Arrange the following compounds in order of increasing boiling point and explain the order: ethanol (CH_3CH_2OH), propane ($CH_3CH_2CH_3$), methoxymethane (CH_3OCH_3). **[4]**

The order is $CH_3CH_2CH_3$, CH_3OCH_3, CH_3CH_2OH

$CH_3CH_2CH_3$ has London forces, CH_3OCH_3 is polar and CH_3CH_2OH has hydrogen bonding. The bonds in CH_3CH_2OH are strongest and those in $CH_3CH_2CH_3$ are weakest, therefore require least energy to break.

...between molecules

2/4

This word makes the answer confusing – it is not clear that intermolecular forces are being considered rather than covalent bonds – boiling points increase as **intermolecular forces** get stronger.

Although it is correct that methoxymethane is polar, the type of intermolecular force is not stated (dipole-dipole interactions).

Correct order although it could have been more clearly stated as $CH_3CH_2CH_3 < CH_3OCH_3 < CH_3CH_2OH$.

TEST YOURSELF 4.6

1 Select the substance from each of the following pairs that has the higher boiling point:

NH_3 or PH_3 CH_3COOH or $CH_3OCH_2CH_3$

2 Which of the following have hydrogen bonding between molecules?

HF CH_3OH CH_3Cl HCN NH_3 CH_3CH_2F

Solubility

Generally a substance will dissolve in a solvent if the intermolecular forces in the solute and solvent are similar, e.g. pentane (London forces) is readily soluble in hexane (London forces) but insoluble in water (hydrogen bonding). Substances that are able to participate in hydrogen bonding will generally be soluble in water as they are able to hydrogen bond to the water. For example, ethanol (C_2H_5OH) is very soluble in water – the ethanol is able to hydrogen bond to the water.

hint

SiO_2 and the allotropes of carbon are mentioned specifically on the syllabus – you should be able to describe their structures as well as understanding how the nature of the bonding affects physical properties.

Giant covalent structures (network solids)

These have no individual molecules – just covalent bonds between atoms throughout the structure.

SiO_2 – each Si joined to 4 O atoms and each O atom is joined to 2 Si atoms. SiO_2 has a high melting point because covalent bonds (strong) must be broken.

Allotropes of carbon

Diamond, graphite, C_{60} (buckminsterfullerene) and graphene are different forms (allotropes) of carbon.

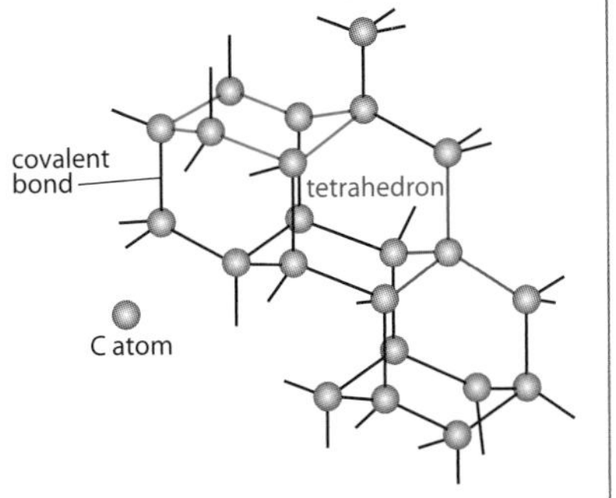

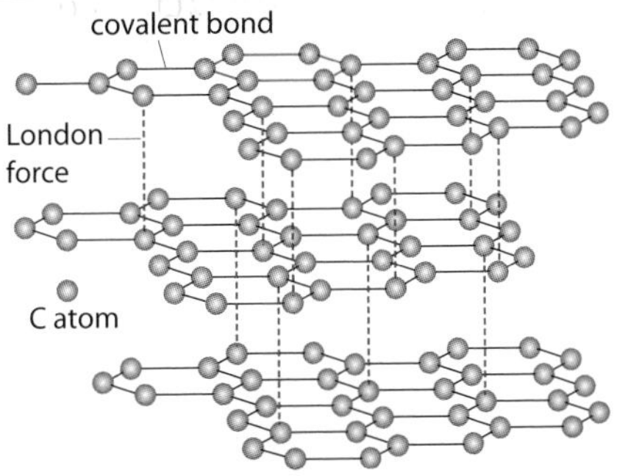

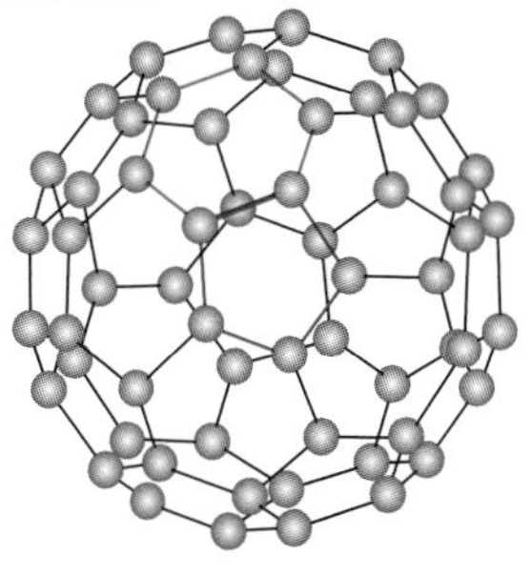

Diamond	Graphite	Buckminsterfullerene
Giant covalent structure. Covalent bonds between all atoms. Tetrahedral arrangement of bonds. Hexagonal rings of C atoms.	Giant covalent structure. Layers of atoms joined by covalent bonds. London forces between layers. Each C forms 3 bonds. Planar hexagonal rings of atoms.	C_{60} molecular structure. Hexagons and pentagons of C atoms. Each C forms 3 bonds. London forces between molecules.
High melting point and hard – strong covalent bonds must be broken. Does not conduct electricity – no free electrons.	High melting point because strong covalent bonds must be broken. Free electrons as C only forms 3 bonds, therefore conducts electricity – 1 electron per C atom free to move along the layers.	Lower melting point than diamond and graphite because only London forces must be broken.

Graphene consists of a single layer of graphite. Graphene has a very high tensile strength and would be expected to have a very high melting point because covalent bonds need to be broken to break the sheet. It is also a good electrical conductor – C forms only three bonds, therefore free electrons are present, which can move around.

hint

Remember that buckminsterfullerene is a covalent molecular substance and not giant covalent.

Annotated exemplar answer 4.2

Explain, in terms of structure and bonding, why carbon dioxide sublimes at −78 °C but silicon dioxide melts at about 1600 °C. **[3]**

When silicon dioxide melts covalent bonds need to be broken but only weak intermolecular forces are broken when carbon dioxide sublimes.

1/3

The question asks about structure and bonding, so these should be stated for each compound:

CO_2 – covalent molecular

SiO_2 – giant covalent.

Compare with covalent bonds – covalent bonds are strong – much more energy is required to break covalent bonds.

State the name of the intermolecular forces.

4.5 Metallic bonding

Metals are made up of a regular lattice arrangement of positive ions surrounded by a 'sea' of delocalised electrons. Metals have a giant structure.

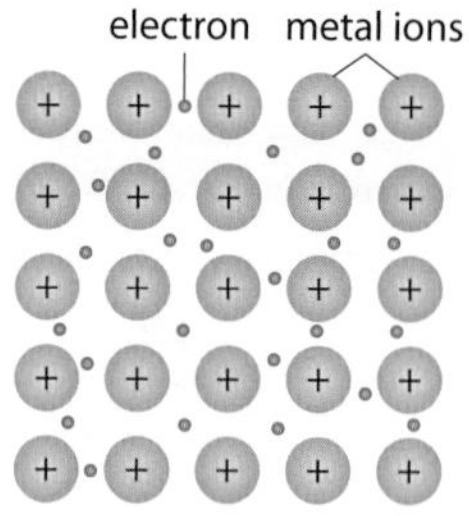

Figure 4.14

DEFINITION

METALLIC BONDING is an electrostatic attraction between the positive ions in the lattice and the delocalised electrons.

☆ Model answer 4.2

Explain why magnesium has a higher melting point than sodium.

- Magnesium forms a 2+ ion compared to sodium, which forms a 1+ ion. This means that the electrostatic attraction between the ions and the delocalised electrons is stronger in magnesium.
- The Mg^{2+} ion is smaller than the Na^{+} ion and therefore the delocalised electrons are closer to the nucleus of the positive ion in magnesium and more strongly attracted.

Some other physical properties of metals:

- Good conductors of electricity – delocalised electrons are free to move around.
- Malleable – may be hammered into shape – layers of metal ions can slide over each other without affecting the bonding – the bonding is non-directional – the metal ions attract the delocalised electrons in all directions.

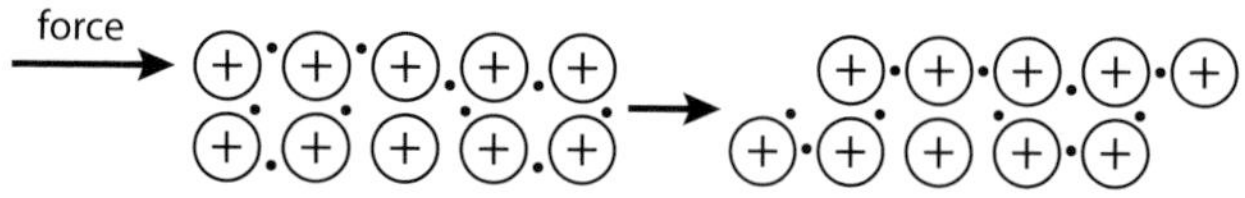

Figure 4.15

Alloys

Alloys are homogeneous mixtures of two or more metals, or of a metal with a non-metal, e.g. steel (Fe, C), brass (Cu, Zn).

Alloys tend to be stronger than pure metals – a different sized atom will prevent planes of metal atoms sliding over each other.

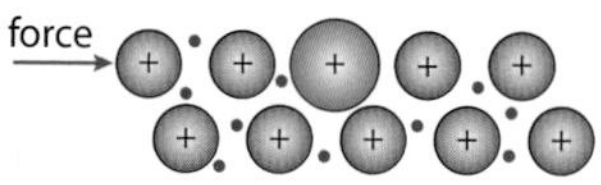

Figure 4.16

4.6 Lewis structures and formal charge (HL only)

Formal charges (FC) on atoms can be used to select the most appropriate Lewis structures for a particular molecule/ion.

hint

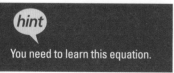

You need to learn this equation.

FC = (number of valence electrons in the uncombined atom) – ½ (number of bonding electrons) – (number of non-bonding electrons).

In general, the preferred Lewis structure is the one in which the formal charges on individual atoms are closest to zero.

Worked example 4.3

Two alternative Lewis structures for SO_4^{2-} are shown here. Use the concept of formal charge to deduce which is the better representation of the actual structure.

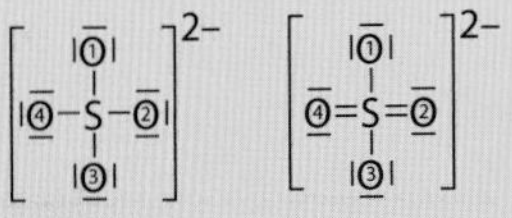

Structure 1 Structure 2

Structure 1		Structure 2	
Formal Charges			
S	$FC = 6 - \frac{1}{2} \times 8 - 0 = 2+$	**S**	$FC = 6 - \frac{1}{2} \times 12 - 0 = 0$
O1	$FC = 6 - \frac{1}{2} \times 2 - 6 = 1-$	**O1**	$FC = 6 - \frac{1}{2} \times 2 - 6 = 1-$
O2	$FC = 6 - \frac{1}{2} \times 2 - 6 = 1-$	**O2**	$FC = 6 - \frac{1}{2} \times 4 - 4 = 0$
O3	$FC = 6 - \frac{1}{2} \times 2 - 6 = 1-$	**O3**	$FC = 6 - \frac{1}{2} \times 2 - 6 = 1-$
O4	$FC = 6 - \frac{1}{2} \times 2 - 6 = 1-$	**O4**	$FC = 6 - \frac{1}{2} \times 4 - 4 = 0$

Structure 2 has lower individual formal charges (S in Structure 1 has a 2+ charge) and therefore it is the preferred Lewis structure.

hint

The sum of the formal charges on all the atoms should add up to the overall charge on the molecule/ion.

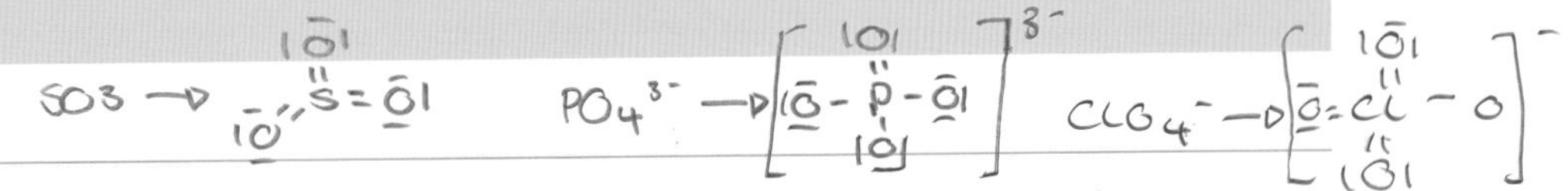

TEST YOURSELF 4.7

Use the concept of formal charge to work out the preferred Lewis structures for:
SO_3 PO_4^{3-} ClO_4^- XeO_4

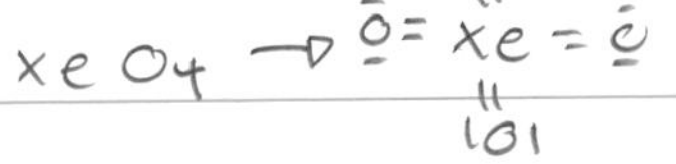

4.7 Sigma and pi bonds (HL only)

Sigma (σ) bonds result from the axial (head-on) overlap of atomic orbitals. The electron density in a σ bond lies mostly along the axis joining the two nuclei.

A pi (π) bond is formed by the sideways overlap of parallel p orbitals. The electron density in the π bond lies above and below the internuclear axis (Figure **4.18**).

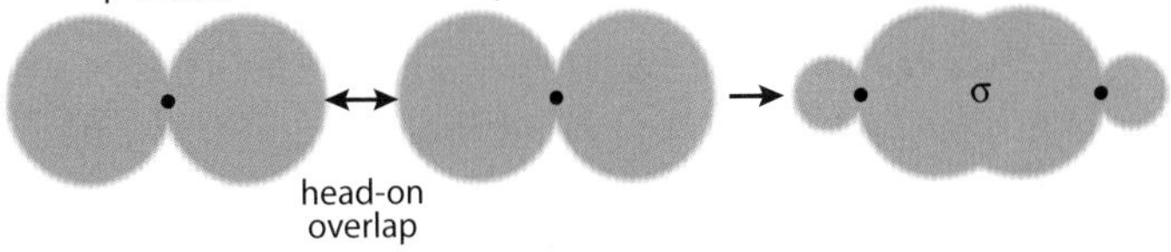

Figure 4.17

Single bond	σ	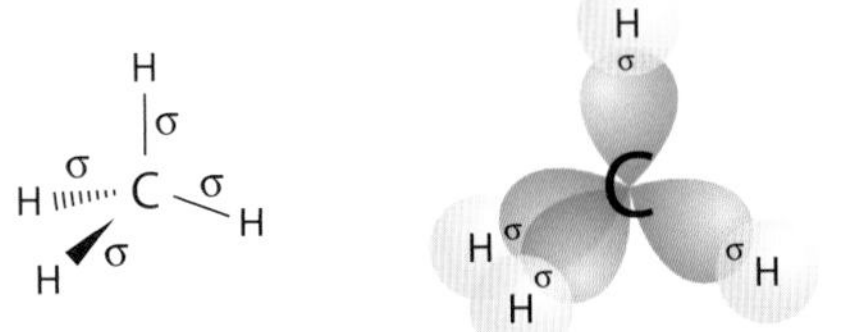
Double bond	σ+π	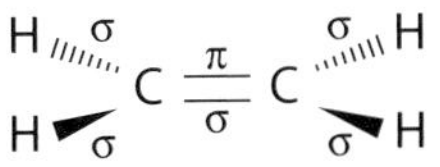
Triple bond	σ+2π	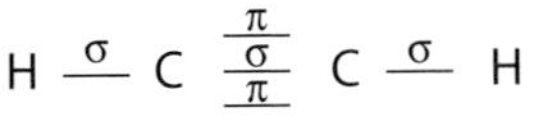

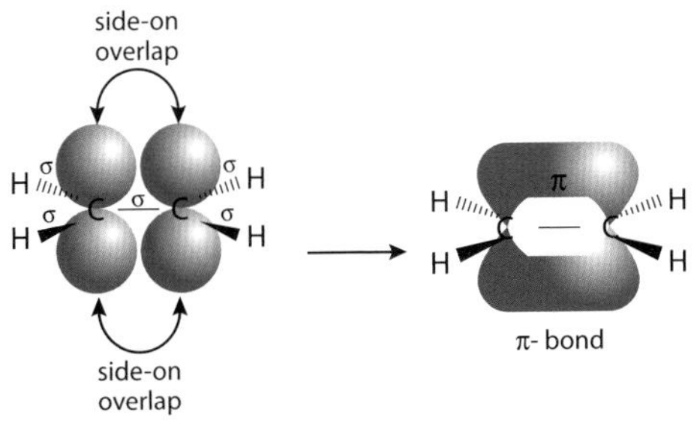

Figure 4.18

TEST YOURSELF 4.8

Work out the number of σ and π bonds in: C_2H_4 HCN C_2H_2.

(handwritten: C_2H_4: 5σ 1π; HCN: 2σ 2π; C_2H_2: 3σ 2π)

4.8 Shapes of molecules with 5 or 6 electron domains (HL only)

VSEPR is used to predict the shapes of molecules/ions with five or six electron domains – the various shapes adopted are:

Total electron domains	Bonding pairs	Lone pairs	Basic shape	Actual shape	Bond angle	Example
5	5	0	trigonal bipyramidal 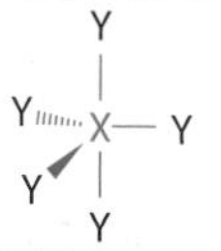	trigonal bipyramidal	90° and 120°	PF_5
5	4	1	trigonal bipyramidal	'see-saw'	88° and 118° (predicted)	SF_4
5	3	2	trigonal bipyramidal	'T-shaped'	88° (predicted)	BrF_3
5	2	3	trigonal bipyramidal	linear	180°	I_3^-, XeF_2
6	6	0	octahedral 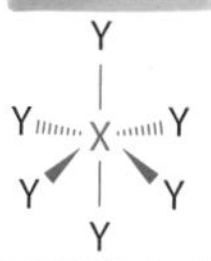	octahedral	90°	SF_6, PF_6^-
6	5	1	octahedral	square pyramidal	88° (predicted)	SF_5^-, BrF_5
6	4	2	octahedral	square planar	90°	XeF_4, SF_4^{2-}

In a trigonal bipyramid the axial and equatorial positions are not equivalent – lone pairs always go in the equatorial position (around the middle). This rule only holds for five electron domains – XeF_4 (six electron domains) is square planar.

Worked example 4.4

Predict the shape and bond angles of SF_4.

The Lewis structure is shown in Figure **4.19a**.

4 bonding pairs of electrons + 1 lone pair = 5 electron domains

5 electron domains therefore the structure is based on trigonal bipyramidal. The lone pair goes around the middle so that the actual shape is 'see-saw' (Figure **4.19b**). The lone pair repels bonding pairs more than they repel each other and so the bond angles close up – bond angles of 88° and 118° could be predicted (a range of values will be accepted).

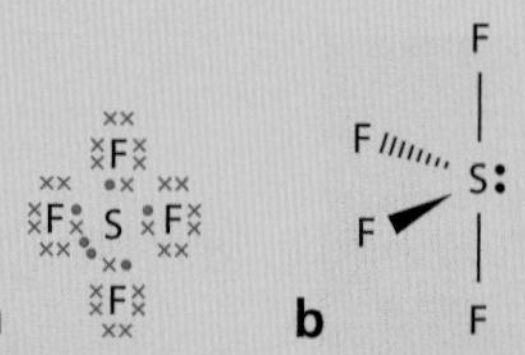

Figure 4.19

TEST YOURSELF 4.9

Predict the shapes and bond angles of: XeF_4 PCl_5 $XeOF_4$ BrF_5 SO_4^{2-} ClF_3 SF_6 XeO_2F_2.

(handwritten: XeF_4 → square planar; PCl_5 → trigonal bipyramidal; $XeOF_4$ → square pyramidal; BrF_5 → square pyramidal; SO_4^{2-} → tetrahedral; ClF_3 → T-shaped; SF_6 → octahedral; XeO_2F_2 → see saw)

4.9 Hybridisation (HL only)

DEFINITION

HYBRIDISATION the mixing of atomic orbitals (s and p) on a particular atom to produce a new set of orbitals (the same number as originally) that have both s and p character and are better arranged in space for covalent bonding

hint

For a 1 mark question on 'what is hybridisation?' all you need is … 'the mixing of atomic orbitals'.

The hybridisation scheme of a particular atom in a molecule is determined by the arrangement of electron domains, i.e. the basic shape of the molecule/ion.

Electron domains	Basic shape	Hybridisation	Examples
2	linear	sp	CO_2, HCN, C_2H_2, NO_2^+
3	trigonal planar	sp^2	BF_3, NO_3^-, CO_3^{2-}, NO_2^-
4	tetrahedral	sp^3	NH_3, PCl_3, H_2O

Methane	CH_4	sp^3	mixing of 1 s and 3 p orbitals	no p orbitals left on C for π bonds
Ethene	$H_2C{=}CH_2$	sp^2	mixing of 1 s and 2 p orbitals on each C	1 p orbital left on each C to form 1 π bond
Ethyne	HC≡CH	sp	mixing of 1 s and 1 p orbital on each C	2 p orbitals left on each C to form 2 π bonds

TEST YOURSELF 4.10

Predict the type of hybridisation present on the central atom in:

CH_4 BF_3 HCN H_2O SO_2 CO_3^{2-} NH_3 CO_2 C_2H_4

Sp3 Sp2 Sp Sp3 Sp2 Sp2 Sp3 Sp Sp2

4.10 Delocalisation (HL only)

DEFINITION

DELOCALISATION is the sharing of a pair of electrons between three or more atoms.

The Lewis structure of O_3 (Figure **4.20**) would suggest one short (O=O) bond and one longer (O–O) bond, but actually the bond lengths in O_3 are equal. The equal bond lengths in O_3 can be explained by delocalisation (or the idea of a resonance hybrid). Instead of the two electrons in the π bond being localised between two atoms, they are delocalised over the three O atoms.

hint

When drawing Lewis structures you must show all lone pairs but must not put in lone pairs when drawing delocalised structures.

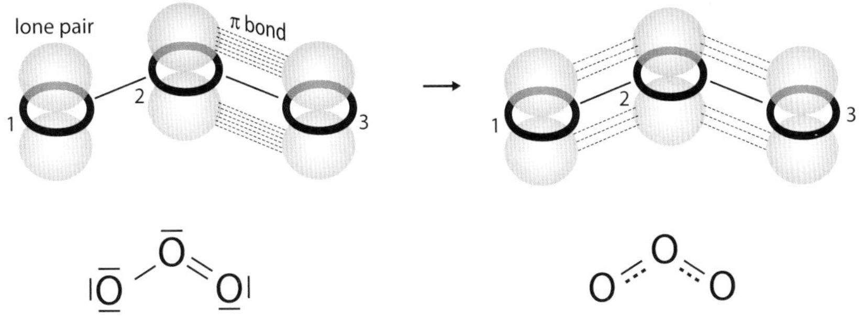

Figure 4.20

Since the π bond in ozone is shared between two single bonds rather than just one, we consider that each bond has a half share of it and we talk about the O–O bond order being 1.5.

hint

Whenever you are asked about bond lengths being equal in a molecule/ion, immediately think that this might be due to resonance hybrids/delocalised structures.

The bond order of a single bond is 1, a double bond has bond order 2 and a triple bond has bond order 3.

If you are asked to decide whether a molecule/ion will have resonance structures/delocalisation you must see whether it is possible to draw exactly equivalent structures where the only difference between them is the position of double bonds (and lone pairs if appropriate), e.g. there is delocalisation in the ethanoate ion (CH_3COO^-) but not in ethanoic acid (CH_3COOH) because moving the double bond results in a molecule that looks different.

TEST YOURSELF 4.11

Which of the following contain delocalised electrons?

CO_2 CO_3^{2-} CO C_4H_8 C_6H_6 CH_3COO^- CH_3COOH H_2SO_4 SO_4^{2-} NO_2^- NO_3^-

Ozone and UV light

The surface of the Earth is protected from UV light by O_2 and O_3 in the atmosphere. UV radiation is absorbed by molecules of O_2 or O_3 as they undergo dissociation (O–O bond broken).

$$O_2 \rightarrow 2O\bullet$$

$$O_3 \rightarrow O\bullet + O_2$$

O_2 absorbs higher energy (higher frequency/shorter wavelength) electromagnetic radiation than O_3 – the bond between the atoms in an O_2 molecule is stronger – it is a double bond (bond order 2), but O_3 has a delocalised structure where the bond order is 1.5 (weaker bond).

Catalysis of ozone depletion by CFCs and NO_x

Chlorofluorocarbons (CFCs) such as CCl_2F_2 are broken down by absorbing UV radiation in the upper atmosphere:

$$CCl_2F_2 \xrightarrow{UV} \underset{\text{free radical}}{\bullet CClF_2} + \underset{\text{free radical}}{Cl\bullet}$$

Ozone molecules are easily destroyed by free radicals (species with unpaired electrons) such as nitrogen oxides or chlorine atoms.

The free radical can take part in a **chain reaction**, which uses up ozone and regenerates the free radical, X•:

$$X\bullet + O_3 \rightarrow XO\bullet + O_2 \qquad X = Cl \text{ or } NO$$

$$XO\bullet + O\bullet \rightarrow O_2 + X\bullet$$

Overall reaction: $O_3 + O\bullet \rightarrow 2O_2$

The free radical (Cl• or NO•) speeds up the decomposition of ozone and is regenerated in these reactions (so it is not used up) – therefore, it acts as a catalyst.

Checklist

At the end of this chapter you should be able to:

- ☐ Identify the type of structure and bonding in a substance.
- ☐ Describe what ions are and work out the formulas of ionic compounds.
- ☐ Explain the physical properties of ionic compounds in terms of structure and bonding.
- ☐ Explain the connection between length and strength of single, double and triple bonds.
- ☐ Draw Lewis structures for a variety of molecules and ions and work out the shapes and bond angles of molecules with up to four electron domains.
- ☐ Predict whether molecules are polar or not.
- ☐ Explain the physical properties of giant covalent substances and allotropes of carbon.
- ☐ Explain the properties of covalent molecular substances in terms of intermolecular forces.
- ☐ Describe metallic bonding and explain the properties of metals and alloys.

Higher Level only

- ☐ Distinguish between different Lewis structures using formal charge.
- ☐ Explain σ and π bonding.
- ☐ Work out the shapes and bond angles of molecules with up to six electron domains.
- ☐ Explain what is meant by hybridisation and predict the hybridisation scheme in a molecule/ion.
- ☐ Explain how delocalisation can be used to explain the structures of molecules/ions.
- ☐ Explain why O_2 and O_3 are dissociated by different wavelengths of light and describe the mechanism of ozone depletion as catalysed by CFCs and NO_x.

5

ENERGETICS

This chapter deals with energy changes in chemical reactions and how we can predict whether a reaction will be spontaneous or not.

This chapter covers the following topics:

- ☐ Measuring energy changes
- ☐ Calculation of enthalpy changes using Hess's law
- ☐ Calculation of enthalpy changes using bond enthalpies
- ☐ Calculation of enthalpy changes using energy cycles (HL only)
- ☐ Entropy (HL only)
- ☐ Predicting the spontaneity of a reaction (HL only)

5.1 Measuring energy changes

Exothermic and endothermic reactions

DEFINITIONS

HEAT a form of energy.

TEMPERATURE a measure of the average kinetic energy of particles.

EXOTHERMIC REACTION heat energy is transferred from a system (chemical reaction) to the surroundings – the surroundings get hotter. The enthalpy change, ΔH, is negative.

ENDOTHERMIC REACTION a system takes in heat energy from the surroundings – the surroundings get cooler. ΔH is positive.

STANDARD CONDITIONS 100 kPa and a specified temperature.

STANDARD STATE the pure substance at 100 kPa and a specified temperature.

hint

Assume that 'standard' refers to 298 K unless another temperature is specified.

Total energy is conserved in a chemical reaction.

Experimental methods for measuring enthalpy changes

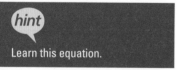

hint

Learn this equation.

We can measure enthalpy changes experimentally by measuring the temperature change of a substance (usually water) and using the equation: $q = mc\Delta T$

q is the amount of heat energy (J) that must be supplied to raise the temperature of mass m (g) by ΔT °C. c is the specific heat capacity ($\text{J g}^{-1}\,°\text{C}^{-1}$) of the substance being heated.

hint

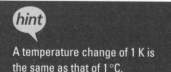

A temperature change of 1 K is the same as that of 1 °C.

Measuring the enthalpy change of combustion

DEFINITION

STANDARD ENTHALPY CHANGE OF COMBUSTION ($\Delta H_c^{\ominus}$) the enthalpy change when one mole of a substance is completely burnt in oxygen under standard conditions.

The heat given out in a combustion reaction can be used to heat another substance of known specific heat capacity, such as water.

Worked example 5.1

A student carried out an experiment to determine the enthalpy change of combustion of methanol, CH_3OH, using the apparatus shown.

Her results were:

volume of water / cm^3	**40.0**
temperature change of the water / °C	**49.5**
mass of methanol burnt / g	**0.52**

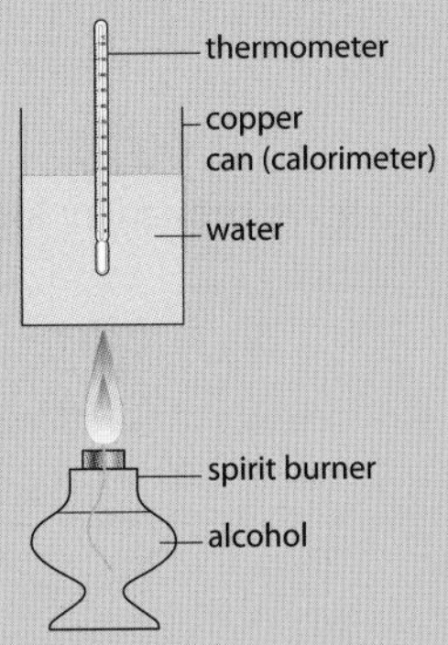

Figure 5.1

Calculate the enthalpy change of combustion of methanol using the data in the table.

The density of water is 1 g cm^{-3} and so the mass of water is 40.0 g.

$q = mc\Delta T = 40.0 \times 4.18 \times 49.5 = 8276.4$ J

Moles of methanol burnt $= \frac{0.52}{32.05} = 0.0162$ mol

Energy given out per mole $= \frac{8276.4\text{ J}}{0.0162\text{ mol}} = 510\,000\text{ J mol}^{-1}$

The enthalpy change of combustion of methanol is therefore $\Delta H = -510\text{ kJ mol}^{-1}$

hint

The water is being heated and it is the mass of water that must be used in this equation – take care, a very common mistake is to use the mass of the substance burnt in $q = mc\Delta T$.

hint

The specific heat capacity of water (4.18 J g^{-1} °C^{-1}) is given in the *IB Data Booklet*.

hint

Don't forget to think about the sign of the final enthalpy change – all combustion reactions are exothermic (ΔH –ve).

The accepted literature value for the enthalpy change of combustion of methanol is -726 kJ mol^{-1} – this experiment does not give a very accurate answer. There are several major flaws (systematic errors) in the experiment, for example:

- heat loss to the surroundings, including heat needed to heat up the calorimeter
- incomplete combustion of the methanol
- evaporation of the methanol
- evaporation of water.

Nature of Science. Scientists must be aware of the limitations of experimental evidence and assess the data obtained in terms of random and systematic errors.

TEST YOURSELF 5.1

Calculate the enthalpy change of combustion of hexane (C_6H_{14}) given that, when 1.20 g of hexane is burnt, the temperature of 250.0 g of water is raised by 56.0 °C.

Enthalpy changes in solution

A general method for measuring enthalpy changes involving solutions in the laboratory is to measure out known amounts of reagents, record their initial temperatures, mix together the reagents in a polystyrene cup (calorimeter) and record the maximum/minimum temperature.

DEFINITION

ENTHALPY CHANGE OF SOLUTION the amount of heat energy given out/taken in when 1 mole of solute dissolves in excess solvent.

Annotated exemplar answer 5.1

A student carried out an experiment to measure the enthalpy change of solution of magnesium chloride. His results are shown in the table.

volume of water ±0.5 / cm^3	50.0
initial temperature of water ±0.02 / °C	21.50
maximum temperature of solution ±0.02 / °C	28.50
mass of $MgCl_2$ ±0.01 / g	1.01

a Calculate the enthalpy change of solution of magnesium chloride. **[3]**

b State any assumptions made in the calculation. **[2]**

c The accepted literature value for the enthalpy change of solution is $-155\,kJ\,mol^{-1}$. Suggest the main source of systematic error in this experiment. **[1]**

Use significant figures as in table

a $q = mc\Delta T = 50 \times 4.18 \times 7 = 1463\,J$

moles of $MgCl_2 = \frac{1.01}{95.21} = 0.0106081294$ mol

Enthalpy change $= \frac{1463}{0.0106081294}$ i.e. 137 913.099

Don't forget to consider whether the reaction is exothermic or endothermic. The correct answer is $\Delta H = -138\,kJ\,mol^{-1}$

Use the appropriate number of significant figures – the minimum number in the data table is 3 so the answer should be given to 3 significant figures.

You need to state units otherwise the answer make no sense – this is in $J\,mol^{-1}$ – the usual units are $kJ\,mol^{-1}$

2/3

b The density and specific heat capacity of the solution are the same as that of water.

2/2

c The main error is heat loss to the surroundings

This is fine – the question says 'state' so no explanation is required.

1/1

150 × 4.18 × 3.59 = x 4.2/86.84 = y

$-(\frac{x}{y})/1000 = -46.5$

TEST YOURSELF 5.2

In an experiment to measure the enthalpy change of solution of lithium bromide ($M_r = 86.84$) 4.20 g of LiBr was dissolved in 150 cm^3 of water and the temperature change measured. If the temperature increased by 3.59 °C calculate the enthalpy change of solution of LiBr.

5.2 Hess's law

The enthalpy change accompanying a chemical reaction is independent of the pathway between the initial and final states.

Hess's law can be used to work out unknown enthalpy changes from known ones.

Worked example 5.2

Given the enthalpy changes below, calculate the enthalpy change for: $2NO_2(g) \rightarrow N_2O_4(g)$

$NO_2(g) \rightarrow \frac{1}{2}N_2(g)+O_2(g)$	**$\Delta H = -34\,kJ\,mol^{-1}$**	**reaction 1**
$N_2O_4(g) \rightarrow N_2(g)+2O_2(g)$	**$\Delta H = -10\,kJ\,mol^{-1}$**	**reaction 2**

We rearrange equations for the data given to produce the equation for the reaction whose enthalpy change we have to find. Reaction 1 is multiplied by 2 so that there are $2NO_2(g)$ on the left-hand side, as in the target equation. Reaction 2 must be reversed to give $N_2O_4(g)$ on the right-hand side as in the target equation.

$2NO_2(g) \rightarrow N_2(g)+2O_2(g)$	$\Delta H = -68\,kJ\,mol^{-1}$	multiplied by 2
$N_2(g)+2O_2(g) \rightarrow N_2O_4(g)$	$\Delta H = +10\,kJ\,mol^{-1}$	reversed so sign changed

Once we have the correct number of each species and on the correct side, the two equations (and the enthalpy changes) are then added together and common terms cancelled from each side, like in a mathematical equation:

$2NO_2(g)+\cancel{N_2}(g)+2\cancel{O_2}(g) \rightarrow \cancel{N_2}(g)+2\cancel{O_2}(g)+N_2O_4(g)$ $\qquad \Delta H = -68+10 = -58\,kJ\,mol^{-1}$

This produces the equation in the question:

$2NO_2(g) \rightarrow N_2O_4(g) \qquad \Delta H = -58\,kJ\,mol^{-1}$

TEST YOURSELF 5.3

1 Calculate the enthalpy change for: $H_2O(l) \rightarrow H_2O(g)$
given the enthalpy changes:

$H_2(g)+\frac{1}{2}O_2(g) \rightarrow H_2O(l)$	$\Delta H = -286\,kJ\,mol^{-1}$
$H_2(g)+\frac{1}{2}O_2(g) \rightarrow H_2O(g)$	$\Delta H = -242\,kJ\,mol^{-1}$

2 Calculate the enthalpy change for: $4NH_3(g)+3O_2(g) \rightarrow 2N_2(g)+6H_2O(g)$
given the enthalpy changes:

$2H_2(g)+O_2(g) \rightarrow 2H_2O(g)$	$\Delta H = -484\,kJ\,mol^{-1}$
$N_2(g)+3H_2(g) \rightarrow 2NH_3(g)$	$\Delta H = -92\,kJ\,mol^{-1}$

Working out enthalpy changes using enthalpy change of formation data

DEFINITION

STANDARD ENTHALPY CHANGE OF FORMATION the enthalpy change when one mole of the substance is formed from its elements in their standard states under standard conditions.

The equation for the enthalpy change of formation of methane is: $C(s)+2H_2(g) \rightarrow CH_4(g)$

TEST YOURSELF 5.4

Write equations to represent the standard enthalpy change of formation of the following:
$H_2O(l)$ $\quad SO_2(g)$ $\quad H_2SO_4(l)$

To solve problems where enthalpy change of formation data has been given use the equation:

$\Delta H_r^{\ominus} = \Sigma\Delta H_f^{\ominus}\text{(products)} - \Sigma\Delta H_f^{\ominus}\text{(reactants)}$

hint

The most stable form of carbon under standard conditions is graphite (solid) and this is therefore the standard state of carbon.

hint

Remember, the enthalpy change of formation of an element in its standard state is zero by definition.

Model answer 5.1

Given the standard enthalpy change of formation data in the table, calculate the enthalpy change for the reaction:
$CF_4(g) + 2H_2O(g) \rightarrow CO_2(g) + 4HF(g)$

	$\Delta H_f^\ominus$/kJ mol^{-1}
$CO_2(g)$	**−394**
$H_2O(g)$	**−242**
$CF_4(g)$	**−680**
HF(g)	**−269**

$\Delta H_r^\ominus = \Sigma\Delta H_f^\ominus(\text{products}) - \Sigma\Delta H_f^\ominus(\text{reactants})$

$\Delta H_r^\ominus = [-394 + (4 \times -269)] - [-680 + (2 \times -242)]$

$\Delta H_r^\ominus = -306\ \text{kJ mol}^{-1}$

TEST YOURSELF 5.5

Use the data in the table to calculate the enthalpy change for the reaction:
$2Pb(NO_3)_2(s) \rightarrow 2PbO(s) + 4NO_2(g) + O_2(g)$

598

	ΔH_f/kJ mol^{-1}
$NO_2(g)$	34
PbO(s)	−218
$Pb(NO_3)_2(s)$	−449

5.3 Bond enthalpies

hint

$HCl(g) \rightarrow \frac{1}{2}H_2(g) + \frac{1}{2}Cl_2(g)$ does **not** represent the bond enthalpy of HCl. $\frac{1}{2}H_2(g)$ represents half a mole of H_2 **molecules** and is not the same as H(g), which is 1 mole of gaseous H **atoms**.

DEFINITION

AVERAGE BOND ENTHALPY (BOND ENERGY) the energy required to break one mole of covalent bonds in a **gaseous** molecule under standard conditions. The value quoted is the average amount of energy to break a particular bond in a range of similar molecules.

The H–Cl bond enthalpy is represented by $HCl(g) \rightarrow H(g) + Cl(g)$ and has a value of 431 kJ mol^{-1}

Bond breaking requires energy – endothermic, ΔH +ve.

Bond making releases energy – exothermic, ΔH −ve.

Worked example 5.3

Use bond enthalpies to work out the enthalpy change for the reaction:
$N_2(g) + 3H_2(g) \rightarrow 2NH_3(g)$

Bond	**Bond enthalpy/kJ mol^{-1}**
N≡N	**945**
H–H	**436**
N–H	**388**

Draw out the molecules showing all the bonds:

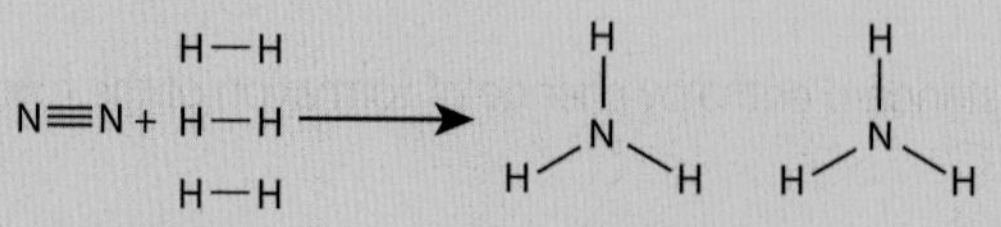

Figure 5.2

Bonds broken				Bonds made			
bond broken	bond energy/ kJ mol^{-1}	number of bonds	total energy/ kJ mol^{-1}	bond made	bond energy/ kJ mol^{-1}	number of bonds	total energy/ kJ mol^{-1}
N≡N	945	1	945	N-H	388	6	2328
H-H	436	3	1308				
energy to break all bonds			2253	energy released when bonds made			2328

$\Delta H_r = \Sigma(\text{bond broken}) - \Sigma(\text{bonds made})$

$\Delta H_r = 2253 - 2328 = -75\ \text{kJ mol}^{-1}$

Because bond energies given in tables are average values this can introduce some inaccuracies into calculations. The enthalpy change for this reaction obtained when enthalpy changes of formation (specific to particular compounds) are used is $-92\ \text{kJ mol}^{-1}$ – this is a more accurate value.

Nature of Science. The validity of models for enthalpy changes based on bond energies can be assessed by comparison with literature values. The models used can be refined by using bond enthalpies that are specific to individual compounds rather than average values.

Bond energies only apply to reactions in the gaseous state. The enthalpy change for the reaction $H_2(g) + \frac{1}{2}O_2(g) \rightarrow H_2O(l)$ could not be worked out using bond enthalpies because water is in the liquid state and intermolecular forces must be taken into account.

A reaction in the **gas phase** will be exothermic if more energy is released when bonds are formed (exothermic) than is required to break bonds (endothermic). This could be the case if stronger/more bonds are made than are broken.

TEST YOURSELF 5.6

Use bond enthalpies to work out the enthalpy change for the reaction:

$Cl_2(g) + H_2(g) \rightarrow 2HCl(g)$

Bond	Bond enthalpy/kJ mol^{-1}
Cl–Cl	242
H–H	436
H–Cl	431

5.4 Potential energy profiles

A potential energy profile shows how the potential energy of the species involved in a chemical reaction changes during that reaction.

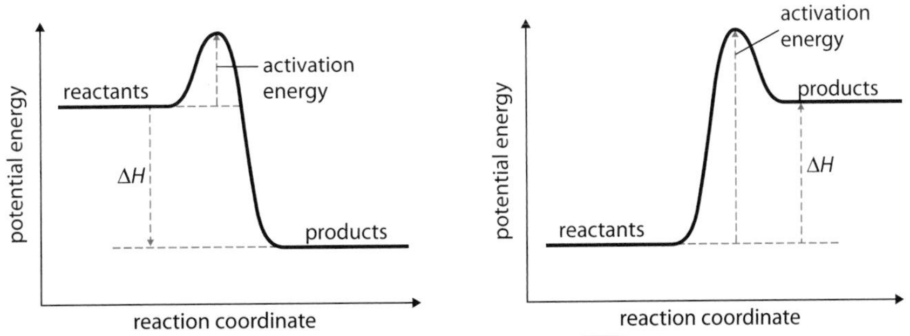

Exothermic reaction
Products have lower energy than reactants and are more stable

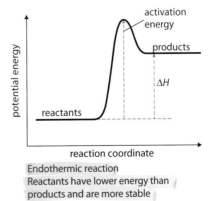

Endothermic reaction
Reactants have lower energy than products and are more stable

Figure 5.3

hint

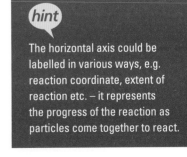

The horizontal axis could be labelled in various ways, e.g. reaction coordinate, extent of reaction etc. – it represents the progress of the reaction as particles come together to react.

Ozone and UV light

The surface of the Earth is protected from UV light by O_2 and O_3 in the atmosphere. UV radiation is absorbed by molecules of O_2 or O_3 as they undergo dissociation (O–O bond broken). O_2 absorbs higher energy (higher frequency/shorter wavelength) UV radiation than O_3 – the bond between the O atoms in an O_2 molecule is stronger (a double bond) than that in O_3 (between a double bond and a single bond).

5.5 Born–Haber cycles (HL only)

DEFINITIONS

	The enthalpy change when ...		
FIRST IONISATION ENERGY	1 electron is removed from each atom in 1 mole of gaseous atoms.	$Na(g) \rightarrow Na^+(g) + e^-$	Endothermic
SECOND IONISATION ENERGY		$Mg^+(g) \rightarrow Mg^{2+}(g) + e^-$	Endothermic
ENTHALPY CHANGE OF ATOMISATION	1 mole of gaseous atoms is formed from an element.	$Na(s) \rightarrow Na(g)$ $\frac{1}{2}Br_2(l) \rightarrow Br(g)$	Endothermic
FIRST ELECTRON AFFINITY	1 electron is added to each atom in 1 mole of gaseous atoms.	$Cl(g) + e^- \rightarrow Cl^-(g)$	Exothermic
SECOND ELECTRON AFFINITY		$O^-(g) + e^- \rightarrow O^{2-}(g)$	Endothermic
LATTICE ENTHALPY	1 mole of ionic compound is broken apart into its constituent gaseous ions.	$NH_4Cl(s) \rightarrow NH_4^+(g) + Cl^-(g)$	Endothermic

It is important to remember that '1 mole' is in each of the definitions and put 'under standard conditions' at the end of each definition if asked about a standard enthalpy change.

A Born–Haber cycle is an enthalpy level diagram that breaks down the formation of an ionic compound into a series of simpler steps.

Worked example 5.4

Draw a Born–Haber cycle for the formation of sodium chloride and use the data given to calculate the lattice enthalpy of sodium chloride.

ΔH_{at} (Na(s))	$109\,kJ\,mol^{-1}$
ΔH_{at} (Cl_2(g))	$121\,kJ\,mol^{-1}$
First ionisation energy (Na)	*$494\,kJ\,mol^{-1}$*
First electron affinity (Cl)	*$-364\,kJ\,mol^{-1}$*
ΔH_f (NaCl(s))	$-411\,kJ\,mol^{-1}$

hint

If you are not given a value for enthalpy change of atomisation for something like Cl, you can use a bond energy instead – the bond energy for Cl_2 [$Cl_2(g) \rightarrow 2Cl(g)$ $\Delta H = 242\,kJ\,mol^{-1}$] is twice the enthalpy change of atomisation [$\frac{1}{2}Cl_2(g) \rightarrow Cl(g)$ $\Delta H = 121\,kJ\,mol^{-1}$]

Upward arrows represent endothermic processes whereas downward arrows represent exothermic processes.

Using Hess's law the enthalpy change for the direct and indirect route are equal. If the indirect route takes you against the direction of an arrow in the cycle (highlighted here in green) then you must reverse the sign of this enthalpy change.

Therefore, $\Delta H_{latt} = 411 + 109 + 494 + 121 - 364$

i.e. $\Delta H_{latt} = 771\,kJ\,mol^{-1}$

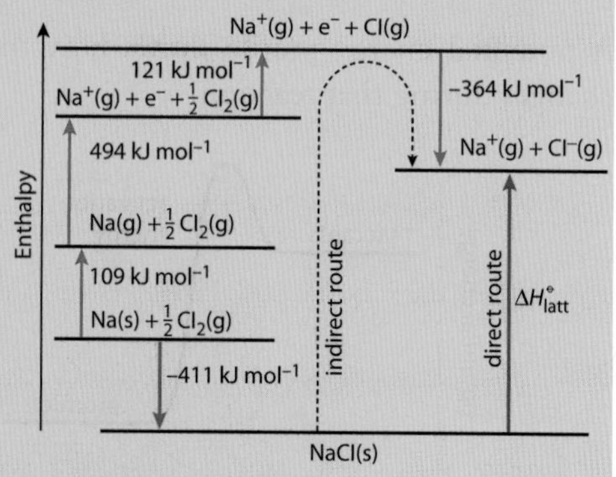

Figure 5.4

TEST YOURSELF 5.7

Use a Born–Haber cycle to calculate the lattice enthalpy of KF.

ΔH_{at} (K(s))	90 kJ mol^{-1}
ΔH_{at} (F_2(g))	79 kJ mol^{-1}
First ionisation energy (K(g))	418 kJ mol^{-1}
First electron affinity (F(g))	−348 kJ mol^{-1}
ΔH_f (KF(s))	−563 kJ mol^{-1}

802 kJ mol^{-1}

hint

State symbols and electrons are essential in the cycle.

hint

Don't forget that the definition of enthalpy change of atomisation is for formation of 1 mole of **atoms**.

5.6 Enthalpy change of solution (HL only)

Hydration enthalpy	Enthalpy change when 1 mole of gaseous ions are surrounded by water molecules to form an infinitely dilute solution	$Na^+(g) \rightarrow Na^+(aq)$ $Cl^-(g) \rightarrow Cl^-(aq)$	exothermic
Enthalpy change of solution	Enthalpy change when one mole of solute is dissolved in excess solvent to form a solution of infinite dilution	$NaCl(s) \rightarrow Na^+(aq)+Cl^-(aq)$ $MgCl_2(s) \rightarrow Mg^{2+}(aq)+2Cl^-(aq)$ $NH_4Cl(s) \rightarrow NH_4^+(aq)+Cl^-(aq)$	endothermic or exothermic

Worked example 5.5

Use the values given in the table to calculate the enthalpy change of solution of calcium chloride.

ΔH_{latt} ($CaCl_2$(s))	**2271 kJ mol^{-1}**
ΔH_{hyd} (Cl^-(g))	**−359 kJ mol^{-1}**
ΔH_{hyd} (Ca^{2+}(g))	**−1616 kJ mol^{-1}**

The first step is to draw a cycle – put what you are trying to find at the top of the cycle and then put in the information from the table – draw the arrows in the same direction as the definitions, e.g. breaking apart the lattice and forming hydrated ions.

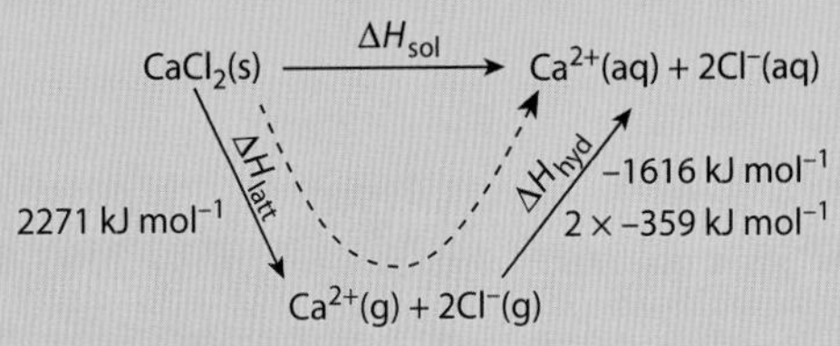

Figure 5.5

$\Delta H_{sol} = \Delta H_{latt}(CaCl_2(s)) + \Delta H_{hyd}(Ca^{2+}(g)) + 2 \times \Delta H_{hyd}(Cl^-(g))$

$\Delta H_{sol} = 2271 + [-1616 + 2 \times -359] = -63\ kJ\ mol^{-1}$

TEST YOURSELF 5.8

Calculate the lattice enthalpy of strontium bromide.

ΔH_{sol} ($SrBr_2$(s))	−72 kJ mol^{-1}
ΔH_{hyd} (Br^-(g))	−328 kJ mol^{-1}
ΔH_{hyd} (Sr^{2+}(g))	−1483 kJ mol^{-1}

2067 kJ mol^{-1}

hint

Lattice enthalpy can also be defined in the opposite direction – forming the lattice – this would be exothermic.

Sizes of lattice enthalpies and enthalpy changes of hydration

Lattice enthalpy is the result of electrostatic attractions between oppositely charged ions in the giant lattice. As such it depends on:

- **Charge** – the higher the charge on the ions, the more strongly they will attract each other and, therefore, the greater the lattice enthalpy: 2+/2− > 2+/1− > 1+/1−.
- **Size** – the smaller the ions the stronger the attraction between them and therefore the larger the lattice enthalpy.

The effect of the charge on the ions is a much larger effect than size variations.

Hydration enthalpies also depend on charge and size of ions:

- The higher the charge the more exothermic the enthalpy change of hydration.
- The smaller the ionic radius the more exothermic the enthalpy change of hydration.

Higher charge and smaller size result in stronger attraction between the ions and water molecules. Again, charge is more important.

TEST YOURSELF 5.9

Arrange the following in order of increasing lattice enthalpy:

KBr 1. CaO 4

CaCl 3 LiF 2

MgO 5.

5.7 Entropy (HL only)

DEFINITION

ENTROPY is a measure of how the available energy is distributed among the particles.

Entropy can also be related to the disorder in a system – the more disordered the system, the more ways there are of distributing the energy and the higher the entropy.

S° is called the **standard entropy.** It is measured in $J\,K^{-1}\,mol^{-1}$ (note: not kJ).

Entropy of a solid < entropy of a liquid < entropy of a gas

The particles in a gas are moving around more randomly and the energy can be distributed in more ways – higher entropy.

It can usually be worked out whether a chemical reaction involves an increase or a decrease in entropy by looking at changes in the numbers of moles of gas:

Increase in number of moles of **gas**: ΔS +ve (entropy increases).

Decrease in number of moles of **gas**: ΔS −ve (entropy decreases).

Worked example 5.6

For which of the following reactions is ΔS positive?

A $2H_2S(g)+SO_2(g) \rightarrow 3S(s)+H_2O(l)$

B $SO_2(g)+\frac{1}{2}O_2(g) \rightarrow SO_3(g)$

C $4NH_3(g)+5O_2(g) \rightarrow 4NO(g)+6H_2O(l)$

D $2Ca(NO_3)_2(s) \rightarrow 2CaO(s)+4NO_2(g)+O_2(g)$

Because the question is looking for a reaction in which ΔS is positive we are looking for an increase in the number of moles of gas from the left to the right – you should work out the change in the number of moles of gas for each reaction:

A 3 moles of gas to 0 moles of gas ⇒ decrease in number of moles of gas: ΔS is negative

B 1½ moles of gas to 1 mole of gas ⇒ decrease in number of moles of gas: ΔS is negative

C 9 moles of gas to 4 moles of gas ⇒ decrease in number of moles of gas: ΔS is negative

D 0 moles of gas to 5 moles of gas ⇒ increase in number of moles of gas: ΔS is positive

The answer is, therefore D.

TEST YOURSELF 5.10

State whether each of the reactions below involves an increase or decrease in entropy.

$NH_3(g)+HCl(g) \rightarrow NH_4Cl(s)$ decrease

$C_2H_5OH(g)+3O_2(g) \rightarrow 2CO_2(g)+3H_2O(l)$ decrease

Calculating an entropy change for a reaction

Values of $\Delta S^\ominus$ may be worked out from standard entropies, $S^\ominus$:

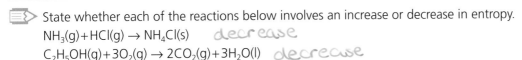

$$\Delta S^\ominus = \Sigma S^\ominus_{products} - \Sigma S^\ominus_{reactants}$$

Model answer 5.2

Calculate the standard entropy change for the following reaction:

$2H_2(g)+O_2(g) \rightarrow 2H_2O(l)$

	$S^\ominus$ / J K^{-1} mol^{-1}
$O_2(g)$	205
$H_2(g)$	131
$H_2O(l)$	70

$$\Delta S^\ominus = \Sigma S^\ominus_{products} - \Sigma S^\ominus_{reactants}$$

$$\Delta S^\ominus = (2\times70)-[(2\times131)+205] = -327\ \text{J K}^{-1}\ \text{mol}^{-1}$$

hint

Don't forget to multiply the entropy value by the coefficient.

hint

Check to see whether your answer makes sense – does the reaction involve an increase or a decrease in entropy?

TEST YOURSELF 5.11

Calculate the standard entropy change for the reaction:

$2C_6H_6(l)+15O_2(g) \rightarrow 12CO_2(g)+6H_2O(l)$

	$S^\ominus$ / J K^{-1} mol^{-1}
$C_6H_6(l)$	173
$CO_2(g)$	214
$H_2O(l)$	70
$O_2(g)$	205

-433 JK⁻ mol⁻¹

5.8 Predicting whether a reaction will be spontaneous (HL only)

DEFINITION

SPONTANEOUS REACTION one that occurs without any outside influence.

For a reaction to be spontaneous ΔG (the change in Gibbs free energy) for the reaction must be **negative**. ΔG can be worked out using the equation:

$\Delta G = \Delta H - T\Delta S$

☆ Model answer 5.3

hint

T must be in K and you must take care with the units – *S* is usually in J but ΔH is in kJ – remember to divide ΔS by 1000.

Consider the following reaction:

$CH_4(g) + 2O_2(g) \rightarrow CO_2(g) + 2H_2O(l)$ $\quad \Delta H^{\ominus} = -890\,kJ\,mol^{-1}$ $\quad \Delta S^{\ominus} = -242\,J\,K^{-1}\,mol^{-1}$

Calculate $\Delta G^{\ominus}$ and hence predict whether the reaction will be spontaneous at 25°C

$\Delta G = \Delta H - T\Delta S$

$\Delta G^{\ominus} = -890 - 298 \times \left(\frac{-242}{1000}\right) = -818\,kJ\,mol^{-1}$

The value of $\Delta G^{\ominus}$ is negative and so the reaction is spontaneous at this temperature.

TEST YOURSELF 5.12

Given that $\Delta H^{\ominus} = -2220\,kJ\,mol^{-1}$ and $\Delta S^{\ominus} = -370\,J\,K^{-1}\,mol^{-1}$, calculate $\Delta G^{\ominus}$ for the following reaction and state whether it is spontaneous or not at 298 K

$C_3H_8(g) + 5O_2(g) \rightarrow 3CO_2(g) + 4H_2O(l)$ -2110 kJ mol⁻¹ → spontaneous

The effect of temperature on the spontaneity of a reaction

Reactions for which ΔS is positive will get more spontaneous as temperature increases. If ΔS is positive, $-T\Delta S$ will be negative and will become more negative as T increases. This means that ΔG will become more negative as T increases.

TEST YOURSELF 5.13

Will the following reaction be spontaneous at high or low temperatures?

$2Pb(NO_3)_2(s) \rightarrow 2PbO(s) + 4NO_2(g) + O_2(g)$

$\Delta H^{\ominus} = +600\,kJ\,mol^{-1}$ $\quad \Delta S^{\ominus} = 900\,J\,K^{-1}\,mol^{-1}$ high temperatures

ΔG and the position of equilibrium

The equilibrium mixture always has a lower Gibbs free energy than either the pure reactants or the pure products. Therefore, the conversion of either pure reactants or pure products into the equilibrium mixture results in a process for which ΔG is negative – i.e. a spontaneous process.

The more negative the value of ΔG for a reaction, the closer the position of equilibrium lies towards the products; the more positive the value, the closer the position of equilibrium lies towards the reactants.

Checklist

At the end of this chapter you should be able to:

- ☐ Explain the difference between temperature and heat.
- ☐ Work out the enthalpy change for a chemical reaction from experimental data.
- ☐ Calculate enthalpy changes using Hess's law.
- ☐ Calculate enthalpy changes from bond energy or enthalpy change of formation data.
- ☐ Understand why O_3 is dissociated by lower energy UV radiation than O_2.

Higher Level only

- ☐ Use Born–Haber cycles to calculate enthalpy changes involving ionic compounds.
- ☐ Use enthalpy cycles to calculate enthalpy changes in solution.
- ☐ Explain what is meant by entropy and work out entropy changes in a reaction.
- ☐ Calculate ΔG values and predict the spontaneity of a reaction.
- ☐ Explain how ΔG varies with temperature and the relationship between ΔG and the position of equilibrium.

6

CHEMICAL KINETICS

It is essential to know not just how favourable a chemical reaction is but also how fast it goes. Chemical kinetics is the study of the rate of chemical reactions.

This chapter covers the following topics:

- ☐ Rates of reaction
- ☐ Collision theory
- ☐ Factors that affect the rate of reaction
- ☐ The rate equation (HL only)
- ☐ Reaction mechanisms (HL only)
- ☐ The Arrhenius equation (HL only)

6.1 Rate of reaction

DEFINITION

RATE OF REACTION $\dfrac{\text{change in concentration of a particular reactant or product}}{\text{time}}$

The units of rate are concentration time^{-1}, e.g. mol dm^{-3} s^{-1}

For the reaction A → B a graph of concentration against time could be:

hint

Take care – other units for time, such as minutes (min) or hours (h) could also be used.

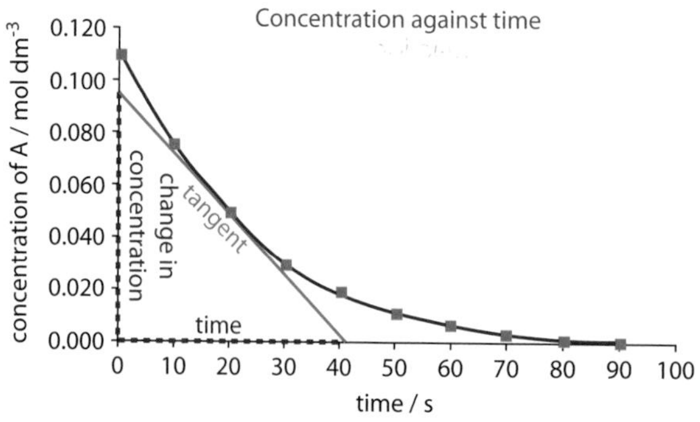

Figure 6.1

The rate of reaction at 20 s can be found by drawing a tangent at that point.

The gradient of the tangent is equal to the rate of reaction

$\text{Rate} = \dfrac{\text{change in concentration}}{\text{time}}$, i.e. rate $= \dfrac{0.095}{42}$ i.e. 2.3×10^{-3} mol dm^{-3} s^{-1}

Measuring rate of reaction experimentally

There are many variations on experiments that you could be faced with and the important thing is to try to work out how the rate of the reaction is being followed – look at the information given – is a gas given off (this could be followed by collecting the gas or monitoring the decrease in mass as the gas is given off), is there something in the reaction that is coloured (common examples are bromine and iodine)? Reactions involving a coloured substance can be followed using a colorimeter.

TEST YOURSELF 6.1

Suggest a method for measuring the rate of each of the following reactions:

$(CH_3)_2CO(aq) + I_2(aq) \xrightarrow{H^+} CH_3COCH_2I(aq) + HI(aq)$ calorimeter

$2H_2O_2(aq) \xrightarrow{MnO_2} 2H_2O(l) + O_2(g)$ gas collection / mass change

6.2 Collision theory

For a reaction to occur, particles must **collide** with more than a **minimum amount of energy** (the activation energy) and with the **correct orientation**.

This means that not every collision will result in a reaction – a collision that results in a reaction is called a **successful** or **effective** collision.

DEFINITION

ACTIVATION ENERGY (E_a) the minimum amount of energy that colliding particles must possess for a collision to result in a reaction.

6.3 Factors that affect the rate of reaction

Concentration of reactants	higher concentration usually results in a faster rate of reaction	with more particles in a certain volume the particles collide more frequently
Pressure for reactions involving gases	higher pressure for gases usually results in a faster rate of reaction	
Surface area of solid reactants (particle size)	increasing the surface area of a solid reactant usually increases the rate of reaction	making a solid more finely divided (e.g. a powder) increases the surface area – with more particles exposed at the surface of the solid there will be more frequent collisions
Temperature	increasing the temperature increases the rate of reaction	increase in the number of particles with energy greater than or equal to the activation energy results in more successful collisions per unit time
Catalyst	the presence of a catalyst increases the rate of reaction	a catalyst provides an alternative reaction pathway of lower activation energy – more particles have energy greater than or equal to the activation energy, therefore there are more successful collisions per unit time

Temperature has a large effect on rate – for most reactions we study in this course a rise in temperature of 10 K causes the reaction rate to be **approximately** doubled.

hint

Any explanations involving the rate of reaction must involve the idea of 'time' – it is not enough to state the particles collide more – you must say that they collide more frequently.

The kinetic theory

The average kinetic energy of the particles in an ideal gas is proportional to the temperature in kelvin.

There are a few important consequences of this:

- at a higher temperature the particles in a gas, on average, move faster
- at temperatures around room temperature (298 K) it takes a large increase in temperature to **significantly** increase the speed at which particles move.
- at the same temperature lighter particles move faster than heavier ones ($KE = \frac{1}{2}mv^2$)

The Maxwell–Boltzmann distribution

This shows the distribution of molecular kinetic energies at a particular temperature for a sample of gas.

Tips for drawing the graph:

- it is **not** symmetrical
- it passes through the origin (no molecules have zero kinetic energy)
- At higher energy the line does not reach the energy axis
- The vertical axis may also be labelled as 'number of particles with a certain amount of energy' or 'proportion of particles with a certain amount of energy'.

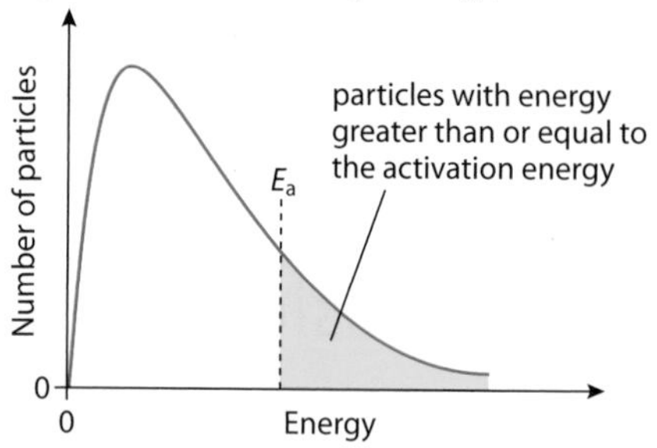

Figure 6.2

The effect of temperature on the rate of reaction

hint

Make sure that you draw the Maxwell–Boltzmann distributions carefully – the maximum for the higher temperature curve must be to the right of that for the lower temperature curve and lower.

The main reason that rate of reaction increases with temperature is:

at the higher temperature more particles have energy greater than or equal to the activation energy – this results in more successful collisions per unit time.

The particles also collide more often at a higher temperature but this is only a small effect.

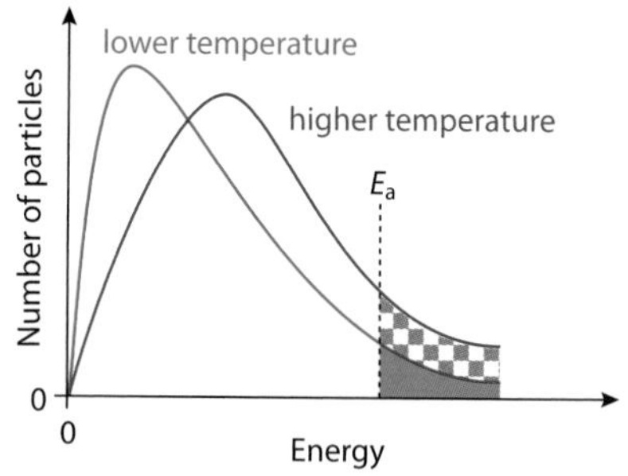

Figure 6.3

Annotated exemplar answer 6.1

Explain, using the Maxwell–Boltzmann distribution, why increasing the temperature causes the rate of reaction to increase. **[4]**

At higher temperatures, particles have more energy so they move faster and collide more. More particles have enough energy to react so there are more effective collisions. On the Maxwell-Boltzmann distribution there are more particles with energy greater than the activation energy.

Yes, but the idea of time must be included – should be 'collide more frequently'

1/4

The increase in collision rate only has a small effect on the rate of the reaction – the increase in the number of particles with energy greater than or equal to the activation energy is more important.

Since the question mentions the Maxwell-Boltzmann distribution it must be drawn

Catalysis

DEFINITION

CATALYST a substance that increases the rate of a chemical reaction without, itself, being used up in the reaction.

A catalyst acts by allowing the reaction to proceed by an alternative pathway of lower activation energy. Therefore, more particles have energy greater than the activation energy and there are more successful collisions per unit time.

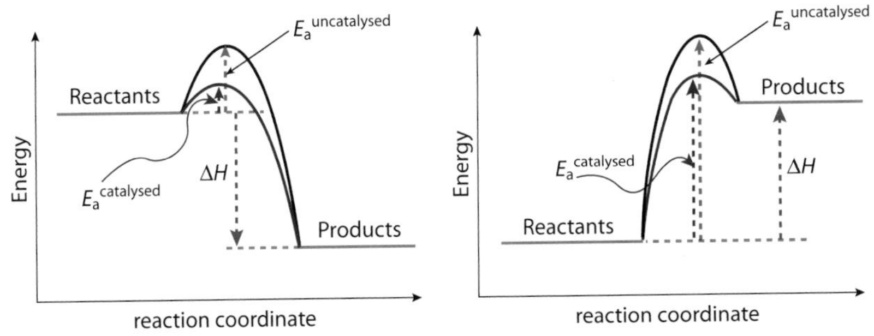

Exothermic reaction | Endothermic reaction

Figure 6.4

hint

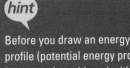

Before you draw an energy profile (potential energy profile) for a reaction with and without a catalyst look to see whether the reaction is exothermic or endothermic

6.4 The rate equation (rate expression) (HL only)

DEFINITIONS

RATE EQUATION an **experimentally-determined** equation that relates the rate of reaction to the concentrations of substances in the reaction mixture.

RATE CONSTANT (k) a constant of proportionality relating the concentrations in the experimentally-determined rate expression to the rate of a chemical reaction. The rate constant is only a constant for a particular reaction at a particular temperature.

ORDER OF REACTION (WITH RESPECT TO A PARTICULAR REACTANT) the power of the reactant's concentration in the experimentally-determined rate equation.

Example of a rate equation: Rate = $k[A]^m[B]^n$

In this equation the order with respect to A is m and that with respect to B is n. The **overall order** of reaction is the **sum** ($m+n$ here) of the powers of the concentration terms in the experimentally-determined rate equation.

hint

When writing a rate equation you must write 'rate =' and not just put the right-hand side. Also, use a small 'k' for the rate constant.

It is important to remember that the rate equation can only be determined from experimental data – i.e. from a series of experiments in which the effect of changing the concentration of the reactants on the rate of reaction is investigated – there is no connection between the chemical equation for a reaction and the rate equation. The reason for this is that we do not know the mechanism for the reaction – the reaction may not occur in a single step.

Worked example 6.1

The data in the table shows the effect of changing the concentrations of A and B on the rate of the reaction: 2A+2B → C+D

Experiment	[A]/mol dm^{-3}	[B]/mol dm^{-3}	Rate/mol dm^{-3} h^{-1}
1	0.10	0.10	0.10
2	0.50	0.10	2.50
3	0.50	0.30	7.50

a Determine the order with respect to A and B and hence the rate equation. Explain your reasoning.

b Determine the value of the rate constant with units.

a To determine the order with respect to A consider experiments 1 and 2 because the only thing that changes between these two experiments is the concentration of A.

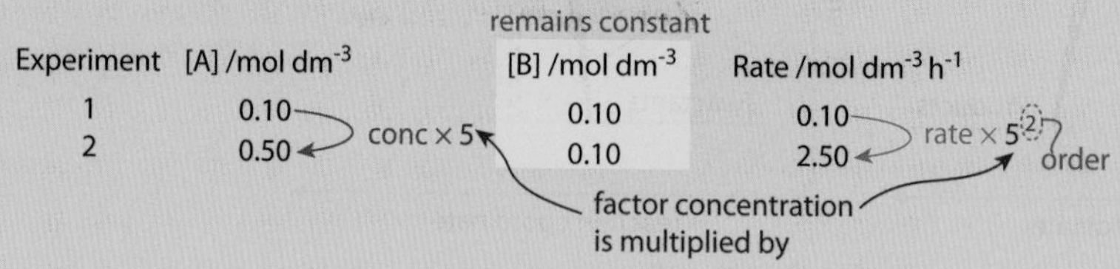

Figure 6.5

From experiment 1 to 2, the concentration of A is multiplied by a factor of 5 and the rate goes up by a factor of 25, i.e. 5^2. This means that the order with respect to A is 2, i.e. the reaction is second order with respect to A.

To determine the order with respect to B consider experiments 2 and 3 because the concentration of A remains constant.

From experiment 2 to 3, the concentration of B is multiplied by a factor of 3 and the rate increases by a factor of 3, i.e. 3^1. This means that the order with respect to B is 1, i.e. the reaction is first order with respect to B.

The rate expression is: Rate = $k[A]^2[B]^1$

b The value of the rate constant can be calculated by substituting values from **one** experiment into the rate equation.

From experiment 2: Rate = $k[A]^2[B]^1$ $2.50 = k \times 0.50^2 \times 0.10$ i.e. $k = 100$

To work out the units for k the units of the concentration and the rate must be substituted into the rate expression: $\text{mol dm}^{-3}\text{ h}^{-1} = k\,(\text{mol dm}^{-3})^2 \times \text{mol dm}^{-3}$

mol dm^{-3} is cancelled from each side: $\cancel{\text{mol dm}^{-3}}\text{ h}^{-1} = k\,(\text{mol dm}^{-3})^2 \times \cancel{\text{mol dm}^{-3}}$

Rearranging: $$\frac{\text{h}^{-1}}{(\text{mol dm}^{-3})^2} = k$$ i.e. $k = \text{mol}^{-2}\text{ dm}^6\text{ h}^{-1}$

The value of the rate constant, k, is 100 mol^{-2} dm^6 h^{-1}

TEST YOURSELF 6.2

1 Determine the rate equation and calculate the rate constant (including units) for the reaction A+B → P+Q given the data in the table: r = k[A][B]^2 k = 2.0×10^{-3}

[A]/mol dm^{-3}	[B]/mol dm^{-3}	Initial rate/mol dm^{-3} s^{-1}
1.0	1.0	2.0×10^{-3}
2.0	1.0	4.0×10^{-3}
2.0	2.0	16×10^{-3}

2 Determine the rate equation and calculate the rate constant (including units) for the reaction 2A+B → D+E given the data in the table: r = k[A^2][k = 2.5

[A]/mol dm^{-3}	[B]/mol dm^{-3}	Initial rate/mol dm^{-3} h^{-1}
0.10	0.20	2.5×10^{-2}
0.40	0.20	4.0×10^{-1}
0.40	0.60	4.0×10^{-1}

Graphs for zero, first and second order reactions (HL only)

You need to be able to recognise/sketch graphs for zero, first and second order reactions:

Zero order reaction

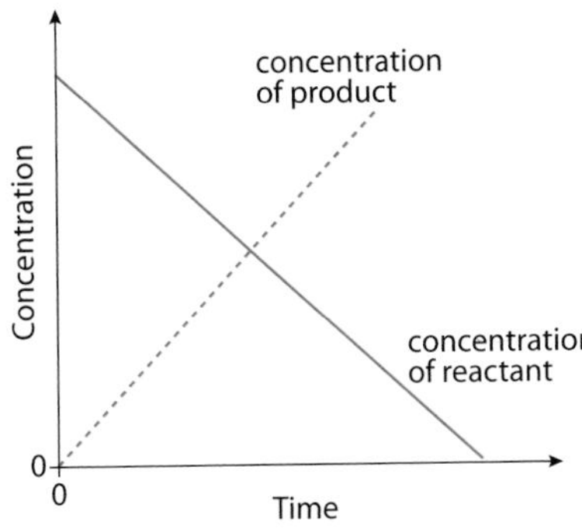

Figure 6.6

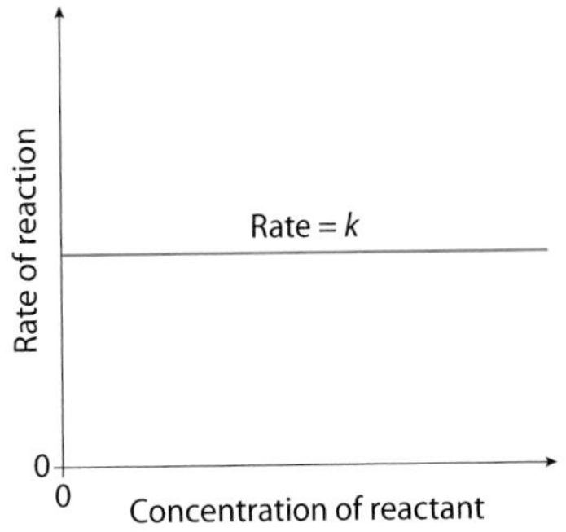

Figure 6.7

First order reaction

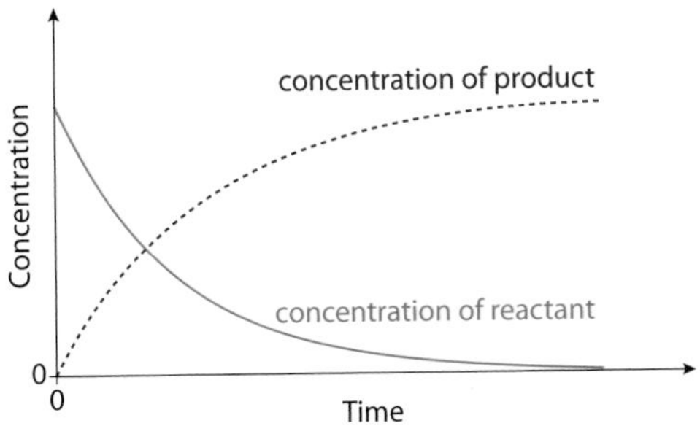

Figure 6.8

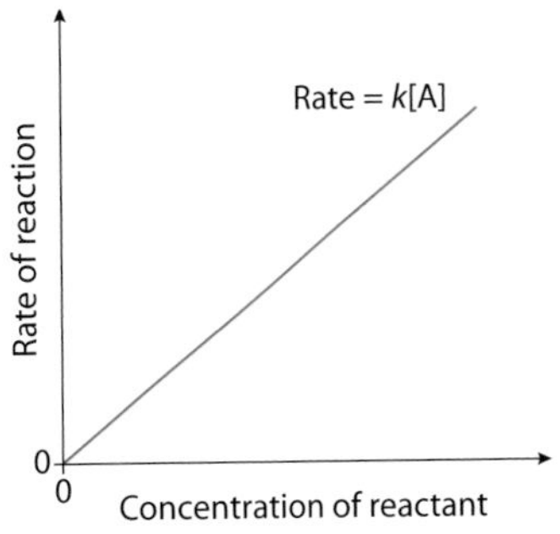

Figure 6.9

Second order reaction

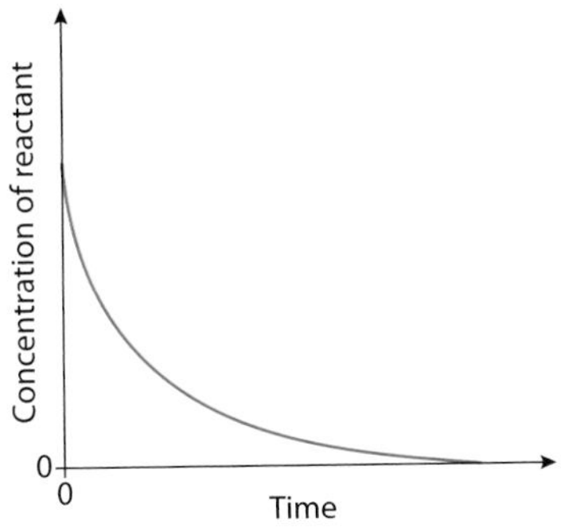

Figure 6.10

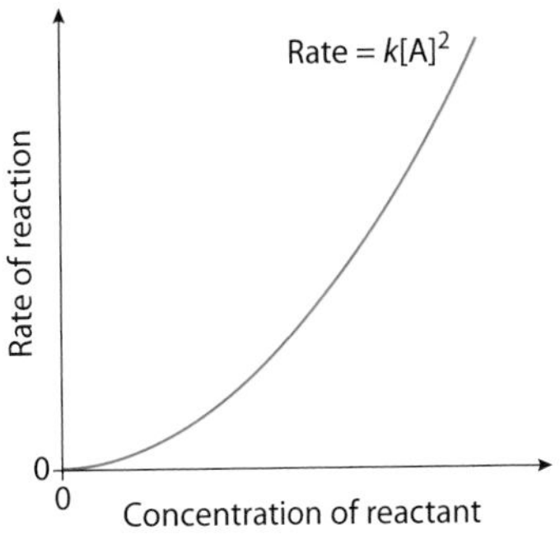

Figure 6.11

6.5 Reaction mechanisms (HL only)

A reaction mechanism shows all the simple steps that make up a more complex reaction.

DEFINITIONS

RATE DETERMINING STEP (RDS) the slowest step (step with highest E_a) in a reaction mechanism.

MOLECULARITY the number of reactant particles taking part in a particular step (usually the RDS) in a reaction mechanism.

You may be asked to work out or evaluate a given reaction mechanism and determine whether it agrees with the rate equation and the stoichiometric equation.

Tips for evaluating mechanisms:

1. Each step must involve a **maximum of two particles** colliding – it is statistically highly unlikely that more than two particles would ever come together at exactly the same time.
2. The mechanism must agree with the stoichiometric equation – when all the individual equations are added together, cancelling out like terms, you must end up with the overall chemical equation.
3. The rate equation contains concentrations of reactants involved in the mechanism **up to and including the rate determining step**.
4. The order of a reactant in the rate equation indicates the number of times the reactant appears in the mechanism up to and including the rate determining step.

Worked example 6.2

Four possible reaction mechanisms for the reaction $Q+2Z \rightarrow E$ are shown here. The rate equation for this reaction is rate = k[Q][Z]. Which mechanism is most likely to be correct?

A	$Z+Z \rightleftharpoons R$	**fast**	**Step 1**
	$R+Q \rightarrow E$	**rate determining step**	**Step 2**
B	$Q+Z \rightarrow X$	**rate determining step**	**Step 1**
	$X+Q \rightarrow E$	**fast**	**Step 2**
C	$Q+Z+Z \rightarrow E$	**rate determining step**	**Step 1**
D	$Q+Z \rightarrow M$	**rate determining step**	**Step 1**
	$M+Z \rightarrow E$	**fast**	**Step 2**

A is not correct because there are two molecules of Z and 1 molecule of Q involved up to and including the rate determining step and so the rate equation rate = $k[Q][Z]^2$ would be expected, which does not agree with the experimental rate equation.

B is not correct because it does not agree with the stoichiometric equation – when the two equations are added together and common terms cancelled, the equation: $2Q+Z \rightarrow E$ is obtained.

C is not correct because it shows three molecules colliding in one step – this is statistically very unlikely.

D is correct – there is one molecule of Q and 1 molecule of Z involved up to and including the rate determining step, which will lead to the rate equation, rate = k[Q][Z], which is the same as the experimental rate equation.

TEST YOURSELF 6.3

1 Work out the rate equation for the reaction $P+2Q \rightarrow R+S$ given the mechanism:

Step 1	$Q+P \rightarrow Y+S$	**rate determining step**
Step 2	$Q+Y \rightarrow R$	**fast**

r = k[Q][P]

2 The rate equation for the reaction $2P+2Q \rightarrow Y+Z$ is Rate = $[Q]^2$. Select the most appropriate mechanism:

A	Step 1	$Q+P \rightarrow X$	**rate determining step**
	Step 2	$X+P \rightarrow Y+Z$	**fast**
B	Step 1	$P+P \rightleftharpoons R$	**fast**
	Step 2	$Q+R \rightleftharpoons Y+L$	**fast**
	Step 3	$L+Q \rightarrow Z$	**rate determining step**

C	Step 1	$Q+Q \rightarrow S$	**rate determining step**
	Step 2	$S+P \rightarrow Y+T$	**fast**
	Step 3	$T+P \rightarrow Z$	**fast**
D	Step 1	$Q+Q \rightarrow B$	**rate determining step**
	Step 2	$B+P+P \rightarrow Z+Y$	**fast**

Nature of Science. The principle of Occam's razor can be used to decide between theories – if two theories have equal explaining power, the simpler is generally chosen as being more likely. This can be useful when deciding between reaction mechanisms.

Catalysts and reaction mechanisms

A catalyst...

- is involved in the rate determining step
- introduces a step with lower activation energy
- allows the reaction to occur by a different mechanism.

6.6 The Arrhenius equation (HL only)

The **Arrhenius equation** shows the variation of the rate constant with temperature. As temperature increases the rate constant increases (exponentially).

$$k = Ae^{\frac{-E_a}{RT}}$$

A is called the frequency factor or pre-exponential factor and is related to the orientation of the collisions and the frequency of collisions. A is essentially a constant for a particular reaction as temperature varies.

Generally, the more complex the molecules, the smaller the value of A – fewer collisions with energy greater than or equal to the activation energy will result in a reaction because it is less likely that complex molecules will collide in the correct orientation, compared to simpler molecules.

R is the gas constant, i.e. 8.31 J K^{-1} mol^{-1}

T is the temperature in kelvin.

The Arrhenius equation may also be used in the form: $\ln k = \frac{-E_a}{R} \times \frac{1}{T} + \ln A$

The Arrhenius equation in this form can be used to work out a value for the activation energy.

hint

Both forms of the Arrhenius equation are shown in the *IB Data Booklet*.

Worked example 6.3

A student conducted a series of experiments at a range of temperatures and calculated a rate constant for each temperature. He then plotted a graph of ln *k* (*y*-axis) against $\frac{1}{T}$ (*x*-axis) – shown here. Determine the activation energy and frequency factor for this reaction.

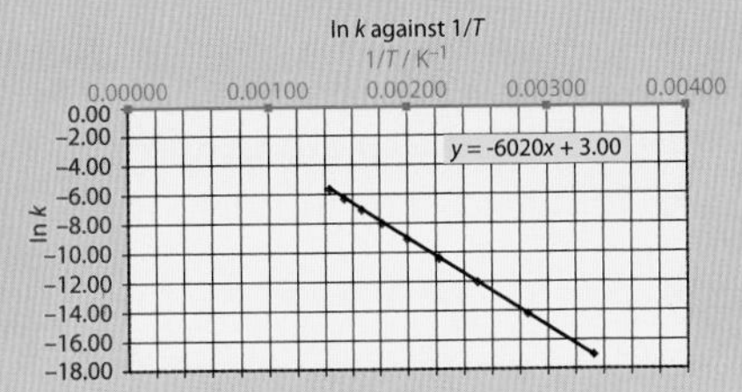

Figure 6.12

The equation for the line is displayed on the graph:

gradient = −6020 K

$$\ln k = \frac{-E_a}{R} \times \frac{1}{T} + \ln A$$

This is the equation of a straight line $y = mx + c$

The gradient of the graph equal to $\frac{-E_a}{R}$

$-6020 = \frac{-E_a}{R}$ $\quad$ $6020 \times 8.31 = E_a$ $\quad$ $E_a = 50\,000\ \text{J mol}^{-1}$ $\quad$ i.e. $E_a = 50.0\ \text{kJ mol}^{-1}$

Determine the intercept of the line on the *y*-axis (ln*k* axis) – this is equal to ln*A*.

From the equation of the line on the graph: $\ln A = 3.00$ therefore $A = 20.1$

The units of *A* would be the same as those of *k*, the rate constant.

A can also be determined by putting known values into the Arrhenius equation.

The activation energy can also be determined using a numerical method, by putting numbers into the equation (given in the *IB Data Booklet*):

$$\ln \frac{k_1}{k_2} = \frac{-E_a}{R}\left(\frac{1}{T_2} - \frac{1}{T_2}\right)$$

TEST YOURSELF 6.4

1 Calculate the activation energy for a reaction given that the gradient of a graph of ln*k* against $\frac{1}{T}$ is –5000 K. 41·6

2 Given that the value of the rate constant for a first order reaction is $2.2 \times 10^{-3}\ \text{s}^{-1}$ at 298 K and the value at 325 K is $9.3 \times 10^{-3}\ \text{s}^{-1}$, calculate the value of the activation energy for the reaction. 43

Checklist

At the end of this chapter you should be able to:

- ☐ **Define rate of reaction and calculate it from graphical or numerical data.**
- ☐ **Suggest methods to measure the rate of reaction.**
- ☐ **Explain collision theory.**
- ☐ **Describe and explain the factors that affect the rate of a chemical reaction.**

Higher Level only

- ☐ **Deduce the rate equation and calculate a value for the rate constant from experimental data.**
- ☐ **Recognise and explain graphs of concentration against time and rate against concentration for zero, first and second order reactions.**
- ☐ **Evaluate reaction mechanisms.**
- ☐ **Work out the activation energy and value of the frequency factor (*A*) from graphical and numerical data.**

7

EQUILIBRIUM

Many chemical reactions are reversible and come to a state of equilibrium. Changing the reaction conditions can affect the position of equilibrium and the amount of product formed in an industrial process.

This chapter covers the following topics:

- ☐ The equilibrium state
- ☐ The equilibrium constant
- ☐ The position of equilibrium and Le Chatelier's principle
- ☐ Calculations involving equilibrium constants (HL only)
- ☐ Explaining Le Chatelier's principle (HL only)
- ☐ Equilibrium constants and ΔG (HL only)

7.1 The equilibrium state

Chemical equilibrium

DEFINITIONS

REVERSIBLE REACTION one that can go in either direction. A reversible reaction can come to a state of equilibrium.

DYNAMIC EQUILIBRIUM macroscopic properties are constant (concentrations of all reactants and products remain constant). The rate of the forward reaction is equal to the rate of the reverse reaction.

CLOSED SYSTEM no exchange of matter with the surroundings – equilibrium can only be attained in a closed system.

hint

When asked to explain what is meant by dynamic equilibrium, don't forget to address both words – 'dynamic' and 'equilibrium'.

hint

Be careful with the word 'equal' – the concentrations of reactants and products are constant at equilibrium, they are not equal.

Physical equilibrium (phase equilibrium)

DEFINITION

PHYSICAL EQUILIBRIUM equilibrium involving a physical process such as melting/freezing or evaporating/condensing.

If a **volatile** liquid, such as bromine, is put in a closed container it will reach a state of equilibrium (macroscopic properties constant) when the rate of evaporation is equal to the rate of condensation.

7.2 The equilibrium constant, K_c

For the reaction: $aA + bB \rightleftharpoons cC + dD$

we can write an expression for the equilibrium constant: $K_c = \dfrac{[C]^c[D]^d}{[A]^a[B]^b}$

where [A] represents the concentration of A **at equilibrium**. The products are always on the top and the reactants are on the bottom. The coefficients of the species in the chemical equation become the powers of the concentrations in the equilibrium constant expression.

e.g. for $N_2(g) + 3H_2(g) \rightleftharpoons 2NH_3(g)$ $\quad K_c = \dfrac{[NH_3(g)]^2}{[N_2(g)][H_2(g)]^3}$

K_c is constant for a particular reaction at a particular temperature.

1. $Kc = \dfrac{[CO_2][H_2]^4}{[CH_4][H_2O]^2}$

2. $Kc = \dfrac{[SO_3]^2}{[O_2][SO_2]^2}$

TEST YOURSELF 7.1

Write expressions for the equilibrium constant, K_c, for each of the following:

1 $CH_4(g) + 2H_2O(g) \rightleftharpoons CO_2(g) + 4H_2(g)$ **2** $2SO_2(g) + O_2(g) \rightleftharpoons 2SO_3(g)$

Different values of the equilibrium constant for the same reaction under the same conditions

The expression (and value) of the equilibrium constant depends on how the chemical equation is written, e.g.

$$2NO_2(g) \rightleftharpoons N_2O_4(g) \quad K_c = \frac{[N_2O_4(g)]}{[NO_2(g)]^2}$$

The value for the equilibrium constant at 400 K is 0.69

However, the reaction could also have been written the other way round, i.e. $N_2O_4(g) \rightleftharpoons 2NO_2(g)$

The expression for the equilibrium constant (K_c') for this reaction is:

$$K_c' = \frac{[NO_2(g)]^2}{[N_2O_4(g)]} \qquad K_c' = \frac{1}{K_c}$$

The value of K_c' at 400 K is $\dfrac{1}{0.69}$ i.e. 1.45

hint

The equilibrium constant expression is only meaningful in conjunction with the stoichiometric equation.

TEST YOURSELF 7.2

The equilibrium constant for the reaction $2SO_2(g) + O_2(g) \rightleftharpoons 2SO_3(g)$ is K_c. The equilibrium constant for the reaction $SO_3(g) \rightleftharpoons SO_2(g) + \frac{1}{2}O_2(g)$ is K_c'. The relationship between K_c and K_c' is:

A $K_c = \sqrt{K_c'}$ **B** $\dfrac{1}{(K_c)^2} = K_c'$ **C** $\dfrac{1}{K_c'} = \sqrt{K_c}$ **D** $\dfrac{1}{K_c} = K_c'$

What use is the equilibrium constant?

The equilibrium constant provides information about how far a reaction proceeds at a particular temperature. In general, you can assume that, for a particular reaction, the larger the value of the equilibrium constant the more the position of equilibrium lies towards the right.

$K_c >> 1$ – the reaction proceeds almost totally towards products.

$K_c << 1$ – the reaction hardly proceeds at all towards products.

7.3 The position of equilibrium

The position of equilibrium refers to the relative amounts of reactants and products present at equilibrium.

e.g. $2NO(g) \rightleftharpoons N_2(g) + O_2(g)$

At 700 K the position of equilibrium lies a long way to the right – there is lots of N_2 and O_2 and very little NO at equilibrium.

Le Chatelier's principle

DEFINITION

LE CHATELIER'S PRINCIPLE if a system at equilibrium is subjected to some change the position of equilibrium will shift in order to minimise the effect of the change.

hint

Remember – only a change in temperature affects the value of the equilibrium constant.

The effect of changing conditions on the position of equilibrium and the value of K_c are summarised in the table.

<table>
<tr><th></th><th>Effect on position of equilibrium</th><th colspan="2">Effect on K_c</th></tr>
<tr><td>increase pressure</td><td>position of equilibrium shifts to side with fewer moles of gas</td><td colspan="2">no change</td></tr>
<tr><td>decrease pressure</td><td>position of equilibrium shifts to side with more moles of gas</td><td colspan="2">no change</td></tr>
<tr><td rowspan="2">increase temperature</td><td rowspan="2">position of equilibrium shifts in endothermic direction</td><td>exothermic reaction</td><td>endothermic reaction</td></tr>
<tr><td>decreases</td><td>increases</td></tr>
<tr><td rowspan="2">decrease temperature</td><td rowspan="2">position of equilibrium shifts in exothermic direction</td><td>exothermic reaction</td><td>endothermic reaction</td></tr>
<tr><td>increases</td><td>decreases</td></tr>
<tr><td>increase concentration of one species in the reaction mixture</td><td>position of equilibrium shifts to use up the added substance – shifts to the side that the added chemical is not on</td><td colspan="2">no change</td></tr>
<tr><td>introduce a catalyst</td><td>no change – the rate of the forward and reverse reactions increase equally (the activation energy for forward and reverse reactions are lowered equally)</td><td colspan="2">no change</td></tr>
</table>

hint

The word **equally** is essential here.

☆ Model answer 7.1

The reaction between nitrogen and hydrogen to form ammonia is reversible:

$$N_2(g) + 3H_2(g) \rightleftharpoons 2NH_3(g) \quad \Delta H = -92\ \text{kJ mol}^{-1}$$

State and explain the effect on the position of equilibrium and the value of the equilibrium constant of:

a increasing temperature

b increasing pressure

a As the temperature increases the position of equilibrium shifts in the endothermic direction, which is to the left.

$$K_c = \frac{[NH_3(g)]^2}{[N_2(g)][H_2(g)]^3}$$

The value of the equilibrium constant decreases because:

- as temperature increases the position of equilibrium shifts to the left,
- at higher temperature the concentration of NH_3 at equilibrium is lower and the concentrations of N_2 and H_2 are higher,
- therefore, the numerator (top of the expression) is smaller and the denominator (bottom of the expression) is larger.

b The position of equilibrium shifts to the right, which is the side with fewer moles of gas.

There is no effect on the value of K_c because this is only affected by a change in temperature.

TEST YOURSELF 7.3

1 Predict the effect of the changes listed on the position of equilibrium for
$CH_4(g)+H_2O(g) \rightleftharpoons CO(g)+3H_2(g)$ $\Delta H = +206\ kJ\ mol^{-1}$
- increasing the pressure left, no change
- decreasing the temperature left, decreases
- adding hydrogen left, no change
- adding a catalyst. no change, no change

2 State how each of the changes in **1** affects the value of the equilibrium constant for the reaction.

The reaction quotient, *Q*

The expression for the reaction quotient is exactly the same as that for K_c except that the concentrations are not equilibrium concentrations:

e.g. for $N_2(g)+3H_2(g) \rightleftharpoons 2NH_3(g)$

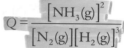

$$Q = \frac{[NH_3(g)]^2}{[N_2(g)][H_2(g)]^3}$$

You could meet the following situations:

$Q>K_c$	the system is not at equilibrium	the reaction must proceed to the left towards equilibrium – the concentrations of the products (on the right of the equation and the top of the expression for Q) are too high and so the products must be used up to reduce the value of Q until it equals K_c (and the system is at equilibrium).
$Q<K_c$	the system is not at equilibrium	the reaction must proceed to the right towards equilibrium – more products must be produced
$Q=K_c$	the system is at equilibrium	

TEST YOURSELF 7.4

The value of K_c for a particular reaction at 300 K is 240. The pressure is changed at constant temperature so that the value of *Q* is now 100. Deduce in which direction (to the left or right) the system must move towards equilibrium. to the right as Q is bigger than K_c

7.4 Calculations involving equilibrium constants (HL only)

It is really important to remember that only equilibrium concentrations must be used to work out an equilibrium constant.

hint

A temperature value is usually included in these questions (because equilibrium constants vary with temperature) but, don't worry, you don't have to do anything with it.

Worked example 7.1

2.00 mol A and 1.00 mol B are mixed together in a vessel of volume 2.00 dm³ and allowed to come to equilibrium at 500 K. At equilibrium there were 1.60 mol of A present in the reaction mixture. Calculate the value of the equilibrium constant at 500 K.

The equation for the reaction is $A(g)+2B(g) \rightleftharpoons 2C(g)$

Write all the information out neatly:

	A(g)	+2B(g)	⇌ 2C(g)
initial number of moles/mol:	2.00	1.00	0.00
equilibrium number of moles/mol:	1.60		

From this we can work out that 0.40 mol of A (2.00 − 1.60) reacted to bring the system to equilibrium.

From the chemical equation we can work out that 0.40 mol A reacts with 2×0.40 = 0.80 mol B to form 0.80 mol C. This leaves 0.20 mol B (1.00 − 0.80).

The concentrations are worked out by dividing the number of moles by the volume of the container, (e.g. $\frac{1.60}{2.00}$ for A).

	A(g)	+ 2B(g)	⇌ 2C(g)
initial number of moles/mol:	**2.00**	**1.00**	**0.00**
equilibrium number of moles/mol:	**1.60**	**0.20**	**0.80**
equilibrium concentrations/mol dm^{-3}:	0.80	0.10	0.40

The expression for K_c is: $K_c = \frac{[C(g)]^2}{[A(g)][B(g)]^2}$ $\quad K_c = \frac{0.40^2}{0.80 \times 0.10^2}$ $\quad K_c = 20$

hint

Note that the number of moles of B at equilibrium is not twice the number of moles of A – the chemical equation only tells us about how many moles of A and B react.

The most tricky type of calculation is when you are given a value for the equilibrium constant and have to work out the equilibrium concentrations.

Worked example 7.2

Consider the reaction: A(g) + B(g) ⇌ 2C(g)

1.00 mol of A and 1.00 mol of B are put into a container of volume 1.00 dm^3 and allowed to come to equilibrium at 600 K. Given that the value of the equilibrium constant at 600 K is 0.25, work out the composition of the equilibrium mixture in terms of concentrations.

	A(g)	+ B(g)	⇌ 2C(g)
initial number of moles/mol:	1.00	1.00	0

We don't know how much A and B react to bring the system to equilibrium, so we will assume that x mol A react with x mol B to form $2x$ mol C

equilibrium number of moles/mol:	$1.00 - x$	$1.00 - x$	$2x$

Divide each of these values by 1.00 (the volume) to get equilibrium concentrations:

equilibrium concentration/mol dm^{-3}:	$\frac{1.00 - x}{1.00}$	$\frac{1.00 - x}{1.00}$	$\frac{2x}{1.00}$

Since we are dividing by 1, we can ignore this – the numbers will not change.

The equilibrium concentrations must be substituted into the expression for K_c

$$K_c = \frac{[C(g)]^2}{[A(g)][B(g)]^2} \quad \text{i.e.} \quad K_c = \frac{(2x)^2}{(1.00 - x)(1.00 - x)}$$

$$\text{i.e.} \quad K_c = \frac{(2x)^2}{(1.00 - x)^2}$$

We know that the value of K_c is 0.25 and so we can put this into the equation:

$$0.25 = \frac{(2x)^2}{(1.00 - x)^2}$$

Now this looks pretty complicated, however, you can solve it fairly easily by just taking the square root of each side:

$$0.25 = \frac{2x}{1.00 - x}$$

which can be rearranged to give $x = 0.2$

You must now remember to substitute this value into the expressions for the equilibrium concentrations:

	A(g)	+	B(g)	$\rightleftharpoons$	2C(g)
equilibrium concentration/mol dm^{-3}	$\frac{1.00-x}{1.00}$		$\frac{1.00-x}{1.00}$		$\frac{2x}{1.00}$

Which gives, [A] = 0.80 mol dm^{-3}, [B] = 0.80 mol dm^{-3} and [C] = 0.40 mol dm^{-3}

TEST YOURSELF 7.5

1 Calculate K_c for the reaction: $2A(g)+X(g) \rightleftharpoons 4Q(g)+Z(g)$

Initial number of moles of A = 0.800 mol	initial number of moles of X = 0.400 mol
initial number of moles of Q = 0.000 mol	initial number of moles of Z = 0.000 mol
number of moles of Z at equilibrium = 0.100 mol	volume of container = 10.0 dm^3

Temperature = 800 K 2.37×10^{-4}

2 Consider the reaction: $A(g)+Z(g) \rightleftharpoons X(g)+Q(g)$. The value for the equilibrium constant at 500 K is 9.00. 0.100 mol of A and 0.100 mol of Z are placed in a container of volume 1.00 dm^3 and allowed to come to equilibrium, at 500 K. Calculate the number of moles of X present at equilibrium.

0.075 mol

7.5 Explaining Le Chatelier's principle (HL only)

The effect of changes of concentration on the position of equilibrium can be explained using the reaction quotient (Q) and K_c.

Consider a reaction: $A(g)+B(g) \rightleftharpoons C(g)+D(g)$

equilibrium concentrations/mol dm^{-3}: 1.0 2.0 4.0 6.0

The equilibrium constant expression is:

$$K_c = \frac{[C(g)][D(g)]}{[A(g)][B(g)]} \qquad K_c = \frac{4.0 \times 6.0}{1.0 \times 2.0} = 12$$

If some more C is added to the reaction mixture so that its concentration increases to 8.0, then, at the moment of addition, all the other concentrations have not changed and we have:

$$Q = \frac{[C(g)][D(g)]}{[A(g)][B(g)]} \qquad Q = \frac{8.0 \times 6.0}{1.0 \times 2.0} = 24$$

Since $Q \neq K_c$ the system is not at equilibrium. Because Q is greater than K_c, the reaction must proceed more to the left to reduce the concentrations of C and D (on the top of the expression) and so reduce the value of Q until it equals K_c. There will be more A and B present when equilibrium is established again, which is in agreement with Le Chatelier's principle – the position of equilibrium shifts to the left to use up the added C.

7.6 Equilibrium constants and Gibbs free energy (HL only)

The position of equilibrium corresponds to the reaction mixture with the **maximum value of entropy** (of the universe) and the **minimum value of Gibbs free energy** (of the system).

A reaction proceeds in the direction that involves a decrease in Gibbs free energy (ΔG negative). Therefore, if substances are mixed together they will react to produce the equilibrium mixture because it has lower Gibbs free energy than any other reaction mixture. Any movement away from the equilibrium mixture will be unfavourable because it will involve an increase in Gibbs free energy (ΔG positive).

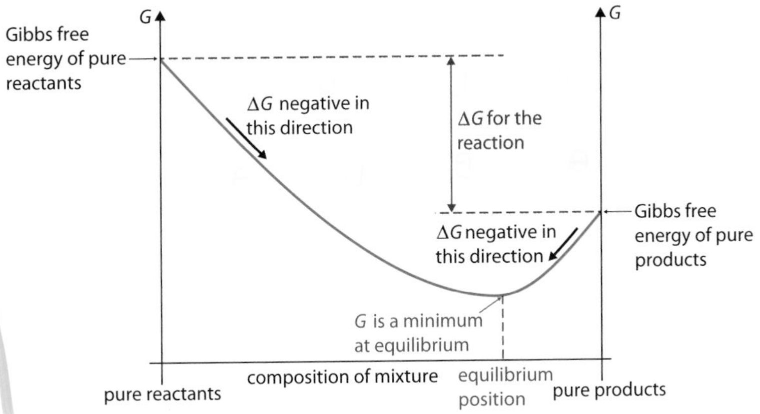

Figure 7.1

Calculations involving ΔG and the equilibrium constant

The equation $\Delta G = -RT \ln K$ relates ΔG and the equilibrium constant (K).

hint

This equation is given in the *IB Data Booklet*.

Worked example 7.3

Consider the reaction:

$A + B \rightleftharpoons C \qquad \Delta G = 100\ kJ\ mol^{-1}$

Calculate the value of the equilibrium constant at 300 K.

Use the equation $\Delta G = -RT \ln K$

Because the units of R are $J\ K^{-1}\ mol^{-1}$ we must convert the value of ΔG to $100\,000\ J\ mol^{-1}$.

Substituting into the equation we get $100\,000 = -8.31 \times 300 \times \ln K$

$\ln K = -40.1$

Use **2ND+ln** key combinations on your calculator to get the value of K.

$K = 3.80 \times 10^{-18}$

hint

You have to be careful with units in these calculations because R is in $J\ K^{-1}\ mol^{-1}$ but ΔG is usually in $kJ\ mol^{-1}$

hint

Some calculators have a 'SHIFT' instead of '2ND' key.

The table shown gives you a summary of how the values of ΔG, K and the position of equilibrium are related:

value of ΔG	value of K	position of equilibrium
negative	greater than 1	to the right
positive	smaller than 1	to the left

TEST YOURSELF 7.6

Consider the reaction: $X(g) + Y(g) \rightleftharpoons Z(g)$.

1 Calculate the value of ΔG at 298 K given that the value of the equilibrium constant at this temperature is 2.2×10^3. $-19\ kJ\ mol^{-1}$

2 Calculate the value of the equilibrium constant at 500 K given that the value of ΔG at this temperature is $56\ kJ\ mol^{-1}$. 1.4×10^{-6}

Checklist

At the end of this chapter you should be able to:

- ☐ Explain what is meant by dynamic equilibrium.
- ☐ Write expressions for the equilibrium constant.
- ☐ Deduce relationships between different equilibrium constants for the same reaction.
- ☐ Write expressions for the reaction quotient.
- ☐ State and explain how changes in conditions affect the position of equilibrium and the value of the equilibrium constant.

Higher Level only

- ☐ Calculate values for equilibrium concentrations and equilibrium constants.
- ☐ Explain Le Chatelier's principle in terms of Q and K_c.
- ☐ Explain the relationship between the equilibrium constant and Gibbs free energy.
- ☐ Carry out calculations using $\Delta G = -RT \ln K$.

8

ACIDS AND BASES

Acids and bases are very important classes of compounds and many acids and bases are familiar from both everyday life and the chemical laboratory.

This chapter covers the following topics:

- ☐ Brønsted–Lowry acids and bases
- ☐ Reactions of acids
- ☐ Strong and weak acids and bases
- ☐ pH
- ☐ Acid deposition
- ☐ Calculations involving acids and bases (HL only)
- ☐ pH (titration) curves (HL only)
- ☐ Buffer solutions (HL only)
- ☐ Lewis acids and bases (HL only)

8.1 Brønsted–Lowry acids and bases

DEFINITIONS

BRØNSTED–LOWRY ACID a proton (H^+) donor.

BRØNSTED–LOWRY BASE a proton (H^+) acceptor.

CONJUGATE ACID–BASE PAIR a pair of species that differ by 1 proton (H^+).

hint

Learn these definitions.

For example, consider the reaction of ethanoic acid with water:

$$\underset{\text{acid 1}}{CH_3COOH(aq)} + \underset{\text{base 2}}{H_2O(l)} \rightleftharpoons \underset{\text{base 1}}{CH_3COO^-(aq)} + \underset{\text{acid 2}}{H_3O^+(aq)}$$

- **Forward direction:** the CH_3COOH donates a proton (H^+) to the H_2O – the CH_3COOH is an acid as it donates a proton. The H_2O accepts a proton and therefore acts as a base.
- **Reverse direction:** H_3O^+ acts as an acid because it donates a proton to CH_3COO^-. CH_3COO^- accepts a proton and is a base.

CH_3COOH acts as an acid and donates a proton to forms its conjugate base, CH_3COO^-. CH_3COOH and CH_3COO^- differ by $1H^+$ and represent a **conjugate acid–base pair**. H_2O and H_3O^+ are also a conjugate acid–base pair.

Amphiprotic and amphoteric

DEFINITIONS

AMPHIPROTIC a species that can donate (acting as an acid) or accept (acting as a base) a proton. (This refers to the Brønsted–Lowry definition of acids and bases.)

AMPHOTERIC any substance that can act as both an acid and a base.

TEST YOURSELF 8.1

1 Identify the Brønsted–Lowry acids and bases in the following:

$$NH_4^+(aq)+H_2O(l) \rightleftharpoons NH_3(aq)+H_3O^+(aq)$$

$$HSO_4^-(aq)+H_2O(l) \rightleftharpoons SO_4^{2-}(aq)+H_3O^+(aq)$$

2 Give the formula of the conjugate acid of OH^-.

Nature of Science. In science, theories are often replaced by new ones as new data comes to light. It was originally believed that all acids contained oxygen but the realisation that hydrochloric acid does not contain oxygen required a new theory to be developed.

8.2 The reactions of acids

General Reaction	Example
acid + metal → salt + H_2	$Mg(s)+2HCl(aq) \rightarrow MgCl_2(aq)+H_2(g)$
acid + carbonate → salt + CO_2 + H_2O	$Na_2CO_3(aq)+H_2SO_4(aq) \rightarrow Na_2SO_4(aq)+H_2O(l)+CO_2(g)$
acid + hydrogencarbonate → salt + CO_2 + H_2O	$NaHCO_3(aq)+HNO_3(aq) \rightarrow NaNO_3(aq)+H_2O(l)+CO_2(g)$
acid + base → salt + water	$CuO(s)+2HNO_3(aq) \rightarrow Cu(NO_3)_2(aq)+H_2O(l)$
acid + alkali → salt + water	$2NaOH(aq) + H_2SO_4(aq) \rightarrow Na_2SO_4(aq) + 2H_2O(l)$

hint

You should learn the general equations – specific equations can be worked out knowing the formulae of ions.

One of the products of all the reactions listed here is a salt – a salt is formed when the acidic protons of an acid are replaced by metal ions or the ammonium ion, e.g. NaCl or $(NH_4)_2SO_4$.

Metal oxides are generally bases (there are a few exceptions that are amphoteric).

Alkalis are bases that dissolve in water and the reaction between an acid and alkali is the same as that between an acid and a base. Alkalis are solutions that contain hydroxide ions.

Ammonia solution is an alkali but the equation for the reaction with acids is usually written differently from the general equation – the water is omitted, e.g. $NH_3(aq)+HCl(aq) \rightarrow NH_4Cl(aq)$.

DEFINITION

NEUTRALISATION REACTION an exothermic reaction between an acid and an alkali or between an acid and a base to produce a salt and water.

TEST YOURSELF 8.2

Complete and balance the following equations:

$CaCO_3+H_2SO_4 \rightarrow$ $\quad$ $CaO+HNO_3 \rightarrow$ $\quad$ $Mg+CH_3COOH \rightarrow$ $\quad$ $NH_3+H_2SO_4 \rightarrow$

$CaSO_4 + H_2O + CO_2$ $\quad$ $Ca(NO_3)_2 + H_2O$ $\quad$ $Mg(CH_3COO)_2 + H_2$ $\quad$ $(NH_4)_2SO_4$

8.3 Strong and weak acids and bases

Strong and weak acids

When an acid reacts with water it **dissociates** or **ionises**. The reaction can be shown with or without H_2O – both representations are correct.

$HA(aq)+H_2O(l) \rightleftharpoons H_3O^+(aq)+A^-(aq)$ $\quad$ or $\quad$ $HA(aq) \rightleftharpoons H^+(aq)+A^-(aq)$

DEFINITION

A STRONG ACID is one that dissociates completely in aqueous solution.

e.g. hydrochloric acid

$HCl(aq) \rightarrow H^+(aq) + Cl^-(aq)$

> **hint**
> The one-way arrow indicates complete dissociation and is essential in an exam answer.

A strong acid is a good proton donor and has a weak conjugate base – the position of equilibrium for the dissociation of the acid lies a long way to the right – there is very little tendency for the conjugate base to pick up a proton to re-form the acid.

> **hint**
> Strong acids that you need to remember are hydrochloric (HCl), sulfuric (H_2SO_4) and nitric (HNO_3) acids.

DEFINITION

A WEAK ACID is one which dissociates partially in aqueous solution.

e.g. ethanoic acid

$CH_3COOH(aq) \rightleftharpoons CH_3COO^-(aq) + H^+(aq)$

> **hint**
> The reversible arrow is essential here.

The stronger the acid, the more it dissociates and the weaker its conjugate base.

Strong and weak bases

DEFINITION

STRONG BASES are hydroxides and ionise completely in aqueous solutions.

> **hint**
> Weak acids that you have to remember are carboxylic acids such as ethanoic acid (CH_3COOH) and carbonic acid (H_2CO_3).

Strong bases are the group 1 hydroxides (LiOH, NaOH etc.) and $Ba(OH)_2$.

e.g. $NaOH(aq) \rightarrow Na^+(aq) + OH^-(aq)$

When a weak base reacts with water it accepts a proton from the water and ionises according to the equation:

$B(aq) + H_2O(l) \rightleftharpoons BH^+(aq) + OH^-(aq)$

The stronger the base, the better it is at accepting a proton and the weaker its conjugate acid.

Weak bases are ammonia (NH_3) and amines (RNH_2, R_2NH, R_3N).

> **hint**
> Remember, the reversible arrow is critical in exam answers about weak bases – you will probably not get the mark if you do not include it.

Distinguishing experimentally between strong and weak acids and between strong and weak bases

It is important to remember that, when trying to distinguish experimentally between solutions of strong and weak acids, you must compare solutions of **equal concentrations**.

Strong acids conduct electricity better than solutions of weak acids.	Strong acids dissociate more, therefore there is a higher concentration of ions in solution.
Strong acids have a lower pH than weak acids.	Strong acids dissociate more, therefore there is a higher concentration of H^+ ions in solution.
Strong acids react more violently with metals or carbonates – more rapid fizzing.	

Solutions of strong bases have a higher pH and are better conductors than solutions of weak bases.

> **hint**
> It is important to remember that titration using an indicator **cannot** be used to distinguish between a weak and a strong acid – if they have the same concentration they will have the same end point.

TEST YOURSELF 8.3

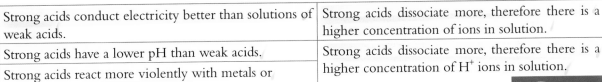

Classify the following as strong or weak acids or bases:

H_2SO_4 →SA $CH_3CH_2NH_2$ →WB CH_3COOH →WA NH_3 →WB CsOH →SB $Ba(OH)_2$ →SB
$H_2CO_3(CO_2(aq))$ →WA HNO_3 →SA HCl →SA LiOH →SB

8.4 pH

DEFINITION

pH = $-\log_{10}[H^+(aq)]$

> **hint**
> The 10 is often omitted in '$\log_{10}$' i.e. pH = $-\log[H^+(aq)]$

pH has no units.

$[H^+(aq)] = 10^{-pH}$

> **hint**
> Remember that, the higher the pH, the lower the $[H^+(aq)]$

☆ Model answer 8.1

What is the pH of a 0.0100 mol dm^{-3} solution of hydrochloric acid?

Hydrochloric acid is a strong acid and so it will dissociate completely to form 0.0100 mol dm^{-3} $H^+(aq)$.

pH = $-\log[H^+(aq)]$ pH = $-\log 0.0100$ pH = $-\log 10^{-2}$ pH = 2

> **hint**
> When $[H^+(aq)]$ is given as a power of 10 you can get the pH by changing the sign of the power, e.g. $[H^+(aq)] = 10^{-2}$ gives pH = 2.

Because pH is a log scale, you need to remember that a 10-fold change in the H^+ concentration is required to change the pH by one unit – every time a solution of a strong acid is diluted by a factor of 10 the pH goes up by 1 unit.

pH values can be used to decide whether a solution is acidic, alkaline or neutral. pH can be measured using a pH meter or universal indicator.

> **hint**
> Remember that $\log_{10} 1 = 0$, so the pH of, for example, a 1.0 mol dm^{-3} solution of hydrochloric acid is 0.0

TEST YOURSELF 8.4

Try to do these questions without using a calculator.

1 State how the pH changes when 10 cm^3 of 0.1 M HCl is diluted with 90 cm^3 of water. + 1 unit
2 Arrange the following in order of increasing pH: 1.0 mol dm^{-3} NH_3, 0.010 mol dm^{-3} HCl, 1.0 mol dm^{-3} NaOH, 0.001 mol dm^{-3} CH_3CH_2COOH, 0.10 mol dm^{-3} H_2SO_4, 0.010 mol dm^{-3} CH_3CH_2COOH, 0.0010 mol dm^{-3} NH_3, 0.10 mol dm^{-3} HCl.
3 State how the concentration of $H^+(aq)$ changes when the pH changes from 2 to 5. decreases by a factor of 1000
4 Calculate the pH of 1.00×10^{-5} M HNO_3. 5
5 Calculate $[H^+(aq)]$ of NaOH(aq) with pH = 9.0. 1×10^{-9}

How to work out the pH of an alkali

In order to work out the pH of an alkali/base you have to know about the dissociation of water.

Water dissociates according to the equation: $H_2O(l) \rightleftharpoons H^+(aq) + OH^-(aq)$

An equilibrium constant, K_w, can be derived for this reaction: $K_w = [H^+(aq)][OH^-(aq)]$

The concentration of water is not included in the expression for K_w because it is the solvent and its concentration is essentially constant.

K_w is called the **ionic product constant** for water and has a value of 1.0×10^{-14} at 25°C.

In any aqueous solution at 25 °C the product of the H^+ concentration and OH^- concentration is equal to 1.0×10^{-14}.

It is important to realise that $[H^+]$ only equals $[OH^-]$ in a neutral solution or pure water, if the solution is acidic $[H^+] > [OH^-]$ and if it is alkaline $[OH^-] > [H^+]$.

☆ Model answer 8.2

Calculate the pH of 0.0500 mol dm^{-3} sodium hydroxide solution at 25 °C.

NaOH is a strong base and ionises completely to form 0.0500 mol dm^{-3} OH^-(aq)

$[OH^-(aq)] = 0.0500$ mol dm^{-3}

$K_w = [H^+(aq)][OH^-(aq)]$

$1.0 \times 10^{-14} = [H^+(aq)] \times 0.0500$

$[H^+(aq)] = 2.00 \times 10^{-13}$ mol dm^{-3}

$pH = -\log[H^+(aq)] = -\log 2.00 \times 10^{-13} = 12.7$

hint

HL students could also use the relationship, pH + pOH = 14 at 25 °C – see the sections that follow.

TEST YOURSELF 8.5

Calculate pH values for the following solutions:

0.010 mol dm^{-3} NaOH(aq) 12

0.200 mol dm^{-3} $Ba(OH)_2$(aq) 13.6

8.5 Acid deposition

DEFINITION

ACID DEPOSITION any process in which acidic substances (particles, gases and precipitation) leave the atmosphere to be deposited on the surface of the Earth.

Rain is naturally acidic due to dissolved carbon dioxide and has a pH of about 5.6.

CO_2 dissolves in water to form carbonic acid (H_2CO_3): $H_2O(l) + CO_2(g) \rightleftharpoons H_2CO_3(aq)$

H_2CO_3 dissociates:

$$H_2CO_3(aq) \rightleftharpoons H^+(aq) + HCO_3^-(aq)$$

$$HCO_3^-(aq) \rightleftharpoons H^+(aq) + CO_3^{2-}(aq)$$

Acid rain is rain with a pH lower than 5.6.

Acid deposition is caused when nitrogen oxides (NO_x) or sulfur oxides (SO_2, SO_3) dissolve in water to form HNO_3 (nitric(V) acid), HNO_2 (nitric(III) acid), H_2SO_4 (sulfuric(VI) acid) and H_2SO_3 (sulfuric(III) acid).

Oxide	Source	Equation	Formation of acid
SO_2	burning coal and other sulfur-containing fuels, smelting of metal ores, volcanic activity	$S(s) + O_2(g) \rightarrow SO_2(g)$	$SO_2(g) + H_2O(l) \rightleftharpoons H_2SO_3(aq)$
SO_3	formed from SO_2 in the atmosphere	$2SO_2(g) + O_2(g) \rightleftharpoons 2SO_3(g)$	$SO_3(g) + H_2O(l) \rightarrow H_2SO_4(aq)$
NO	internal combustion engine, coal, oil and gas-fuelled power stations	$N_2(g) + O_2(g) \rightarrow 2NO(g)$	converted to NO_2
NO_2	oxidation of NO in the atmosphere	$2NO(g) + O_2(g) \rightarrow 2NO_2(g)$	$2NO_2(g) + H_2O(l) \rightarrow HNO_2(aq) + HNO_3(aq)$

hint

You need to learn the information in the table.

Some effects of acid deposition are:

- damages/kill trees and plants
- kills fish in lakes
- damages limestone buildings [$CaCO_3(s) + H_2SO_4(aq) \rightarrow CaSO_4(s) + H_2O(l) + CO_2(g)$].

Pre- and post-combustion methods for reducing sulfur oxide emissions

Sulfur is present in fossil fuels such as coal and oil and you should be able to explain the pre- and post-combustion methods for removing it.

pre-combustion	the sulfur is removed from the fuel before it is burnt – used for producing petrol/diesel for vehicles	e.g. heating the sulfur-containing fuel with hydrogen in the presence of a catalyst converts S to H_2S
post-combustion	in a power station the sulfur oxides are removed after the fuel has been burnt	The exhaust gases are passed through a base such as calcium oxide or calcium carbonate, e.g. $CaCO_3(s)+SO_2(g) \rightarrow CaSO_3(s)+CO_2(g)$

Nature of Science. The advancement of science and technology can cause problems but also solve them.

8.6 Calculations involving acids and bases (HL only)

The acid and base dissociation/ionisation constants

The acid and base dissociation constants give us information about the strength of acids and bases – these are essentially equilibrium constants. The types of constant that you will encounter are summarised in the table. You will be expected to be able to solve problems using these constants.

	Acids		
K_a	$HA(aq) \rightleftharpoons H^+(aq)+A^-(aq)$	$K_a = \frac{[A^-(aq)][H^+(aq)]}{[HA(aq)]}$	larger value indicates a stronger acid
pK_a		$pK_a = -\log_{10}K_a$ $K_a = 10^{-pK_a}$	smaller value indicates a stronger acid
	Bases		
K_b	$B(aq)+H_2O(l) \rightleftharpoons BH^+(aq)+OH^-(aq)$	$K_b = \frac{[BH^+(aq)][OH^-(aq)]}{[B(aq)]}$	larger value indicates a stronger base
pK_b		$pK_b = -\log_{10}K_b$ $K_b = 10^{-pK_b}$	smaller value indicates a stronger base

hint

This is a standard type of question and it is important to learn the method.

hint

The pK_a value will not always be given in the question but there is a table of them in the *IB Data Booklet*.

hint

The assumption is essential in exam answers

Worked example 8.1

The pK_a of HCN(aq) is 9.40. Calculate the pH of a 0.200 mol dm^{-3} solution.

Calculate K_a using $K_a = 10^{-pK_a}$ $\quad K_a = 10^{-9.40} = 3.98 \times 10^{-10}$

Write the expression for dissociation of the acid: $HCN(aq) \rightleftharpoons H^+(aq)+CN^-(aq)$

Write the expression for K_a:

$$K_a = \frac{[CN^-(aq)][H^+(aq)]}{[HCN(aq)]}$$

When 1 HCN dissociates equal numbers of H^+ and CN^- are produced, therefore $[CN^-(aq)] = [H^+(aq)]$

We will make the assumption that the dissociation of HCN is negligible compared to its concentration, therefore, we can take [HCN(aq)] as 0.200 mol dm^{-3}.

Substitute known values into the K_a expression: $\quad 3.98 \times 10^{-10} = \frac{[H^+(aq)]^2}{0.200}$

Rearrange the expression to get $[H^+(aq)]$: $\quad [H^+(aq)] = 8.92 \times 10^{-6}$ mol dm^{-3}

Work out pH using $pH = -\log[H^+(aq)]$: $\quad pH = -\log 8.92 \times 10^{-6} = 5.05$

TEST YOURSELF 8.6

1 Arrange the following acids in order of acid strength (strongest first):
HA ($pK_a = 4.8$) HB ($K_a = 1.3\times10^{-5}$) HC ($K_a = 4.7\times10^{-4}$) HD ($pK_a = 4.2$)

2 Calculate pH of a 0.100 mol dm^{-3} solution of propanoic acid (CH_3CH_2COOH) given that its pK_a is 4.87.

The relationship between K_a and K_b

We have already met the idea that the stronger the acid, the weaker its conjugate base - for a conjugate acid–base pair the values of K_a and K_b are related:

$K_a\times K_b = K_w$	$pK_a+pK_b = pK_w$	At 25 °C: $pK_a+pK_b = 14$

hint

It is important to remember that these relationships **only work for a conjugate acid base pair,** e.g. for NH_3 and NH_4^+, not for any random acid and base.

TEST YOURSELF 8.7

Calculate K_a for NH_4^+ given that pK_b for NH_3 is 4.75.

Working out the pH of a weak base

pOH can be used to simplify the calculation of the pH of a weak base. pOH is basically the same as pH except it refers to the concentration of OH^- ions rather than H^+ ions.

$$pOH = -\log[OH^-(aq)] \quad \text{or} \quad [OH^-(aq)] = 10^{-pOH}$$

The ionic product constant expression, $K_w = [H^+(aq)][OH^-(aq)]$, can then be converted to $pK_w = pH+pOH$, or, at 25 °C, $pH+pOH = 14$.

In order to calculate the pH of a 0.100 mol dm^{-3} solution of ammonia ($pK_b = 4.75$):

- Use the expression for K_b to calculate $[OH^-(aq)]$
- Calculate pOH
- Use $pOH+pH = 14$ to work out pH (the answer here is 11.1).

TEST YOURSELF 8.8

1 Calculate pH, pOH, $[H^+(aq)]$ and $[OH^-(aq)]$ for each of the following solutions:
0.1 M HCl 0.1 M NaOH 0.0200 M $Ba(OH)_2$

2 Calculate the pH of a 0.0100 mol dm^{-3} solution of methylamine (CH_3NH_2) at 25 °C given that $pK_b = 3.34$.

The variation of K_w with temperature

The ionisation of water [$H_2O(l) \rightleftharpoons H^+(aq)+OH^-(aq)$] is endothermic and therefore the degree of ionisation increases as temperature increases (Le Chatelier's principle). This means that the pH of water decreases as temperature increases (higher concentration of $H^+(aq)$ at higher temperature).

Worked example 8.2

Calculate the pH of water at 50 °C given that $pK_w = 13.26$ at this temperature.

Water dissociates according to the equation: $H_2O(l) \rightleftharpoons H^+(aq)+OH^-(aq)$

$pK_w = pH+pOH$ $pK_w = 13.26$ therefore: $13.26 = pH+pOH$

In pure water $[H^+] = [OH^-]$, therefore: pH = pOH

$13.26 = 2\times pH$ therefore: pH = 6.63

Although the pH is less than 7 the water is not acidic, because $[H^+(aq)] = [OH^-(aq)]$ – pure water is neutral at all temperatures.

TEST YOURSELF 8.9

Calculate the $[H^+]$, $[OH^-]$ and pH for water at 313 K given that K_w at 313 K is 2.92×10^{-14}.

pH = 6.77 $[H^+] = 1.71 \times 10^{-7}$ $[OH^-] = 1.71 \times 10^{-7}$

8.7 pH (titration) curves (HL only)

You will need to be able to draw pH curves for titrating the different combinations of strong/weak acids and bases with each other.

When drawing these curves you should pay special attention to:

- the general shape
- the initial pH (this will just be the pH of the acid/alkali you started with)
- the volume of acid/base required for an exact reaction (position of mid-point of steep part)
- whether the steep part is more in the acidic/alkaline region
- the pH at the equivalence point
- the final pH (this will just be the pH of the acid/alkali that you are adding).

DEFINITION

EQUIVALENCE POINT the point at which equivalent numbers of moles of acid and alkali have been added.

Figure **8.1** compares the pH curves for the addition of 0.10 mol dm^{-3} sodium hydroxide solution to 25 cm^3 of 0.10 mol dm^{-3} hydrochloric acid (red line) or to 25 cm^3 of 0.10 mol dm^{-3} ethanoic acid (blue line).

The pH at the equivalence point is the pH at the mid-point of the steep part of the curve (point of inflexion).

The table summarises the similarities and differences between these curves:

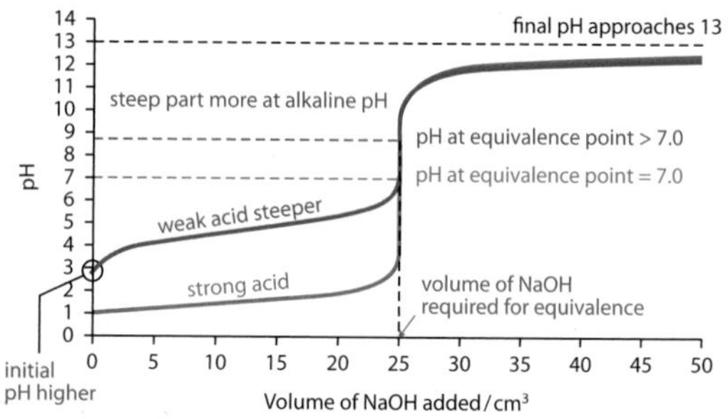

Figure 8.1

Titration with a strong base	
weak acid	**strong acid**
same basic shape	
initial pH higher because it only dissociates partially	initial pH lower because it fully dissociates
Volume of NaOH required for exact reaction (mid-point of steep part of curve) the same because the same volume of and concentration of acids is used.	
steep part more in alkaline region – the base is stronger than the acid	steep part equally in acidic and alkaline regions
pH at equivalence point > 7 because the base is stronger than the acid	pH at equivalence point = 7 because the base and the acid are both strong
final pH the same because the same alkali was used	

For the four combinations of acid–base titrations you need to know whether the pH at the equivalence point is equal to, less than or greater than 7.

Acid	Base	pH at equivalence point
strong	strong	7
strong	weak	below 7
weak	strong	above 7
weak	weak	depends on the relative strength of the acid and base

Note: the titration curve for a weak acid-weak base titration has a slightly different form and there is no really steep part.

hint

In a titration where one component is weak and the other strong, the pH at the equivalence point will be acidic if the acid is strong and alkaline if the base is strong – you can think about this as if the stronger component dominates (don't write this in the exam!).

TEST YOURSELF 8.10

Sketch a pH curve for the addition of 0.100 mol dm^{-3} HCl(aq) to 25.0 cm^3 of NH_3(aq) [pK_b = 4.75]. Mark the initial pH, final pH and an approximate pH at the equivalence point on your sketch.

Determination of pK_a or pK_b from a titration curve

For a weak acid-strong base titration the pK_a of the acid is given by the pH at the half equivalence point.

Consider adding 0.10 mol dm^{-3} NaOH(aq) to 25 cm^3 of 0.10 mol dm^{-3} CH_3COOH. 25.0 cm^3 of NaOH would have to be added to get to the equivalence point – when 12.5 cm^3 of NaOH is added the pH of the mixture is 4.76, which is the pK_a of ethanoic acid.

For a titration where a strong acid is added to a weak base pK_b = pOH at the half equivalence point. Since at 25 °C, pH+pOH = 14,
14 - (pH at the half equivalence point) = pK_b for the weak base.

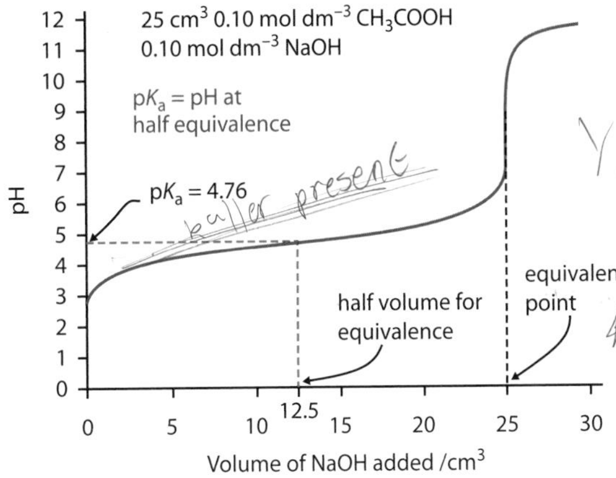

Figure 8.2

Indicators

DEFINITION

INDICATORS are **weak acids** or **weak bases** which exhibit different colours depending on the pH – the ionised and unionised forms have **different colours**.

For an indicator that is a weak acid:

$$HIn(aq) \rightleftharpoons H^+(aq) + In^-(aq)$$

colour I   colour II

If acid (H^+) is added to the indicator the position of equilibrium will shift to the left to use up, as far as possible, the H^+ that has been added – the indicator will appear as colour I.

If we add some alkali to this solution the indicator changes to colour II. When we add alkali the OH^- from the alkali reacts with the H^+ on the right-hand side of the equilibrium to produce water. The position of equilibrium thus shifts to the right to replace the H^+ as far as possible.

Annotated exemplar answer 8.1

Explain how an acid–base indicator works. **[3]**

An indicator has two colours. When acid is added it goes to colour I but when alkali is added it goes to colour 2. This is because of Le Chatelier's principle.

Vague – it must be stressed that ionised and unionised forms have different colours.

0/3

Explain the effect of adding acid/alkali on the colour – If acid (H^+) is added the position of equilibrium shifts to the left to use up the H^+ – the indicator appears colour I. If alkali (OH^-) is added the H^+ is used up and the position of equilibrium shift to the right – colour II.

This only makes sense if an equation is given for the ionisation of the indicator, e.g.

$HIn(aq) \rightleftharpoons H^+(aq) + In^-(aq)$

colour I colour II

Selecting an indicator for a titration

DEFINITION

pH RANGE OF AN INDICATOR the pH range over which intermediate colours of the indicator can be seen.

As a rough guide, the pH range can be taken as $pK_a \pm 1$.

DEFINITION

THE END POINT OF A TITRATION is the point at which the indicator changes colour.

hint

The pK_a values and ranges of indicators are given in the *IB Data Booklet*.

An indicator should be chosen so that the pH range of the indicator occurs entirely within the very steep part of the pH curve.

If you are given a pH curve then you just have to fit the range into the steep part of the curve, e.g. bromothymol blue (pH range 6.0–7.6) is suitable for the strong acid–weak base titration shown in Figure **8.3**

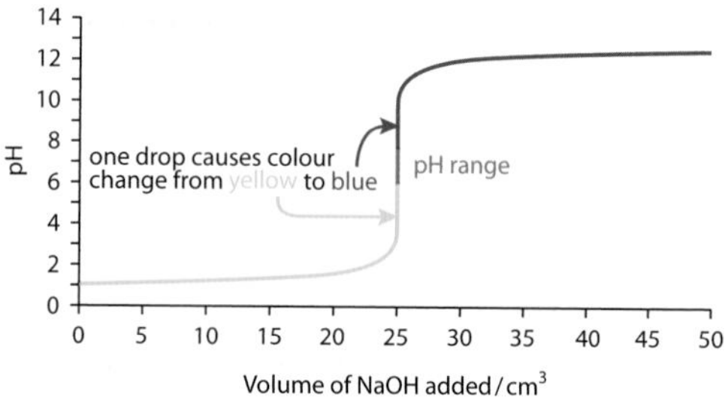

Figure 8.3

When a pH curve has not been given use the pH at the equivalence point as a guide.

acid	base	pH at equivalence point	indicator
strong	strong	7	most common indicators, e.g. methyl red
strong	weak	below 7	indicators with pK_a less than 7, e.g. bromocresol green
weak	strong	above 7	indicators with pK_a greater than 7, e.g. phenolphthalein
weak	weak	depends on the relative strength of the acid and base	indicators cannot be used – the pH curve does not show a really steep part and a gradual change in colour will be seen.

Salt hydrolysis

You need to be able to predict and explain whether a solution of a particular salt will be acidic, alkaline or neutral. Assume that the salt has been made from an acid and base (alkali) – if the acid is stronger the pH will be acidic (less than 7) but if the base is stronger the pH will be basic (greater than 7).

Acid	Base	pH of salt solution	Example	Equation
strong	strong	7	NaCl	
strong	weak	less than 7	NH_4Cl	$NH_4^+(aq) \rightleftharpoons NH_3(aq)+H^+(aq)$
weak	strong	greater than 7	CH_3COONa	$CH_3COO^-(aq)+H_2O(l) \rightleftharpoons CH_3COOH(aq)+OH^-(aq)$
weak	weak	depends on acid and base	CH_3COONH_4	$CH_3COO^-(aq)+H_2O(l) \rightleftharpoons CH_3COOH(aq)+OH^-(aq)$ and $NH_4^+(aq) \rightleftharpoons NH_3(aq)+H^+(aq)$

A solution of CH_3COONa is alkaline because CH_3COO^- is the conjugate base of a weak acid and therefore a reasonably strong base.

A solution of NH_4Cl is acidic because NH_4^+ is the conjugate acid of a weak base and therefore a reasonably strong acid.

hint

CH_3COONH_4 has a pH very close to 7 – it is the salt of a weak acid and a weak base of very similar strengths.

TEST YOURSELF 8.11

State whether solutions of the following salts will be acidic, basic or neutral:

NaCl CH_3COONa NH_4Cl KNO_3

neutral   basic   acidic   neutral

The acidity of positive ions in solution

All 3+ ions in solution are acidic, e.g. a 0.10 mol dm^{-3} solution of iron(III) will have a pH of less than 2. The high charge density (or charge:radius ratio) of the ion causes the water molecule to be sufficiently **polarised** for H^+ to dissociate.

$[Fe(H_2O)_6]^{3+}(aq) \rightleftharpoons [Fe(H_2O)_5(OH)]^{2+}(aq)+H^+(aq)$

8.8 Buffer solutions (HL only)

DEFINITION

A BUFFER SOLUTION is one which resists changes in pH when **small amounts** of acid or alkali are added.

A buffer solution consists of two components and must always contain something to react with any acid added and something to react with any base added.

An acidic buffer solution (e.g. CH_3COOH and CH_3COONa) is a solution of a weak acid (HA) and the salt/conjugate base of that weak acid (NaA is in the buffer but the A^- is the important part).

The equilibrium that exists in this buffer is: $HA(aq) \rightleftharpoons H^+(aq)+A^-(aq)$

If some hydrochloric acid is added to this solution the extra H^+ added reacts with the A^- in the solution:

$$A^-(aq)+H^+(aq) \rightarrow HA(aq)$$

If some sodium hydroxide is added to the solution the extra OH^- added reacts with the HA in the solution:

$$HA(aq)+OH^-(aq) \rightarrow A^-(aq)+H_2O(l)$$

Because large amounts of HA and A^- are present in the buffer solution, small changes in their concentration do not change the position of the original equilibrium very much and hence the pH remains approximately constant.

A basic buffer solution contains a weak base (reacts with any acid added) and the salt of that weak base (reacts with any base added). For example, a solution containing ammonia (NH_3) and ammonium chloride (NH_4Cl).

Worked example 8.3

Which of the following would act as a buffer solution?

A 50.0 cm^3 0.100 mol dm^{-3} NaOH(aq) and 50.0 cm^3 0.100 mol dm^{-3} CH_3COOH(aq)

B 50.0 cm^3 0.100 mol dm^{-3} NaOH(aq) and 20.0 cm^3 0.100 mol dm^{-3} CH_3COOH(aq)

C 20.0 cm^3 0.100 mol dm^{-3} NaOH(aq) and 50.0 cm^3 0.100 mol dm^{-3} CH_3COOH(aq)

D 50.0 cm^3 0.100 mol dm^{-3} NaOH(aq) and 50.0 cm^3 0.100 mol dm^{-3} HCl(aq)

In the previous section we discussed that a buffer is made from a weak acid and its salt or a weak base and its salt. None of the options appear to contain these things; however, when a strong base reacts with a weak acid the salt of the acid will be formed:

$NaOH(aq) + CH_3COOH(aq) \rightarrow CH_3COONa(aq) + H_2O(l)$

For a buffer to be formed there must be some weak acid left after the strong base has reacted with the acid, so we must look at the relative amounts of each substance added.

Option A is not correct because there are equal numbers of moles of NaOH and CH_3COOH so the NaOH will react with **all** the CH_3COOH – there will be no weak acid left and the solution cannot function as a buffer.

Option B is not correct – there are more moles of NaOH than CH_3COOH so, although the salt will be formed, there will be no weak acid left and this is not a buffer.

Option C is the **correct answer** – there are more moles of CH_3COOH than NaOH. The NaOH will react with some of the CH_3COOH to form the salt but there will still be some CH_3COOH left over. The solution will thus contain a weak acid and its salt and be a buffer solution.

Option D is not correct because it involves a strong acid and base – there are no weak acids/bases.

TEST YOURSELF 8.12

Which of the following combinations will produce a buffer solution?

A 25.0 cm^3 0.100 mol dm^{-3} NH_3(aq) + 25.0 cm^3 0.100 mol dm^{-3} HNO_3
B 25.0 cm^3 0.100 mol dm^{-3} NH_3(aq) + 25.0 cm^3 0.050 mol dm^{-3} HCl
C 25.0 cm^3 0.200 mol dm^{-3} NH_3(aq) + 25.0 cm^3 0.100 mol dm^{-3} HNO_3
D 25.0 cm^3 0.200 mol dm^{-3} NH_3(aq) + 25.0 cm^3 0.200 mol dm^{-3} NaOH

8.9 Lewis acids and bases (HL only)

Lewis acid	Lewis base
electron pair **acceptor**	electron pair **donor**
must have a lone pair of electrons	must have space in its outer shell to accept a pair of electrons

A coordinate covalent bond is always formed in a Lewis acid–base reaction when the base donates a pair of electrons to the acid.

e.g. $NH_3 + H^+ \rightarrow NH_4^+$

NH_3 is the Lewis base as it donates an electron pair to H^+, which is the electron pair acceptor, i.e. a Lewis acid.

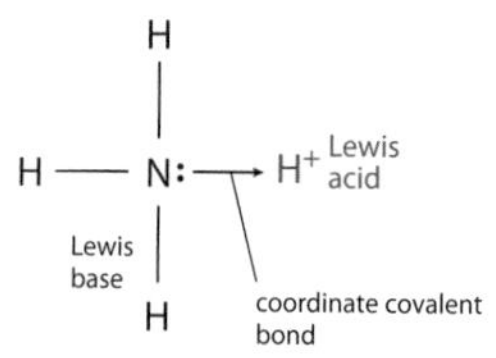

Figure 8.4

Formation of a complex ion by a transition metal ion is a Lewis acid–base reaction. The transition metal ion is the Lewis acid and the ligand is the Lewis base.

e.g. $Fe^{2+} + 6H_2O \rightleftharpoons [Fe(H_2O)_6]^{2+}$

This is only an acid–base reaction according to the Lewis theory and not according to the Brønsted–Lowry theory – there is no transfer of a proton.

Many reactions in organic chemistry can also be described using the Lewis theory – a nucleophile is a Lewis base and an electrophile is a Lewis acid.

TEST YOURSELF 8.13

1 Explain whether BCl_3 can act as a Lewis acid or base. acid, has space to accept more e^-

2 Classify each of the following as a Lewis acid or Lewis base:

OH^- Cl^- $(CH_3)_3C^+$

Base Base Acid

Checklist

At the end of this chapter you should be able to:

- ☐ Explain the Brønsted–Lowry definition of acids and bases and identify conjugate acid–base pairs.
- ☐ Write equations for reactions of acids.
- ☐ Explain the differences between strong and weak acids and bases.
- ☐ Define pH and solve problems involving pH and $[H^+(aq)]$.
- ☐ Describe and explain the causes of acid deposition

Higher Level only

- ☐ Solve problems involving K_a, pK_a, K_b, pK_b, K_w, pK_w, pH and pOH.
- ☐ Sketch titration curves and identify the key features for titrations involving all the combinations of weak/strong acids/bases.
- ☐ Explain how indicators work and select suitable indicators for titrations.
- ☐ Explain whether salt solutions are acidic, neutral or basic.
- ☐ Explain what is meant by a buffer solution and how they are made.
- ☐ Explain the Lewis definition of acids and bases and give examples of each.

9 REDOX PROCESSES

Redox reactions are an extremely important class of reactions – they power our world – combustion reactions of fuel, producing electricity from batteries, reactions in the body that allow us to get the energy from food – these are all redox reactions.

This chapter covers the following topics:

- ☐ Oxidation state
- ☐ Oxidation and reduction
- ☐ Biochemical oxygen demand
- ☐ Electrochemical cells
- ☐ Standard electrode potentials (HL only)
- ☐ Electrolysis of aqueous solutions (HL only)

9.1 Oxidation state

hint

The terms 'oxidation state' and 'oxidation number' may be used interchangeably in the exam.

You need to be able to assign oxidation states to atoms in molecules/ions.
Tips for working out oxidation states:

- If the compound is ionic then the charges on the ions are the oxidation states.
- In most cases you can assume that O has an oxidation number of −2 and H of +1.
- The most electronegative atom in a molecule is assigned a negative oxidation state according to how many electrons it needs to gain to have a full outer shell.
- The sum of the oxidation states must equal the overall charge on the molecule/ion.
- The oxidation state of atoms in an element is zero.

Worked example 9.1

hint

It is important to remember that oxidation state is written with the sign first, e.g. −2 but a charge is written with the number first, e.g. 2−.

Deduce the oxidation state of sulfur in Na_2SO_4.

Na_2SO_4 is an ionic compound and can be separated into Na^+ and SO_4^{2-} ions.

In SO_4^{2-} O is more electronegative – oxidation state −2.

The total oxidation state of 4 O atoms in SO_4^{2-} is 4×−2, i.e. −8.

Since the overall charge on the SO_4^{2-} ion is 2− the oxidation state of S must be +6 to cancel out all but 2 of the total oxidation state of the 4 O atoms.

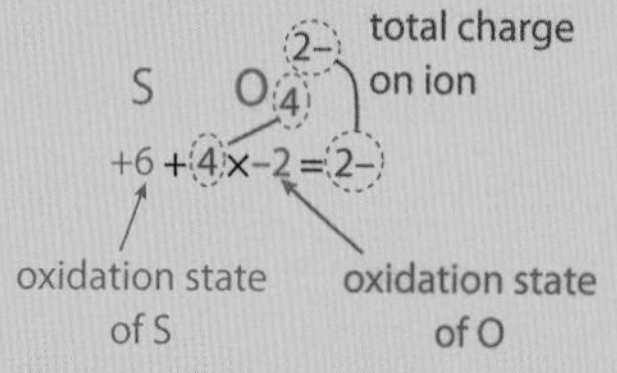

Figure 9.1

hint

The term 'oxidation number' may be used in questions about transition metal complex ions - this is written using Roman numerals, eg II for +2 and −III for −3.

Oxidation states are written as Roman numerals in names of ions/compounds. The SO_4^{2-} ion can be named as the sulfate(VI) ion and Na_2SO_4 as sodium sulfate(VI).

A couple of odd cases that are important to remember

- H_2O_2 – hydrogen peroxide (and other peroxides) – the oxidation state of O is −1.
- In metal hydrides such as NaH the oxidation state of H is −1.

Variable oxidation states

Transition metals and main group non-metals can exhibit variable oxidation states, e.g. $FeCl_2$ [iron(II) chloride], $FeCl_3$ [iron(III) chloride], SO_2 [sulfur(IV) oxide] and SO_3 [sulfur(VI) oxide].

TEST YOURSELF 9.1

1 Determine the oxidation states of all elements in the following species:

NO_2 NO_3^- HCl $HClO_3$ ClO_4^- CrO_4^{2-} $Cr_2O_7^{2-}$ Na_2O_2 LiH

2 Name the following molecules/ions using oxidation states:

NO_2 N_2O $Cr_2O_7^{2-}$ MnO_4^- ClO_4^- ClO^-

nitrogen (IV) oxide nitrogen (I) oxide dichromate (IV) Manganate (VII) clorate (VII) Chlorate (I)

9.2 Oxidation and reduction

DEFINITIONS

OXIDATION	loss of hydrogen	gain of oxygen	loss of electrons	increase in oxidation state
REDUCTION	gain of hydrogen	loss of oxygen	gain of electrons	decrease in oxidation state

In the equation: $Zn(s) + Cu^{2+}(aq) \rightarrow Zn^{2+}(aq) + Cu(s)$ the Zn has been oxidised to Zn^{2+} as it has lost electrons ($Zn \rightarrow Zn^{2+} + 2e^-$) and the Cu^{2+} has been reduced to Cu as it has gained electrons ($Cu^{2+} + 2e^- \rightarrow Cu$).

Oxidation
Is
Loss of electrons
Reduction
Is
Gain of electrons

Figure 9.2

DEFINITION

A REDOX REACTION is one that involves both oxidation and reduction.

If something loses electrons something else must gain them, therefore **oxidation** and reduction always occur together. A redox reaction involves a change in oxidation state.

The reaction shown is a redox reaction. Br_2 is reduced as the oxidation state decreases from 0 to −1 and the S in SO_2 is oxidised as its oxidation state increases from +4 to +6. We say that the SO_2 is oxidised.

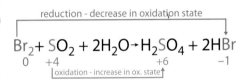

reduction - decrease in oxidation state

$$Br_2 + SO_2 + 2H_2O \rightarrow H_2SO_4 + 2HBr$$

0 +4 +6 −1

oxidation - increase in ox. state

Figure 9.3

Oxidising and reducing agents

DEFINITIONS

OXIDISING AGENT species that oxidises other species and is, in the process, reduced. Oxidising agents remove electrons from something.

REDUCING AGENT species that reduces other species and is, in the process, oxidised. Reducing agents give electrons to something.

Worked example 9.2

Explain which species is the oxidising agent in the reaction: $2KBr(aq) + Cl_2(aq) \rightarrow 2KCl(aq) + Br_2(aq)$

This can be written as an ionic equation by leaving out the K^+ (called a spectator ion):

$2Br^- + Cl_2 \rightarrow 2Cl^- + Br_2$

The Cl_2 has gained electrons and been reduced, therefore it must be the oxidising agent.

Cl_2 is the oxidising agent as it oxidises the Br^- to Br_2. The Cl_2 removes electrons from the bromide ions thus oxidising them.

TEST YOURSELF 9.2

Work out which of the following equations represent redox reactions and identify the oxidising and reducing agents:

$I_2(s) + 2OH^-(aq) \rightarrow I^-(aq) + OI^-(aq) + H_2O(l)$

$SO_4^{2-}(aq) + H_2O(l) \rightarrow HSO_4^-(aq) + OH^-(aq)$

$Fe + Cu^{2+} \rightarrow Fe^{2+} + Cu$

$C_2H_2 + 2H_2 \rightarrow C_2H_6$

hint

Remember the oxidising agent is the species that gets reduced and the reducing agent is the species that gets oxidised.

The activity series

Any group of metals may be arranged in activity series such as:

magnesium	most reactive – strongest reducing agent. Mg will reduce Zn^{2+}, Fe^{2+}, Cu^{2+}
zinc	Zn will reduce Fe^{2+}, Cu^{2+} $\quad Zn + Cu^{2+} \rightarrow Zn^{2+} + Cu$
iron	Fe will reduce Cu^{2+}
hydrogen	
copper	least reactive – weakest reducing agent. Cu^{2+} is the strongest oxidising agent.

Metals higher in the activity series are stronger reducing agents than metals lower in the activity series.

Metals above hydrogen in the activity series are stronger reducing agents than hydrogen and should displace hydrogen from a solution of its ions (an acid). Thus, the reaction between magnesium and hydrogen ions is $Mg(s) + 2H^+(aq) \rightarrow Mg^{2+}(aq) + H_2(g)$.

Half-equations

A redox equation may be broken down into two **half-equations**. These half-equations show the oxidation and reduction processes separately.

Redox equation	Half equations	
$2Br^-(aq) + Cl_2(aq) \rightarrow 2Cl^-(aq) + Br_2(aq)$	$2Br^-(aq) \rightarrow Br_2(aq) + 2e^-$	oxidation
	$Cl_2(aq) + 2e^- \rightarrow 2Cl^-(aq)$	reduction

Remember that each half-equation must balance in terms of the number of atoms on both sides and in terms of the total charge on both sides.

Balancing half-equations in acidic solution

Balancing half-equations in neutral solution is fairly straightforward but it is best to learn a set of rules for balancing half-equations in acidic solution.

Acidic solutions:

1 balance all atoms except H and O,

2 add H_2O to side with fewer O atoms to balance O,

3 add H^+ to side with fewer H atoms to balance H,

4 add e^- to side deficient in negative charge to balance charge.

Worked example 9.3

Balance the half-equation $Cr_2O_7^{2-} \rightarrow Cr^{3+}$ in acidic solution.

The Cr atoms must be balanced: $Cr_2O_7^{2-} \rightarrow \mathbf{2}Cr^{3+}$

O atoms are balanced: $Cr_2O_7^{2-} \rightarrow 2Cr^{3+} + \mathbf{7}H_2O$

H atoms are balanced: $Cr_2O_7^{2-} + \mathbf{14}H^+ \rightarrow 2Cr^{3+} + \mathbf{7}H_2O$

The total charge on LHS is $2- + 14+ = 12+$

The total charge on the RHS is $2 \times 3+ = 6+$

$6e^-$ must be added to LHS to balance the charges: $Cr_2O_7^{2-} + 14H^+ + 6e^- \rightarrow 2Cr^{3+} + 7H_2O$.

hint

As a final check, make sure that all the atoms and charges are balanced on each side.

If you are asked to balance an overall redox reaction it is usually easiest to split it into half-equations, balance each one individually and then combine them to give the overall equation.

When an oxidation half-equation is combined with a reduction half-equation to produce an overall redox equation the number of electrons lost in the oxidation reaction must be the same as the number gained in the reduction reaction. You must multiply the half-equations by the appropriate numbers to balance electrons.

Worked example 9.4

Balance the following redox equation in acid solution:

$MnO_4^-(aq) + H^+(aq) + Fe^{2+}(aq) \rightarrow Mn^{2+}(aq) + H_2O(l) + Fe^{3+}(aq)$

The equation is split into half-equations:

$MnO_4^-(aq) + H^+(aq) + 5e^- \rightarrow Mn^{2+}(aq) + 4H_2O(l)$ **reduction**

$Fe^{2+}(aq) \rightarrow Fe^{3+}(aq) + e^-$ **oxidation**

The Fe^{2+}/Fe^{3+} half-equation is multiplied by 5 to balance the electrons in both half-equations:

$MnO_4^-(aq) + 8H^+(aq) + 5e^- \rightarrow Mn^{2+}(aq) + 4H_2O(l)$

$5Fe^{2+}(aq) \rightarrow 5Fe^{3+}(aq) + 5e^-$

The half-equations are added together and the electrons cancelled:

$MnO_4^-(aq) + 8H^+(aq) + 5Fe^{2+}(aq) \rightarrow Mn^{2+}(aq) + 4H_2O(l) + 5Fe^{3+}(aq)$

hint

These half-equations are provided ready balanced in the *IB Data Booklet*.

TEST YOURSELF 9.3

1 Balance the following half-equations in neutral solution:

$Fe^{2+} \rightarrow Fe$ $\quad Br^- \rightarrow Br_2$

2 Balance the following half-equations in acidic solution:

$Cr_2O_7^{2-} + H^+ + e^- \rightarrow Cr^{3+} + H_2O$ $\quad H_3PO_3 \rightarrow H_3PO_4$

3 Balance the following redox equation:

$MnO_4^-(aq) + H^+(aq) + I^-(aq) \rightarrow Mn^{2+}(aq) + H_2O(l) + I_2(aq)$

hint

Remember, half-equations have e^- in them but overall redox equations do not.

9.3 Biochemical oxygen demand (BOD)

DEFINITION

BOD the amount of oxygen used by the aerobic microorganisms in water to decompose the organic matter in the water over a fixed period of time (usually 5 days) at a fixed temperature (usually 20 °C).

Biochemical oxygen demand (BOD) is used as a measure of the quality of water. The lower the BOD, the less polluted the water.

The Winkler method can be used to measure the concentration of dissolved oxygen in a sample of water – it involves a redox titration. The equations involved are:

$$2Mn(OH)_2(s) + O_2(aq) \rightarrow 2MnO(OH)_2(s)$$

$$MnO(OH)_2(s) + 4H^+(aq) \rightarrow Mn^{4+}(aq) + 3H_2O(l)$$

$$Mn^{4+}(aq) + 2I^-(aq) \rightarrow Mn^{2+}(aq) + I_2(aq)$$

$$I_2(aq) + 2S_2O_3^{2-}(aq) \rightarrow S_4O_6^{2-}(aq) + 2I^-(aq)$$

hint

A different set of equations could be given in the exam but the overall stoichiometry will be the same.

The various steps in the calculation involve a series of moles calculations – the end result is that the number of moles of oxygen in the sample is ¼ the number of moles of sodium thiosulfate used in the titration.

Model answer 9.1

The Winkler method was used to calculate the concentration of dissolved oxygen in 50.0 cm^3 of water. The iodine that was formed was titrated against a sodium thiosulfate solution. It was found that 25.60 cm^3 of sodium thiosulfate of concentration 2.00×10^{-3} mol dm^{-3} was required in the titration.

a Calculate the concentration of dissolved oxygen in ppm.

b After 5 days incubation at 20 °C another sample of water from the same source had a dissolved oxygen concentration of 3.50 ppm. Calculate the BOD.

a Number of moles of sodium thiosulfate = volume (in dm^3) × concentration

$$= \frac{25.60}{1000} \times 2.00 \times 10^{-3} = 5.12 \times 10^{-5} \text{ mol}$$

Number of moles of oxygen in water sample $= \frac{5.12 \times 10^{-5}}{4} = 1.28 \times 10^{-5}$ mol

Mass of dissolved oxygen in 50.0 cm^3 $= 1.28 \times 10^{-5} \times 32.00 = 4.096 \times 10^{-4}$ g

The concentration in g dm^{-3} is $4.096 \times 10^{-4} / \left(\frac{50.0}{1000}\right) = 8.19 \times 10^{-3}$ g dm^{-3}

This is equivalent to 8.19 ppm.

b The biochemical oxygen demand is determined by subtracting the final concentration of oxygen from its initial concentration.

BOD = 8.19 − 3.50 = 4.69 ppm.

TEST YOURSELF 9.4

The Winkler method was used to determine the concentration of dissolved oxygen in 100.0 cm^3 of water from a lake. 25.80 cm^3 of sodium thiosulfate of concentration 3.20×10^{-3} mol dm^{-3} was required for the titration.

1 Calculate the concentration of dissolved oxygen in ppm. 6.60

2 After 5 days' incubation at 20 °C another sample of water from the same source had a dissolved oxygen concentration of 2.50 ppm. Calculate the BOD. 4.10

9.4 Electrochemical cells

Electrochemical cells can be classified as voltaic (galvanic) cells and electrolytic cells.

DEFINITIONS

ELECTROLYSIS is the breaking down of a substance (in molten state or solution) by the passage of electricity through it.

ANODE electrode at which oxidation occurs.

CATHODE electrode at which reduction occurs.

How voltaic and electrolytic cells work and some similarities/differences are summarised in the table below. Some key differences are in bold.

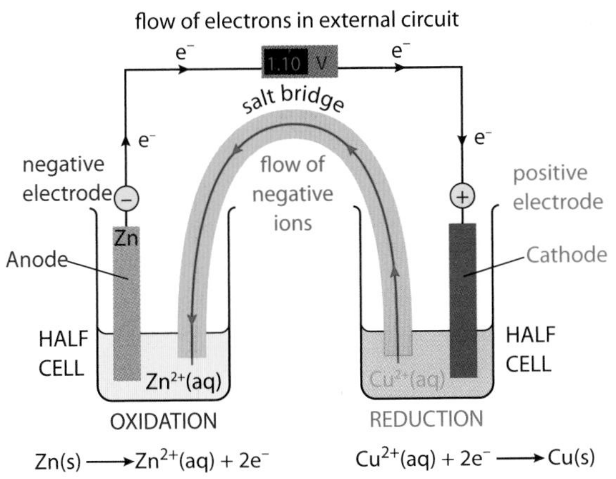

Figure 9.4

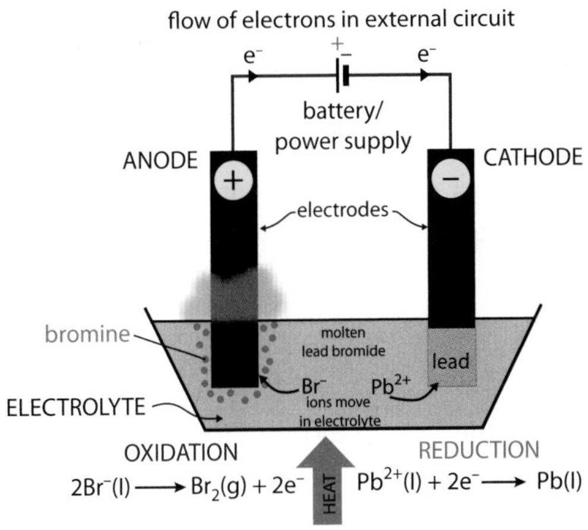

Figure 9.5

Voltaic cell	Electrolytic cell
Overall reaction: $Zn(s)+Cu^{2+}(aq) \rightarrow Zn^{2+}(aq)+Cu(s)$ Cell convention: $Zn(s)\vert Zn^{2+}(aq)\Vert Cu^{2+}(aq)\vert Cu(s)$	Overall reaction: $PbBr_2(l) \rightarrow Pb(l)+Br_2(g)$
Converts chemical energy from a spontaneous exothermic process to electrical energy.	**Converts electrical energy to chemical energy by forcing a non-spontaneous reaction to occur.**
Anode is negative and cathode is positive.	**Anode is positive and cathode is negative.**
The more reactive metal is always oxidised (at the anode).	The negative ion is oxidised at the anode.
Electrons flow from the more reactive metal through the external circuit to the ion of the less reactive metal. Electrons flow from the anode to the cathode.	Electrons flow from the battery to the negative electrode (cathode) and from the positive electrode (anode) to the battery.
The ion of the less reactive metal is reduced at the cathode.	The positive ion is reduced at the cathode.
The salt bridge completes the circuit and allows ions to flow into/out of the half-cells. The salt bridge can be made from a piece of paper soaked in a solution of an ionic salt such as KCl(aq).	No salt bridge.
Ions move in the solution and the salt bridge. No electrons flow in the salt bridge or the solutions.	Ions move in the electrolyte and electrons flow in the external circuit.
The bigger the difference in reactivity between the metals, the bigger the voltage.	The salt must be molten for electricity to flow – when molten the ions are free to move – if it is solid the ions are not free to move.
	The products of electrolysis of a molten salt are simply the elements of which it is made up, so, for example Al_2O_3 will produce aluminium and oxygen and $MgCl_2$ will produce magnesium and chlorine.

The cell convention for voltaic cells [e.g. $Zn(s)\,|\,Zn^{2+}(aq)\,||\,Cu^{2+}(aq)\,|\,Cu(s)$] describes the overall reaction. A single vertical line represents a phase boundary and the double line is the salt bridge. The reaction at the anode (oxidation) is shown on the left and that at the cathode (reduction) on the right.

9.5 Standard electrode potentials (HL only)

DEFINITION

STANDARD ELECTRODE POTENTIAL the EMF (voltage) of a half-cell connected to a standard hydrogen electrode (SHE), measured under standard conditions (solutions of concentration 1 mol dm^{-3}, pressure = 100 kPa and temperature usually 298 K).

You need to learn the details of the **standard hydrogen electrode**.

$H^+(aq) + e^- \rightleftharpoons \frac{1}{2}H_2(g) \qquad E^\ominus = 0.00\,V$

The standard electrode potential for the SHE is assigned a value of 0.00 V.

The standard electrode potential ($E^\ominus$) of a half-cell is measured by connecting it to the SHE. Standard electrode potentials are always written as **reduction** reactions and show the favourability of a particular reduction reaction relative to the SHE.

e.g. for $Zn|Zn^{2+}$ $\qquad Zn^{2+}(aq) + 2e^- \rightleftharpoons Zn(s) \qquad E^\ominus = -0.76\,V$

This reduction reaction is unfavourable relative to the SHE.

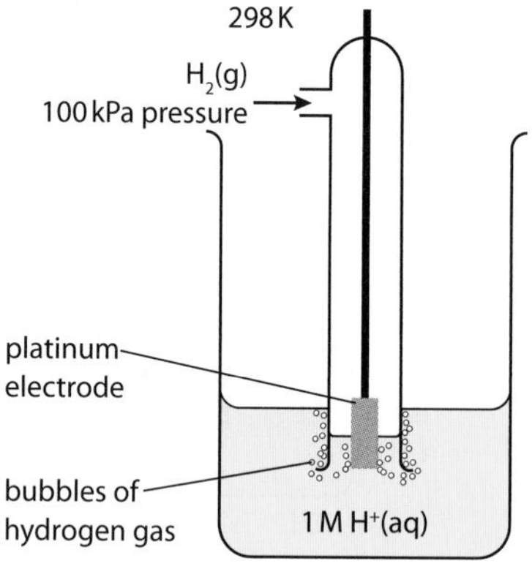

Figure 9.6

hint

Make sure that you look up the correct E° value – for instance, there is more than one half-equation involving Fe^{2+} and Cu^{2+}. You will always be given the oxidised and the reduced species.

Working out cell potentials

The cell potential is the potential (voltage) of a cell formed when two half-cells are connected.

Method for working out a cell potential:

- Write down the half-equations and standard electrode potentials for the two reactions.
- Change the sign of the more negative (less positive) $E^\ominus$ and add it to the other $E^\ominus$.

We are always looking for a cell potential that is **positive** overall as a positive value indicates a **spontaneous** reaction.

Worked example 9.5

a Deduce the overall reaction and calculate the cell potential for a cell consisting of an Fe^{3+}/Fe^{2+} half-cell and a $Zn|Zn^{2+}$ half-cell.

b State the polarity of the zinc electrode in the cell.

c State which is the anode and cathode in the cell and the direction of electron flow in the external circuit.

hint

Because there is no metal in the Fe^{3+}/Fe^{2+} half-cell, a platinum (inert) electrode is used to provide an electrical connection to the half-cell.

a Half-equations are given in the *IB Data Booklet*: $Zn^{2+}(aq) + 2e^- \rightleftharpoons Zn(s) \qquad E^\ominus = -0.76\,V$

$Fe^{3+}(aq) + e^- \rightleftharpoons Fe^{2+}(aq) \qquad E^\ominus = +0.77\,V$

The more negative value is −0.76 V and this half-equation is reversed:

$Zn(s) \rightarrow Zn^{2+}(aq) + 2e^- \qquad E^\ominus = +0.76\,V$ oxidation

$Fe^{3+}(aq) + e^- \rightarrow Fe^{2+}(aq) \qquad E^\ominus = +0.77\,V$ reduction

Add these to give the cell potential: $E^{\ominus}_{cell} = 0.76 + 0.77$ i.e. $\boldsymbol{E^{\ominus}_{cell} = +1.53}$ **V**

This is a positive value and so the reaction is spontaneous.

In order to combine the two half-equations to produce the overall redox equation the Fe^{3+}/Fe^{2+} half-equation must be multiplied by **2** so that the electrons balance:

$Zn(s) + 2Fe^{3+}(aq) \rightarrow Zn^{2+}(aq) + 2Fe^{2+}(aq)$

b Which is the negative electrode can be worked out from the original $E^{\ominus}$ values – the more negative value gives the negative electrode. Therefore, the Zn electrode is negative.

c The anode is the electrode at which oxidation occurs – the zinc electrode. Reduction occurs at the cathode and so the $Pt|Fe^{3+}$ electrode is the cathode.

Electrons always flow from the anode (negative electrode) to the cathode (positive electrode), that is from Zn to the Fe^{3+}/Fe^{2+} half-cell.

Note: negative charge always flows in the same continuous direction in the circuit and so negative ions in the salt bridge flow from the Fe^{3+}/Fe^{2+} half-cell to the Zn/Zn^{2+} half-cell.

hint

Do not use reversible arrows here.

hint

The standard electrode potentials in the *IB Data Booklet* are arranged from most negative to most positive – it is always the equation that is higher up the page that is reversed.

TEST YOURSELF 9.5

1 Calculate the cell potential for the following system and write an overall equation for the reaction that goes on in the cell.

$Fe^{3+}(aq) + e^- \rightleftharpoons Fe^{2+}(aq)$ $E^{\ominus} = 0.77$ V 1·37 − 0·77 = 0·60V

$Cl_2(g) + 2e^- \rightleftharpoons 2Cl^-(aq)$ $E^{\ominus} = 1.36$ V $2Fe^{2+} + Cl_2 \rightarrow 2Cl^- + 2Fe^{3+}$

2 State which is the positive electrode, which way electrons flow in the external circuit and which way ions flow in the salt bridge for the cell in **1**. Cl_2 is the positive electrode, e⁻ flow from Fe to Cl, salt bridge – negative flow to Fe^{3+}, positive flow to Cl^-

3 Explain which electrode is the anode and which the cathode in the cell in **1**.

Pt electrode is anode as Fe is oxidised

Oxidising and reducing agents

A very **positive** value for $E^{\ominus}$ means that the reduction reaction is very favourable (relative to the SHE) and the substance has a very strong tendency to pick up electrons from other species, i.e. it is a strong oxidising agent. A very negative $E^{\ominus}$ value indicates a strong reducing agent.

In the *IB Data Booklet* table species further up the table are stronger reducing agents (the '**reduced species**' is the reducing agent – on the right) and will reduce species below them. Species lower down the table are stronger oxidising agents (the '**oxidised species**' is the oxidising agent – on the left) and will oxidise species above them.

oxidised species		reduced species	$E^{\ominus}$/V	
$K^+(aq) + e^-$	$\rightleftharpoons$	K(s)	−2.92	K very strong reducing agent
$Zn^{2+}(aq) + 2e^-$	$\rightleftharpoons$	Zn(s)	−0.76	K will reduce Zn^{2+}
$2BrO_3^-(aq) + 12H^+(aq) + 10e^-$	$\rightleftharpoons$	$Br_2(aq) + 6H_2O(l)$	+1.52	F_2 will oxidise Br_2 to BrO_3^-
$F_2(g) + 2e^-$	$\rightleftharpoons$	$2F^-(aq)$	+2.87	F_2 very strong oxidising agent

A substance with a more positive electrode potential will oxidise a substance with a less positive electrode potential.

Standard cell potentials can also be used to predict whether redox reactions occur.

Worked example 9.6

Explain which of the halide ions (F^- or Br^-) will be oxidised by acidified potassium dichromate(VI), deduce a redox equation and calculate the cell potential for the reaction that occurs.

The half-equations are:

$Cr_2O_7^{2-}(aq) + 14H^+(aq) + 6e^- \rightleftharpoons 2Cr^{3+}(aq) + 7H_2O(l)$ $E^\ominus = +1.33$ V

$F_2(g) + 2e^- \rightleftharpoons 2F^-(aq)$ $E^\ominus = +2.87$ V

$Br_2(aq) + 2e^- \rightleftharpoons 2Br^-(aq)$ $E^\ominus = +1.09$ V

From this we can see that $Cr_2O_7^{2-}$ has a more positive standard electrode potential than Br_2 and so is a stronger oxidising agent than Br_2 but a weaker oxidising agent than F_2. This means that $Cr_2O_7^{2-}$ will oxidise Br^- to Br_2 but not F^- to F_2.

Alternatively – work out the cell potential for each reaction – the positive value indicates the spontaneous reaction.

Br^- is oxidised and therefore the half-equation must be reversed:

$2Br^-(aq) \rightarrow Br_2(aq) + 2e^-$ $E^\ominus = -1.09$ V **oxidation**

$Cr_2O_7^{2-}(aq) + 14H^+(aq) + 6e^- \rightarrow 2Cr^{3+}(aq) + 7H_2O(l)$ $E^\ominus = +1.33$ V **reduction**

$E^\ominus_{cell} = -1.09 + 1.33 = +0.24$ V

A positive $E^\ominus_{cell}$ value indicates that the reaction is spontaneous.

The $Br^-|Br_2$ half-equation is multiplied by 3 to balance electrons, then the half-equations are added together to give the overall equation:

$Cr_2O_7^{2-}(aq) + 14H^+(aq) + 6Br^-(aq) \rightarrow 2Cr^{3+}(aq) + 7H_2O(l) + 3Br_2(aq)$

hint

Remember to use single arrows, not reversible ones, in the final redox equation – you could lose marks otherwise.

TEST YOURSELF 9.6

1 Using standard electrode potentials given previously, explain whether the following reaction will be spontaneous or not: yes as E° is positive

$Cr_2O_7^{2-}(aq) + 14H^+(aq) + 6Fe^{2+}(aq) \rightarrow 2Cr^{3+}(aq) + 7H_2O(l) + 6Fe^{3+}(aq)$

2 Select the strongest reducing agent from these species:

$U^{4+} + e^- \rightleftharpoons U^{3+}$ $E^\ominus = -0.61$ V

$Eu^{3+} + e^- \rightleftharpoons Eu^{2+}$ $E^\ominus = -0.43$ V

hint

Positive $E^\ominus_{cell}$ value indicates that a reaction will be spontaneous; however, the fact that a reaction is spontaneous does not tell us anything about the speed of the reaction.

Standard cell potentials and ΔG

The standard cell potential and the Gibbs free energy change (ΔG) are related by the equation: $\Delta G^\ominus = -nFE^\ominus$

where n is the number of electrons transferred in a particular redox reaction and F is the Faraday constant (96 500 C mol^{-1}).

hint

Remember that ΔG will come out in J mol^{-1} using this equation.

Annotated exemplar answer 9.1

Calculate $\Delta G^{\ominus}$ (in kJ mol^{-1}) for the reaction:

$Cr_2O_7^{2-}(aq) + 14H^+(aq) + 6Br^-(aq) \rightarrow 2Cr^{3+}(aq) + 7H_2O(l) + 3Br_2(aq)$ $E^{\ominus}_{cell} = +0.24\,V$ **[2]**

0/2

$\Delta G^{\ominus} = -2 \times 96\,500 \times 0.24 = -46\,320$

It is good practice to state the equation you are using:

$\Delta G^{\ominus} = -nFE^{\ominus}$

Correct answer is $\Delta G^{\ominus} = -140$ kJ mol^{-1}

The half equations must be written out to work out how many electrons are transferred:

$6Br^-(aq) \rightarrow Br_2(aq) + 6e^-$

$Cr_2O_7^{2-}(aq) + 14H^+(aq) + 6e^- \rightarrow 2Cr^{3+}(aq) + 7H_2O(l)$

6 electrons are transferred, therefore $n = 6$.

Units needed – this answer would be in J mol^{-1} but an answer in kJ mol^{-1} is required. The answer should be given to 2 significant figures.

TEST YOURSELF 9.7

Using standard electrode potentials from the *IB Data Booklet*, calculate ΔG for the following reaction: -357 KJ mol⁻¹

$MnO_4^-(aq) + 8H^+(aq) + 5Fe^{2+}(aq) \rightarrow Mn^{2+}(aq) + 4H_2O(l) + 5Fe^{3+}(aq)$

hint

The Faraday constant is given in the *IB Data Booklet*.

9.6 Electrolysis of aqueous solutions (HL only)

The products of electrolysis of an aqueous solution can be different from those of a molten salt. The difference arises because water can also be oxidised/reduced at the electrodes as well as the individual ions. Water can be oxidised to oxygen at the anode and reduced to hydrogen at the cathode.

In general the products of electrolysis of an aqueous solution are:

Electrode	Product
Positive (**anode**)	Oxygen or a halogen
Negative (**cathode**)	A metal or hydrogen

The three main factors that affect the products formed at the electrodes are:

- The standard electrode potentials of the species in solution.
- The concentration of the electrolyte.
- The material from which the electrodes are made.

Using the standard electrode potential to predict the product of electrolysis

Consider the electrolysis of sodium chloride solution. There are two possible reduction reactions at the cathode

$Na^+(aq) + e^- \rightarrow Na(s)$ $E^{\ominus} = -2.71\,V$

$H_2O(l) + e^- \rightarrow \frac{1}{2}H_2(g) + OH^-(aq)$ $E^{\ominus} = -0.83\,V$

The reduction of water at the cathode is more favourable (more positive $E^{\ominus}$) than the reduction of sodium ions therefore water is reduced to form H_2 and sodium is not produced.

hint

For metals with electrode potentials between –0.83 V and 0.00 V the situation is more complicated and the products are more difficult to predict.

In the electrolysis of copper sulfate solution the reduction of copper ions to copper ($E^\ominus$ = +0.34 V) is more favourable than the reduction of water and therefore copper is produced at the cathode.

In general, if a metal has a very negative electrode potential hydrogen will be discharged at the cathode. But, if the metal has a positive electrode potential the metal itself will be discharged at the cathode.

The effect of concentration of the electrolyte

The concentration of ions in solution can also affect the product obtained. You will probably only meet this for solutions containing chloride ions – at very low chloride concentrations the product of electrolysis is mainly oxygen (oxidation of water) but with more concentrated solutions chlorine is the major product.

The effect of nature of the electrodes

What the electrodes are made of can also affect the product obtained – you will probably only meet this for solutions of copper sulfate with inert (graphite or platinum) electrodes or copper electrodes.

Electrolysis of copper sulfate solution:

Electrode material	Electrode	Product	Half-equation	Observations
graphite/Pt	anode (+)	O_2	$2H_2O(l) \rightarrow O_2(g) + 4H^+(aq) + 4e^-$	cathode (–) coated in copper, bubbles of O_2 given off at the anode (+), blue colour of the solution fades ($[Cu^{2+}]$ decreases), the solution gets more acidic (H^+ ions produced)
	cathode (–)	Cu	$Cu^{2+}(aq) + 2e^- \rightarrow Cu(s)$	
copper	anode (+)	electrode dissolves	$Cu(s) \rightarrow Cu^{2+}(aq) + 2e^-$	anode gets smaller, cathode coated with copper and the solution remains the same colour ($[Cu^{2+}]$ does not change)
	cathode (–)	Cu	$Cu^{2+}(aq) + 2e^- \rightarrow Cu(s)$	

At the copper anode the oxidation of copper is more favourable than the oxidation of water:

$Cu(s) \rightarrow Cu^{2+}(aq) + 2e^-$ $\qquad E_{ox} = -0.34\,V$

$2H_2O(l) \rightarrow O_2(g) + 4H^+(aq) + 4e^-$ $\qquad E_{ox} = -1.23\,V$

hint

To help you with working out a suitable substance for the electrolyte remember that all nitrates are soluble in water.

Electroplating

This is the process of coating an object with a thin layer of a metal using electrolysis. The object to be coated should be used as the cathode (since metal ions are positive), the anode should be made of the metal with which the object is to be plated and the electrolyte is a solution containing the ions of the coating metal.

Electrolysis of water

Distilled water is a very poor conductor of electricity and so a small amount of sulfuric acid is usually added to increase the concentration of ions present.

Twice as much hydrogen as oxygen should be collected. This can be seen if the half-equations for the oxidation and reduction of water are written with equal numbers of electrons:

$2H_2O(l) \rightarrow O_2(g) + 4H^+(aq) + \mathbf{4e^-}$ $\qquad$ **anode**

$4H_2O(l) + \mathbf{4e^-} \rightarrow 2H_2(g) + 4OH^-(aq)$ $\qquad$ **cathode**

When four electrons are lost at the anode one molecule of O_2 is formed but when four electrons are gained at the cathode two molecules of H_2 are formed, therefore two molecules of H_2 are formed for every one molecule of O_2 formed.

Nature of Science. Scientists must be aware of systematic errors in an experiment – the volume of hydrogen collected is more than twice the volume of oxygen because oxygen is more soluble in water than hydrogen.

TEST YOURSELF 9.8

Predict the products formed at the electrodes when the following solutions are electrolysed:

- concentrated $MgCl_2$(aq) anode - chlorine cathode - hydrogen
- $Cu(NO_3)_2$(aq) with platinum electrodes anode - oxygen cathode - copper
- H_2SO_4(aq) anode - oxygen cathode - hydrogen

Quantitative electrolysis

The factors which affect the amount of product formed at the electrodes are:

Current	Time of electrolysis	Charge on the ion
proportional to number of moles of product		inversely proportional to number of moles of product
double current or time – twice as much charge will pass and twice as many electrons will be transferred to/from ions, therefore twice as many moles of substance will be produced		the higher the charge on the ion the more electrons will be required to produce 1 mole of atoms $Na^+ + e^- \rightarrow Na$ $Mg^{2+} + 2e^- \rightarrow Mg$ 2 moles of electrons produce 2 moles of Na but only 1 mole of Mg

You need to be able to work out the relative amounts of products formed during electrolysis – to do this look at the number of electrons that must flow to produce 1 mole of substance by writing out the half-equations. Remember that, if two cells are connected in series, the current is the same in each – the same number of electrons will flow through each in a certain amount of time.

☆ Model answer 9.2

During the electrolysis of copper sulfate solution using platinum electrodes 0.636 g of copper is formed at the cathode. Calculate the volume of oxygen (measured at STP) produced at the anode.

moles of copper formed $= \frac{0.636}{63.55} = 0.0100$ mol

Half-equations: cathode: $Cu^{2+}(aq) + 2e^- \rightarrow Cu(s)$

anode: $2H_2O(l) \rightarrow O_2(g) + 4H^+(aq) + 4e^-$

There are $4e^-$ in the second equation but only $2e^-$ in the first, therefore twice as many electrons must flow to form 1 mole of O_2 compared to 1 mole of Cu.

Therefore number of moles of O_2 is half the number of moles of copper: i.e. 0.00500 mol.

Volume of $O_2 = 0.00500 \times 22.7 = 0.114$ dm^3.

TEST YOURSELF 9.9

An electrolytic cell containing Na_2SO_4(aq) was connected to one containing $CuSO_4$(aq). Both cells had platinum electrodes. 100.0 cm^3 of hydrogen (measured at STP) was produced in the first cell. Calculate the volume of oxygen and mass of copper produced in the second cell.

volume of oxygen = 50 cm^3
mass of copper = 0.28 g

Checklist

At the end of this chapter you should be able to:

- ☐ Work out the oxidation state of an atom in a compound/ion.
- ☐ Work out whether a reaction involves oxidation/reduction and identify oxidising and reducing agents.
- ☐ Balance half-equations for reactions in neutral/acidic solution and write balanced redox equations.
- ☐ Carry out calculations involving biochemical oxygen demand.
- ☐ Describe the different types of electrochemical cells including similarities and differences between them.
- ☐ Predict the products of electrolysis of molten salts and write equations/half-equations for reactions occurring.

Higher Level only

- ☐ Describe the standard hydrogen electrode and define standard electrode potential.
- ☐ Carry out calculations involving standard electrode potentials.
- ☐ Use standard electrode potentials to predict the feasibility of redox reactions.
- ☐ Explain the products of electrolysis of aqueous solutions.
- ☐ Describe the factors that affect the amount of product formed in electrolysis and determine the relative amounts of products formed.

ORGANIC CHEMISTRY

There are more compounds of carbon than of all other elements put togeth[er]. Organic chemistry deals with the chemistry of carbon compounds and the rea[c]tions of functional groups.

This chapter covers the following topics:

- ☐ Fundamentals of organic chemistry
- ☐ Functional group chemistry
- ☐ Types of organic reactions (HL only)
- ☐ Synthetic routes (HL only)
- ☐ Stereoisomerism (HL only)

10.1 Fundamentals of organic chemistry

Homologous series and functional groups

DEFINITIONS

HOMOLOGOUS SERIES series of compounds with the same functional group, in which each member differs from the next by a common structural unit (usually $-CH_2-$).

FUNCTIONAL GROUP atom/group of atoms in a molecule that gives it its characteristic chemical properties – the reactive part of a molecule.

HYDROCARBON a compound containing carbon and hydrogen only.

The members of a particular homologous series have similar **chemical properties** as they have the same functional group. There is, however, a gradation in physical properties, e.g. boiling point increases as the carbon chain gets longer – this is due to an increase in the strength of London forces between molecules as the relative molecular mass of the compound increases.

Various types of structural formulae are used to describe organic molecules, e.g. for propan-1-ol:

Name	Structural formulae		
	Condensed	Full	Skeletal
propan-1-ol	$CH_3CH_2CH_2OH$	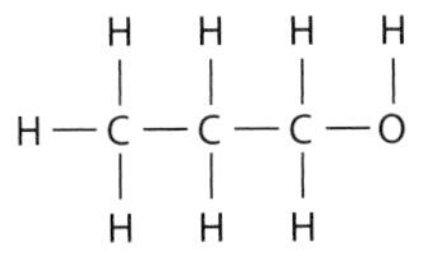	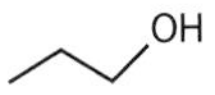

The names of the homologous series and functional groups you need to know are summarised the table on the next page.

hint

For a full structural formula you must show all the bonds between atoms – don't forget the O–H bond.

hint

If presented with a skeletal structure it is usually easiest to draw in all the C and H atoms – don't forget that C forms a maximum of four bonds.

Organic chemistry

Class name	Functional group	Functional group name	Example	General formula	How to name
alkane		alkyl	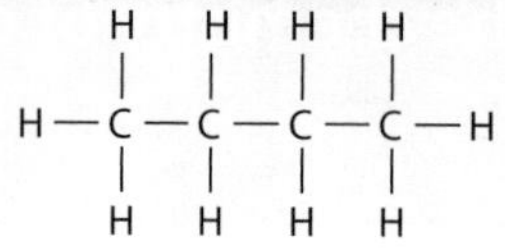	C_nH_{2n+2}	alkane
alkene	C═C	alkenyl	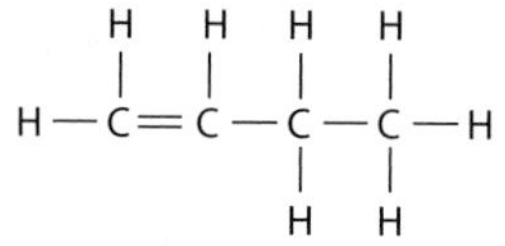	C_nH_{2n}	alk-*x*-**ene** (or *x*-alkene)
alkyne	C≡C	alkynyl	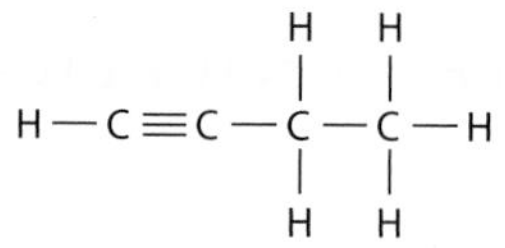	C_nH_{2n-2}	alk-*x*-**yne** (or *x*-alkyne)
alcohol	–OH	hydroxyl	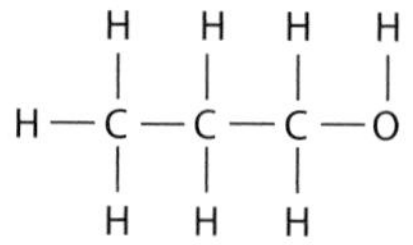	$C_nH_{2n+2}O$ or $C_nH_{2n+1}OH$	alkan-*x*-**ol** (or *x*-alkanol)
ether	C—O—C	ether	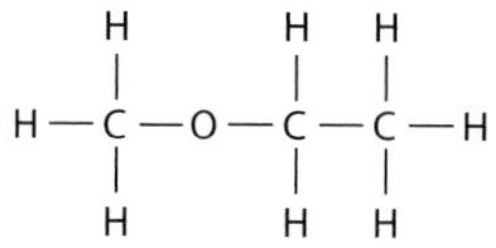	$C_nH_{2n+2}O$	*x*-alkoxyalkane
aldehyde	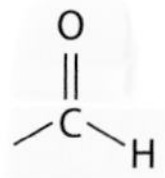	carbonyl	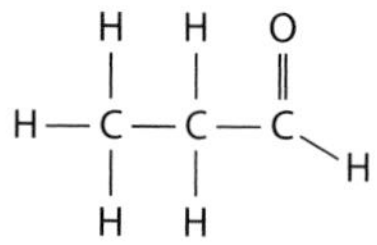	$C_nH_{2n}O$	alkanal
ketone	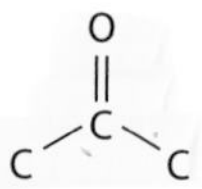		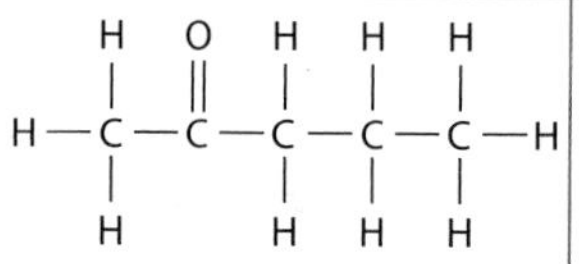	$C_nH_{2n}O$	alkan-*x*-**one** (or *x*-alkanone)
carboxylic acid	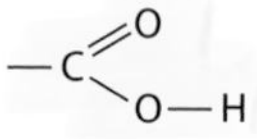	carboxyl	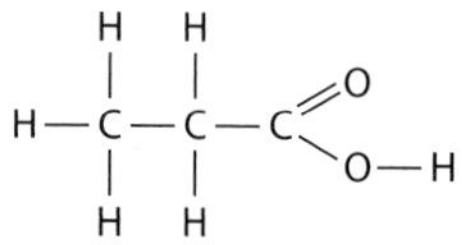	$C_nH_{2n}O_2$ or $C_nH_{2n+1}COOH$	alkanoic acid
halogeno–alkane	—X X=Cl/Br/I	halo	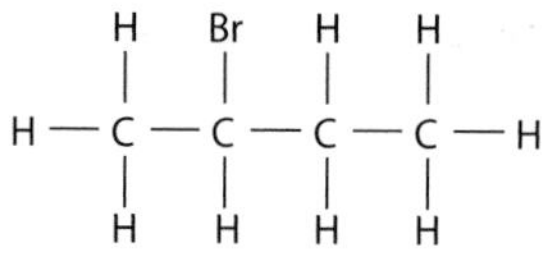		*x*-haloalkane
amine	$-NH_2$ –NHR $-NR_2$	amine	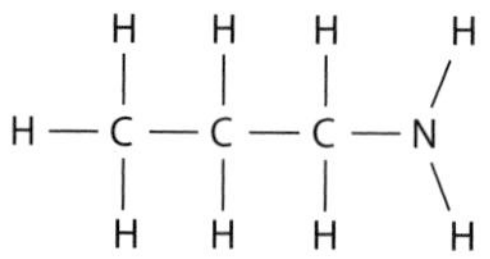		alkylamine or *x*-aminoalkane or alkan-*x*-amine
ester	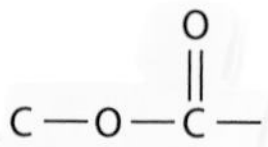	ester	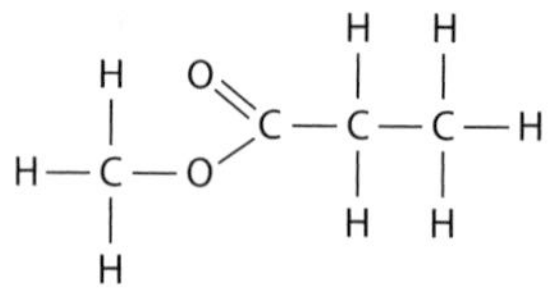		alkyl alkanoate

nitrile	$—C≡N$	nitrile	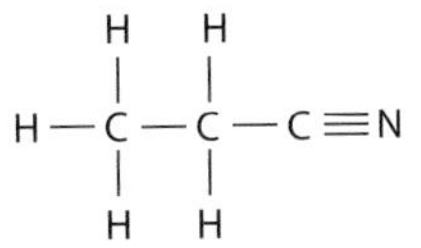		alkanenitrile (C of C≡N included in chain)
amide	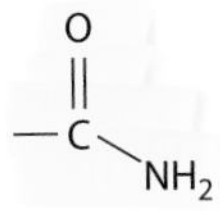	carboxamide	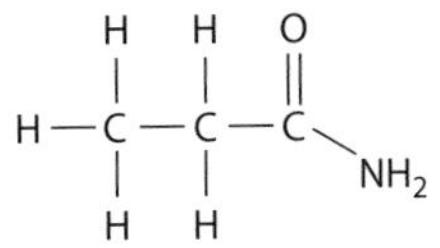		alkanamide

☆ Model answer 10.1

Name the functional groups present in the following molecule:

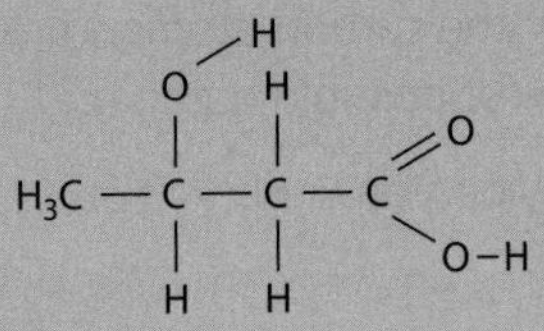

Figure 10.1

The functional groups are: hydroxyl and carboxyl.

hint

Remember that the names of the functional groups are not the same as the class names – you must use the functional group names when asked in the exam.

TEST YOURSELF 10.1

State the names of the functional groups present in the following molecules:

1, 2

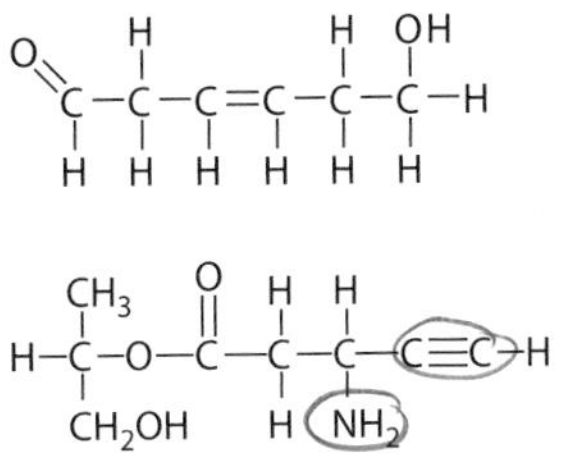

Figure 10.2

1: alkene, alcohol, aldehyde

2: alcohol, ether, alkyne, amine, ketone

Saturated and unsaturated compounds

DEFINITIONS

UNSATURATED COMPOUNDS contain multiple bonds between carbon atoms (C=C or C≡C).

SATURATED COMPOUNDS contain only single bonds between carbon atoms (C–C).

Alkenes and alkynes are described as unsaturated but alkanes are saturated.

hint

The term saturated or unsaturated refers to the carbon skeleton and does not usually include the other groups (e.g. C=O or C≡N) in a molecule – thus it is possible to have a saturated or unsaturated carboxylic acid (referring to whether the hydrocarbon chain contains C=C or not).

Naming organic compounds

Organic compounds are named according to the **IUPAC** system.
Tips for naming compounds:

- Find the longest **continuous carbon chain** in the molecule – don't forget that this may go around a corner.
- Number the carbons so that the functional group (e.g. alkenyl, hydroxyl, etc.) in the molecule has the lowest possible number, i.e. start numbering from the end closest to the functional group.

- Number the positions of alkyl groups (branches) using the combination that includes the lowest individual numbers (not the sum) but bearing in mind the rule that other functional groups take priority over alkyl groups.
- Arrange the names of the groups in alphabetical order.
- When more than one group of a particular type is present use the prefixes di (2), tri (3), tetra (4).
- Use a comma between numbers and a dash between numbers and letters, e.g. 2, 2-dimethylbutane.

Worked example 10.1

State the IUPAC name of the compound shown in Figure 10.3.

Start by highlighting the longest continuous carbon chain in the molecule. Number the carbon atoms so that the OH group has the lowest possible number.

This molecule is called 3-methylpentan-1-ol.

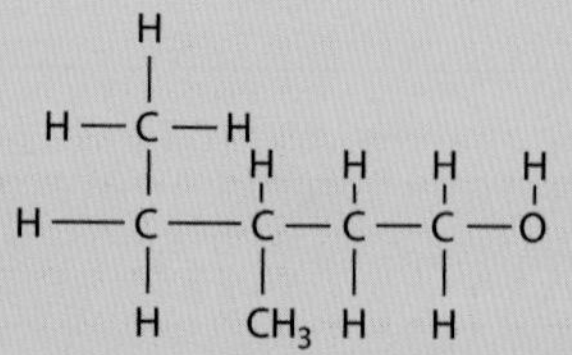

Figure 10.3

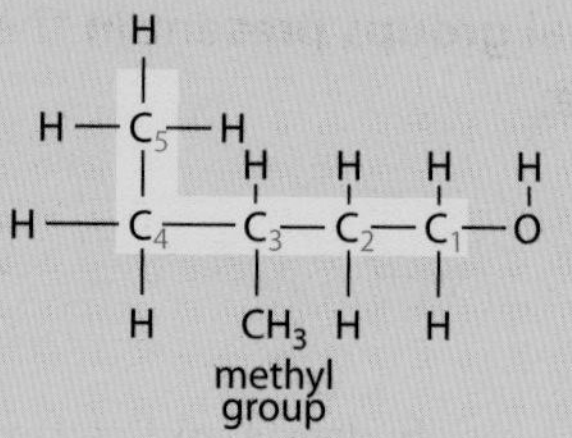

Figure 10.4

TEST YOURSELF 10.2

Name the following molecules.

1

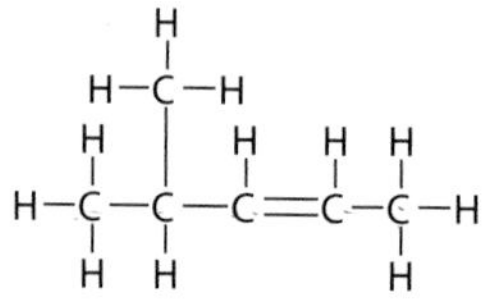

2

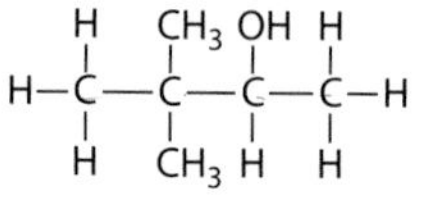

3

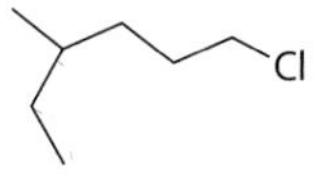

2 methyl pentene

3,3 di methyl butanol

1 chloro 4 methylhexane

Figure 10.5

Naming ethers

Ethers contain the C–O–C group – the shorter carbon chain and the O atom together are named as an alkoxy substituent on the main alkane chain.

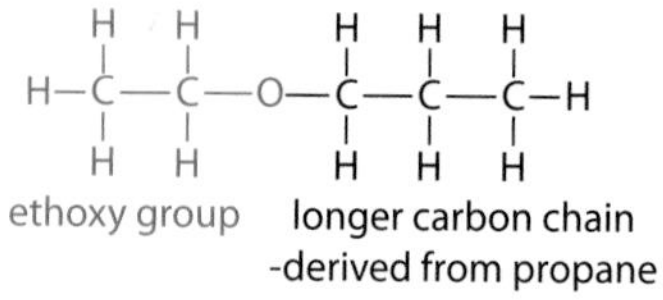

Figure 10.6

Here, the ethoxy group is on Carbon 1 of the longer hydrocarbon chain and so the molecule is 1-ethoxypropane.

TEST YOURSELF 10.3

Name the following molecule.

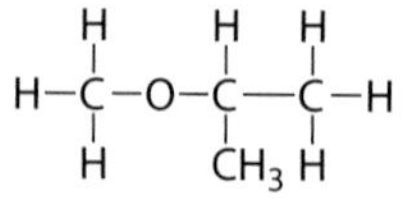

Figure 10.7

2 - Methoxypropane

Isomers

DEFINITION

STRUCTURAL ISOMERS two or more compounds which have the same molecular formula but different structural formulae, i.e. a different arrangement of atoms.

The two structural isomers with the molecular formula C_4H_{10} are:

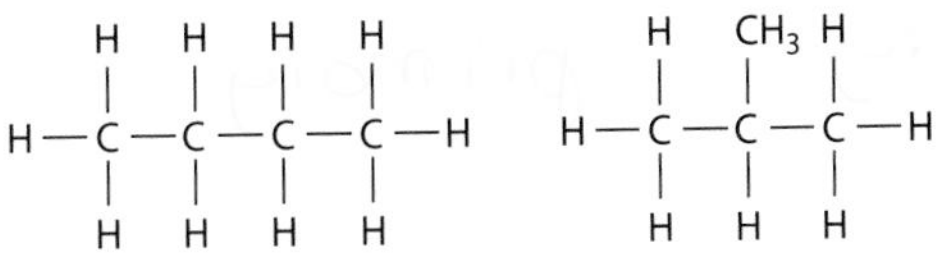

Figure 10.8

TEST YOURSELF 10.4

State the number of isomers of:

1 C_5H_{12} 3
2 C_4H_8 that are alkenes 3
3 $C_4H_{10}O$ that are ethers 3
4 $C_4H_{10}O$ that are alcohols 4

hint

Remember that structural isomers can be generated by just moving functional groups around, e.g. propan-1-ol and propan-2-ol are isomers of each other.

Primary, secondary and tertiary halogenoalkanes and alcohols

Alcohols (and halogenoalkanes) may be described as primary, secondary or tertiary.

Locate the C attached to the OH group (or the halogen atom). Count the number of C atoms attached to this C:

- 1 C atom attached – primary
- 2 C atoms attached – secondary
- 3 C atoms attached – tertiary

hint

Methanol, is also a primary alcohol.

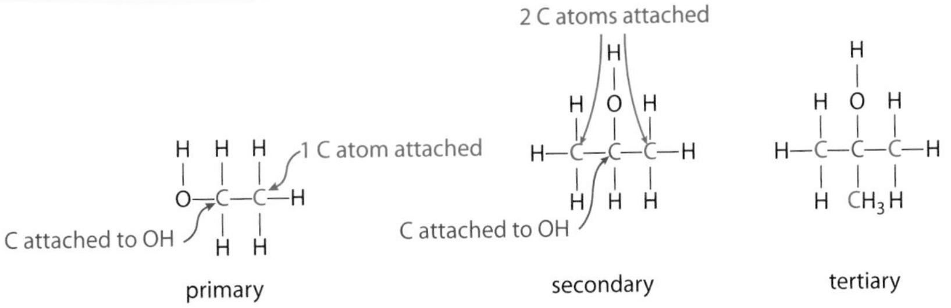

Figure 10.9

Primary, secondary and tertiary amines

Amines are also classified as primary, secondary and tertiary according to the number of alkyl groups attached to the N.

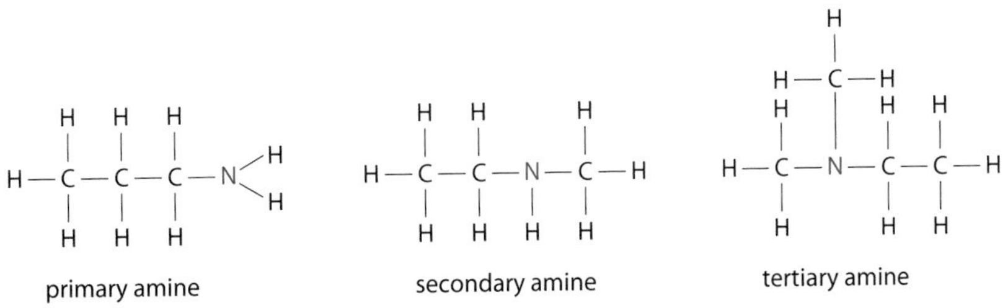

Figure 10.10

TEST YOURSELF 10.5

Classify each of the following as primary, secondary or tertiary.

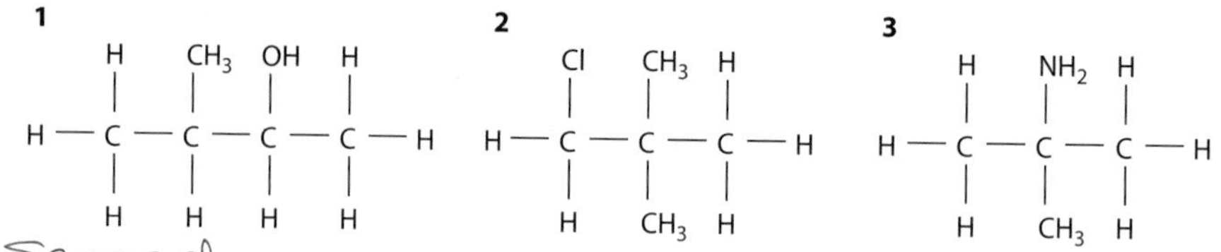

1 secondary 2 primary 3 primary

Figure 10.11

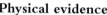

Benzene and aromatic compounds

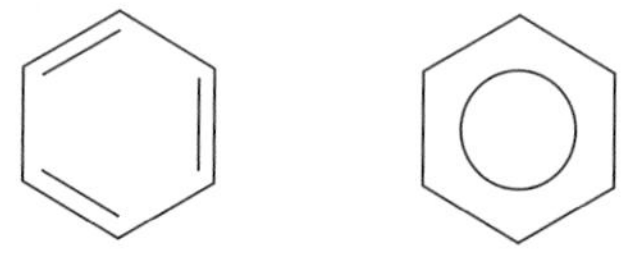

cyclohexa-1,3,5-triene (Kekulé benzene)

benzene (delocalised ring of electrons)

Figure 10.12

Benzene has the molecular formula C_6H_6 and consists of a planar hexagonal ring of carbon atoms. The structure can be drawn with alternating double and single bonds between carbon atoms or with a delocalised ring of six electrons.

hint

HL students should remember that the C-C bond order is 1.5 – equivalent to each π bond being shared over 2 C-C bonds.

Evidence for the delocalised structure rather than the structure with alternating double and single bonds:

Physical evidence

- All the C–C bonds in benzene are the same length but if the structure had double and single bonds there would be shorter (C=C) and longer (C–C) bonds between C atoms.
- Only three isomers exist for $C_6H_4Cl_2$ but if there were three double bonds in the ring, four isomers would be predicted (the 2 Cl atoms could be either side of a C–C or a C=C).

Chemical evidence

- When cyclohexene (6-membered ring with 1 C=C) reacts with hydrogen 120 kJ mol^{-1} are given out but when benzene reacts with hydrogen considerably less than three times as much energy is given out.
- Compounds with C=C undergo **addition reactions** (with e.g. Cl_2) but benzene undergoes substitution reactions.

hint

Technically a phenyl group is C_6H_5 but the name is also used to describe other more substituted derivatives present in molecules, e.g. C_6H_4, C_6H_3 etc.

Benzene is described as an **aromatic unsaturated hydrocarbon**. When a benzene ring is present in a compound it is called a **phenyl** group and compounds containing this are called **aromatic compounds**.

10.2 Functional group chemistry

Alkanes

Alkanes undergo combustion reactions:

- complete combustion produces CO_2 and H_2O
- incomplete combustion (limited supply of oxygen) produces CO/C and H_2O

Alkanes are unreactive because:

- of the high strength of C–C and C–H bonds
- the C–C and C–H bonds and the overall molecules are essentially non-polar

TEST YOURSELF 10.6

Write equations for:

1 The complete combustion of ethane. $C_2H_6 + \frac{7}{2}O_2 \rightarrow 2CO_2 + 3H_2O$

2 The incomplete combustion of propane producing CO. $C_3H_8 + 2O_2 \rightarrow CO + 3H_2O + 2C$

Reaction of alkanes with halogens

DEFINITIONS

SUBSTITUTION REACTION one in which one atom or group is replaced by another atom or group.

FREE RADICALS species (atoms or groups of atoms) with an unpaired electron, e.g. Cl•.

HOMOLYTIC FISSION a covalent bond breaks so that one electron goes back to each atom making up the original covalent bond.

Homolytic fission can be shown using 'fish-hooks', which show the movement of one electron:

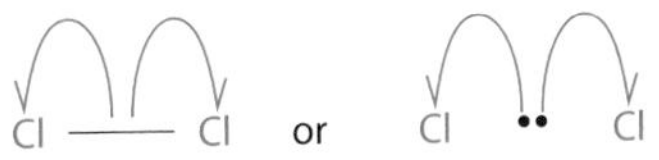

Figure 10.13

In the presence of UV light (sunlight) alkanes undergo substitution reactions with halogens.

The reaction between methane and chlorine: $CH_4 + Cl_2 \xrightarrow{UV} CH_3Cl + HCl$

methane + chlorine ⟶ chloromethane + hydrogen chloride

The mechanism for these reactions is **free radical substitution.**

The first stage is photochemical homolytic fission of the bond between the halogen atoms to produce two free radicals:

$Cl_2 \xrightarrow{UV} 2Cl\bullet$	**initiation**
$Cl\bullet + CH_4 \rightarrow \bullet CH_3 + HCl$	**propagation**
$\bullet CH_3 + Cl_2 \rightarrow CH_3Cl + Cl\bullet$	**propagation**
$Cl\bullet + Cl\bullet \rightarrow Cl_2$	**termination**
$Cl\bullet + \bullet CH_3 \rightarrow CH_3Cl$	**termination**
$H_3C\bullet + \bullet CH_3 \rightarrow C_2H_6$	**termination**

hint

You usually only need to show one of the termination steps – the easiest one to learn is just the first one as it is simply the reverse of the initiation step.

The Cl• generated in the second propagation step can go on to react further with another methane molecule so that the cycle of propagation steps starts again – **a chain reaction**.

hint

When propane reacts with chlorine in the presence of UV light two products are possible: 1-chloropropane and 2-chloropropane depending on which H is substituted.

Reactions of alkenes

Alkenes are more reactive than alkanes. This is because:

- One component (the π bond) of the C=C is weaker than a normal C–C single bond.
- The double bond (four electrons) attracts electrophiles (reagents attracted to electrons).

Alkenes undergo **addition reactions** as shown in Figure **10.14**.

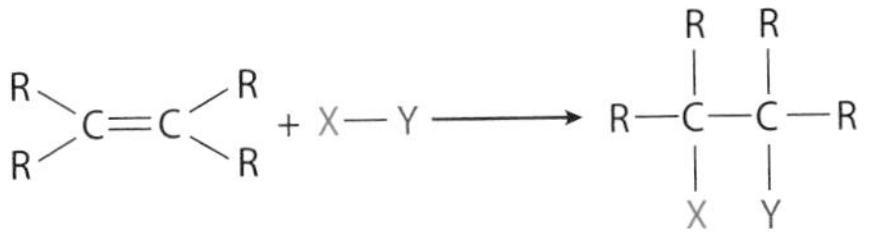

Figure 10.14

You need to learn the following reactions of alkenes:

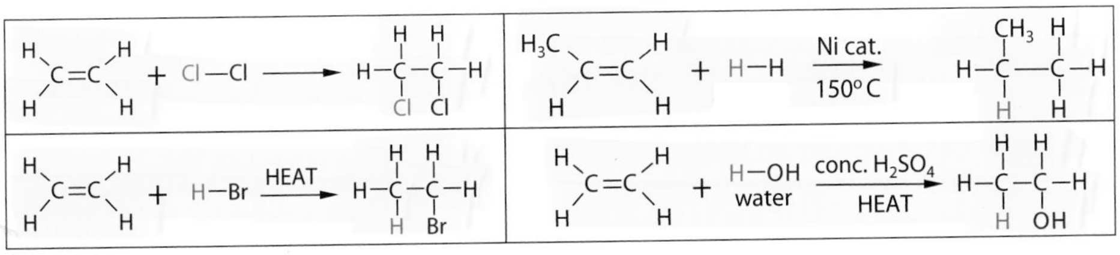

Figure 10.15

TEST YOURSELF 10.7

Give the structure of the product of the following reactions:

1 Propene + Br_2

2 $CH_3CH_2CHCHCH_2CH_3 + H_2O$

Distinguishing between alkanes and alkenes

Alkenes (unsaturated) may be distinguished from alkanes (saturated) by shaking them with bromine water, $Br_2(aq)$:

Alkene: bromine water – orange to colourless

Alkane: bromine water – no colour change

The product of the reaction of ethene with bromine water is CH_2BrCH_2OH.

Addition polymerisation

Alkenes undergo **addition polymerisation**. Reactions often require high temperature and pressure (a catalyst/initiator may also be required).

If all Rs are H we have ethene and poly(ethene); if one R is Cl and the rest are Hs, we have chloroethene and poly(chloroethene) (PVC) – these are important plastics and the addition polymerisation reaction forms the basis of the plastics industry.

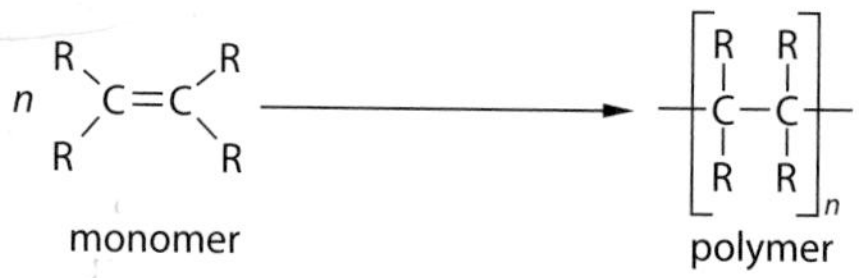

Figure 10.16

The structure in brackets in Figure **10.16** is called the **repeating unit** of the polymer – this is the basic unit from which the whole polymer can be made up.

Addition polymers are named according to the **monomer** from which they were made.

It is important to remember that **only the C=C group reacts** when the polymerisation reaction occurs – all the other groups attached to the C=C are unaffected. To make sure that you do not make a mistake when drawing the structure of a polymer, draw out the alkene monomer with the C=C at the centre and the four groups attached at 90°. Now all you have to do is open out the double bond to get the repeating unit.

Worked example 10.2

Draw the repeating unit of the addition polymer formed by propene.

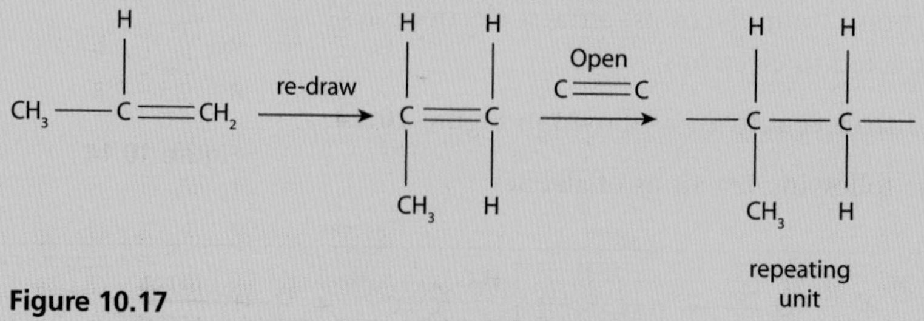

Figure 10.17

If you are given part of the polymer chain, take any two adjacent Cs in the polymer chain to give the repeating unit. Put a double bond between the 2Cs to give the original monomer.

TEST YOURSELF 10.8

1 Draw three repeating units for the polymer formed from but-1-ene.
2 Draw the repeating unit and monomer for the polymer shown.

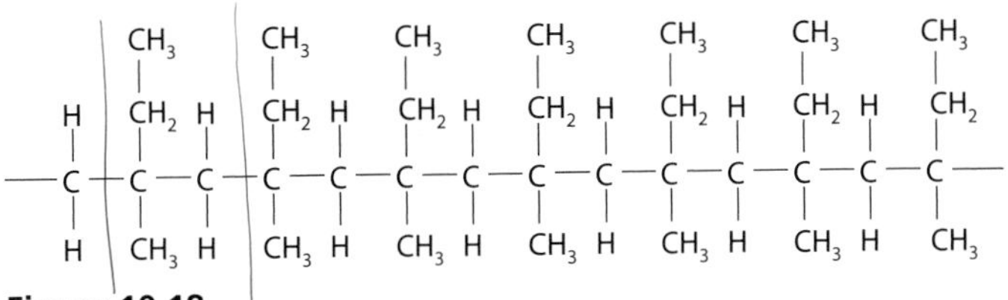

Figure 10.18

Nature of Science. Serendipity (chance) can play a role in scientific discoveries, e.g. the discovery of PTFE [poly(tetrafluoroethene)], which is used as a non-stick coating.

Reactions of alcohols

Combustion of alcohols

In a plentiful supply of O_2 alcohols burn to produce CO_2 and H_2O:

e.g. $C_2H_5OH + 3O_2 \rightarrow 2CO_2 + 3H_2O$

hint
Don't forget the O in the alcohol when balancing.

Oxidation

Primary alcohols are oxidised first of all to an aldehyde (partial oxidation), which can be oxidised further to a carboxylic acid (complete oxidation).

Oxidising agents that can be used are:

- Acidified dichromate(VI) $[Cr_2O_7^{2-}/H^+]$ – colour change: **orange → green**
- Acidified manganate(VII) $[MnO_4^-/H^+]$ – colour change: **purple → colourless**

e.g.

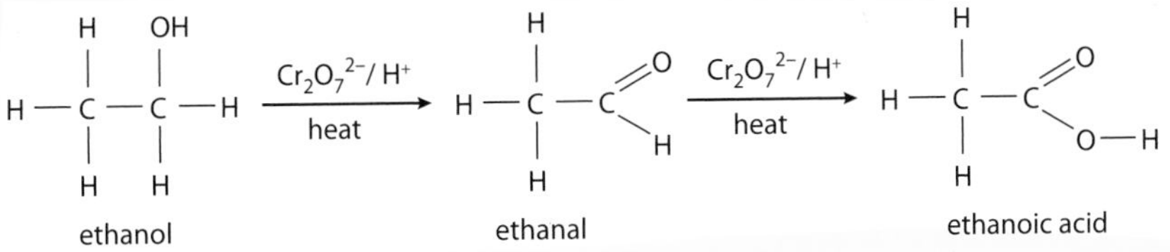

Figure 10.19

It is important to remember, when writing organic reactions, that only the functional groups change and not the rest of the carbon skeleton – focus on the functional groups:

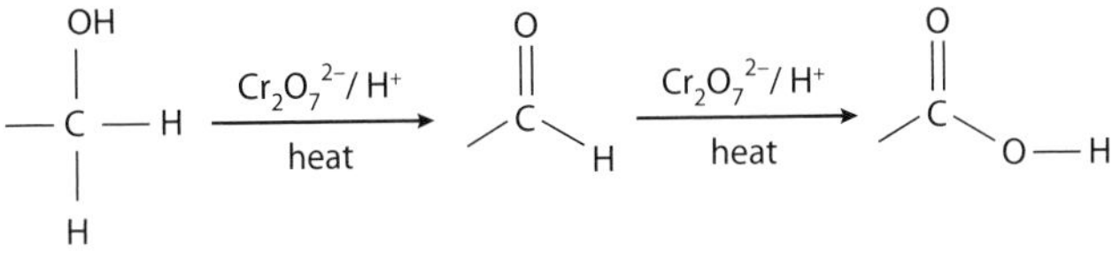

Figure 10.20

To obtain the **carboxylic acid** as the main product (**complete oxidation**) the reaction mixture is heated under **reflux**. To obtain the **aldehyde** (**partial oxidation**) the apparatus must be set up for **distillation** so that the aldehyde distils off as soon as it is formed (the aldehyde has a lower boiling point than the alcohol – no H-bonding between aldehyde molecules).

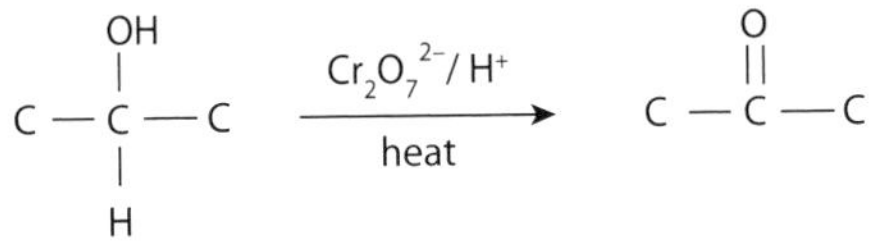

Figure 10.21

Secondary alcohols are oxidised to ketones, which cannot be oxidised any further (Figure **10.21**).

Tertiary alcohols are resistant to oxidation.

TEST YOURSELF 10.9

butan-1-ol, butan-2-ol,

1 Draw and name all the isomers of $C_4H_{10}O$ that are alcohols. 2 methyl propan-1-ol, 2 methyl propan-2-ol
2 Draw out and name the products of complete and partial oxidation (if any) of the alcohols in the previous question.

Esterification

When an alcohol is heated with a carboxylic acid, in the presence of a small amount of concentrated sulfuric acid as a catalyst, an ester is formed. This is a **condensation** reaction – two molecules join together with the elimination of water:

e.g.

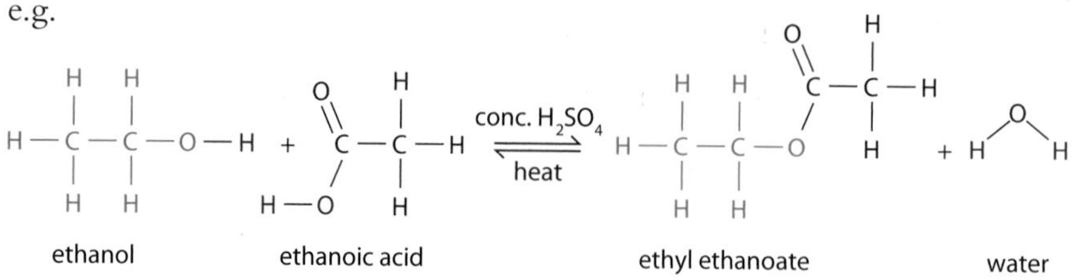

Figure 10.22

To work out the structure of an ester put the OH groups of the alcohol and carboxylic acid together, join the O of the alcohol to the C bearing the C=O and then remove water (Figure **10.23**).

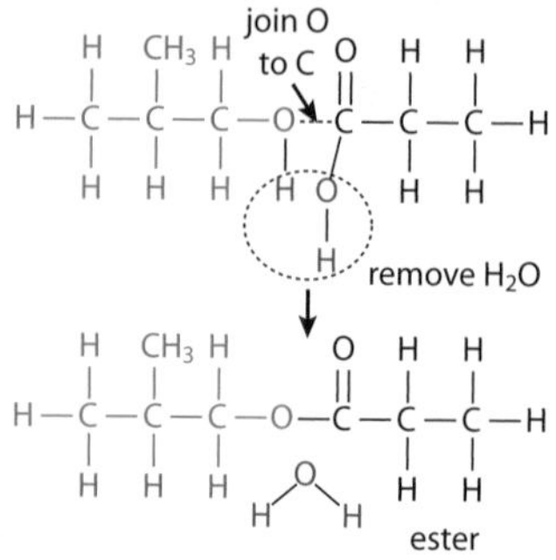

Figure 10.23

Naming esters

The carbon chain containing the COO group is named as *alkanoate* and the other carbon chain as an alkyl group – see Figure **10.24**.

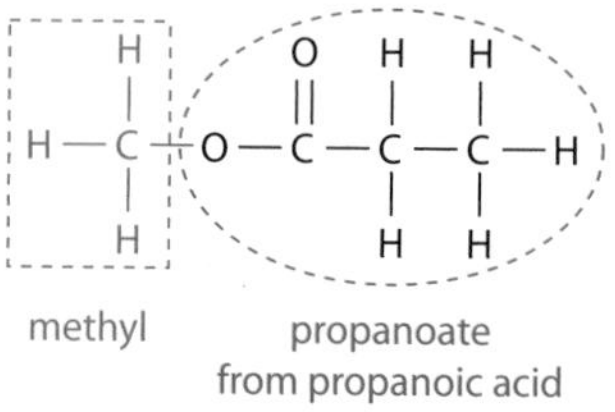

Figure 10.24

TEST YOURSELF 10.10

Draw the full structural formula of and name the ester formed when:

1 propanoic acid reacts with methanol, methyl propanoate

2 ethanoic acid reacts with propan-2-ol. propyl ethanoate

Reactions of halogenoalkanes

DEFINITION

NUCLEOPHILE species possessing a lone pair of electrons, which is attracted to an electron-deficient region in a molecule and donates a lone pair of electrons to form a coordinate covalent bond.

Halogenoalkanes are more reactive than alkanes – the very electronegative halogen atom makes the C that it is attached to δ+ so that it attracts nucleophiles.

Halogenoalkanes can undergo **nucleophilic substitution** reactions, e.g. when 1-bromopropane is heated with **aqueous** NaOH, the Br is replaced by an OH.

$CH_3CH_2CH_2Br + NaOH \rightarrow CH_3CH_2CH_2OH + NaBr$

TEST YOURSELF 10.11

Write an equation for the reaction of 1-bromo-3-methylbutane with NaOH(aq).

Reactions of benzene

Benzene undergoes electrophilic substitution reactions. A hydrogen atom in the ring is replaced by another atom/group.

e.g. $C_6H_6 + Cl_2 \rightarrow C_6H_5Cl + HCl$

10.3 Types of organic reactions (HL only)

Nucleophilic substitution reactions of halogenoalkanes

There are two possible mechanisms for nucleophilic substitution – S_N1 and S_N2; which mechanism occurs depends on whether the halogenoalkane is primary, secondary or tertiary.

S_N2 Substitution **N**ucleophilic **bimolecular** (two species involved in the rate determining step).

S_N1 Substitution **N**ucleophilic **unimolecular** (one molecule involved in the rate determining step).

Primary halogenoalkanes – S_N2 mechanism

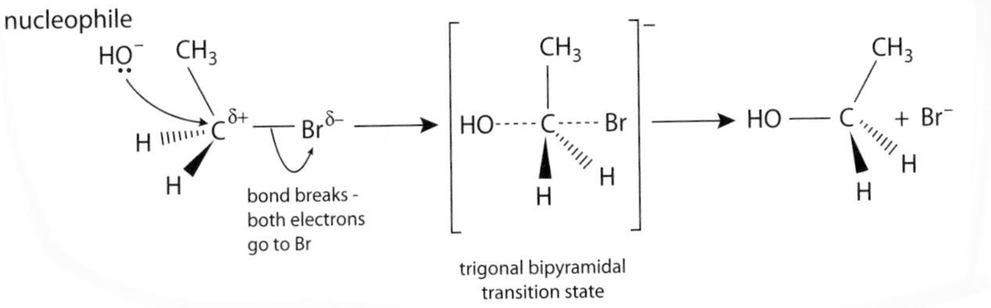

Figure 10.25

hint

The transition state (with negative charge) must be shown in the exam.

In mechanisms a curly arrow represents the movement of a **pair** of electrons. Compare this with using a 'fish hook' to show the movement of one electron in the free-radical substitution reaction.

The molecule inverts (like an umbrella turning inside-out) as the nucleophile attacks so that, if the original halogenoalkane is chiral, the reaction will occur with inversion of the configuration at the C attached to the halogen. This type of reaction is termed stereospecific (the stereochemistry can be predicted from that of the starting material).

Tertiary halogenoalkanes – S_N1 mechanism

This is a two-step mechanism (see Figure **10.26**):

The first step involves heterolytic fission of the C–Br bond – the bond breaks so that both electrons go back to the same atom. Compare with homolytic fission in the free radical substitution reaction – the bond breaks so that one electron goes back to each atom.

Because the carbocation formed is planar, the nucleophile can attack from either side in the second step and the reaction is not stereospecific – if we start with an optically active halogenoalkane a racemic mixture will be formed.

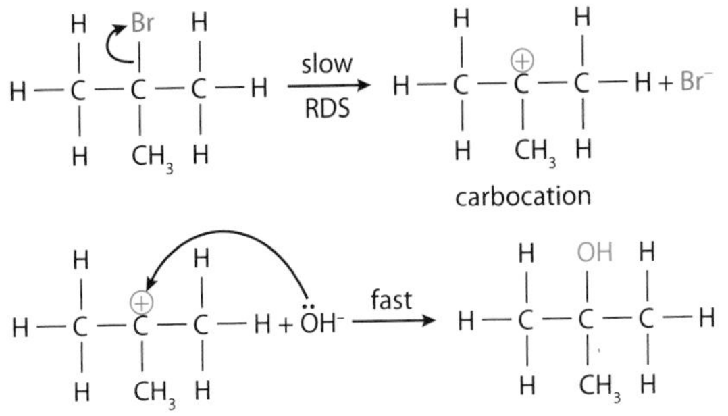

Figure 10.26

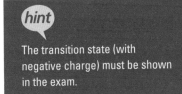

The S_N1 reaction is first order overall and the S_N2 reaction is second order:

	Type of halogenoalkane	Rate	Effect of doubling concentration of OH^-
S_N1	Tertiary	Rate = $k[(CH_3)_3CBr]$	No effect as OH^- is not in the rate equation – it is only involved in a fast step after the RDS.
S_N2	Primary	Rate = $k[CH_3CH_2Br][OH^-]$	Rate of reaction doubles as the reaction is first order with respect to OH^- – OH^- is involved in the RDS

Secondary halogenoalkanes undergo nucleophilic substitution via a mixture of mechanisms (i.e. both mechanisms); the more dominant mechanism will depend on the conditions.

The effect of structure on the rate of nucleophilic substitution

The rate of reaction for S_N2 is: **primary > secondary > tertiary**

> **hint**
> In the exam it is important to remember that, for the reaction of aqueous sodium hydroxide with halogenoalkanes, the S_N1 reactions of tertiary halogenoalkanes are faster than S_N2 reactions of primary halogenoalkanes.

This is mainly due to **steric effects** – more alkyl groups surrounding the central C in a halogenoalkane make it more difficult for the nucleophile to get in to attack the central C.

The rate of reaction for S_N1 is: **tertiary > secondary > primary**

This is due to the stability of the intermediate carbocation. Alkyl groups have an electron-releasing effect (positive inductive effect) so that they are able to stabilise a positively-charged carbon atom to which they are bonded – more alkyl groups results in a greater electron-releasing effect and greater stabilisation of the carbocation.

The effect of the halogen (leaving group) on the rate of nucleophilic substitution

R–I > R–Br > R–Cl > R–F

For both mechanisms the rate determining step involves breaking the carbon-halogen bond. The C–I bond is weakest, therefore the reaction will be fastest for an iodoalkane.

The effect of the nucleophile on the rate of nucleophilic substitution

OH^- is a better nucleophile than water because OH^- is negatively charged and will be more strongly attracted to the δ+ C. Therefore the rate of an S_N2 reaction is slower when OH^- is replaced with H_2O as a nucleophile. Changing the nucleophile will have no effect on the rate of an S_N1 reaction as the nucleophile is only involved in the mechanism after the RDS (the nucleophile is not in the rate equation).

The effect of the solvent on the rate of nucleophilic substitution

Solvents may be polar (e.g. H_2O) or non-polar (e.g. CCl_4). Polar solvents can be further classified as:

- Protic solvents – have H joined to N or O and can participate in hydrogen bonding, e.g. water, ethanol (CH_3CH_2OH) and ammonia.
- Aprotic solvents – do not have H joined to N or O and cannot participate in hydrogen bonding. These solvents do not act as proton donors, e.g. propanone ($(CH_3)_2CO$).

S_N1 reactions are favoured by protic polar solvents because solvent molecules are able to solvate both positive and negative ions, which are generated in the reaction.

S_N2 reactions are favoured by aprotic polar solvents because these are not very good at solvating the negatively-charged nucleophile – this means that the nucleophile is not surrounded by solvent molecules and is better able to get in and attack the δ+ carbon atom. For the same reason, S_N2 reactions are favoured by non-polar solvents.

Electrophilic addition reactions

DEFINITION

ELECTROPHILE species which is attracted to regions of high electron density and accepts a pair of electrons to form a covalent bond.

Electrophiles are Lewis acids.

Alkenes react in an electrophilic addition mechanism (Figure **10.27**).

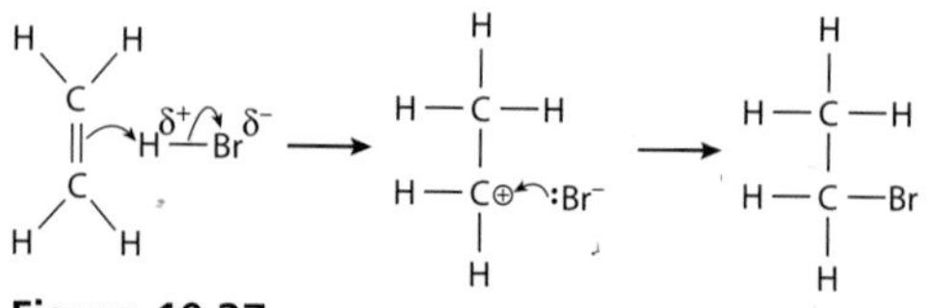

Figure 10.27

A halogen, such as bromine, is non-polar but as it approaches the C=C, which is a region of high electron density, a dipole is induced in the halogen (δ+ closer to the C=C). With interhalogen compounds, such as I–Cl, consider the difference in electronegativity to work out which is δ+ and attacks the C=C.

When hydrogen halides (or interhalogen compounds) react with unsymmetrical alkenes more than one product can be formed depending on which way the atoms add across the double bond:

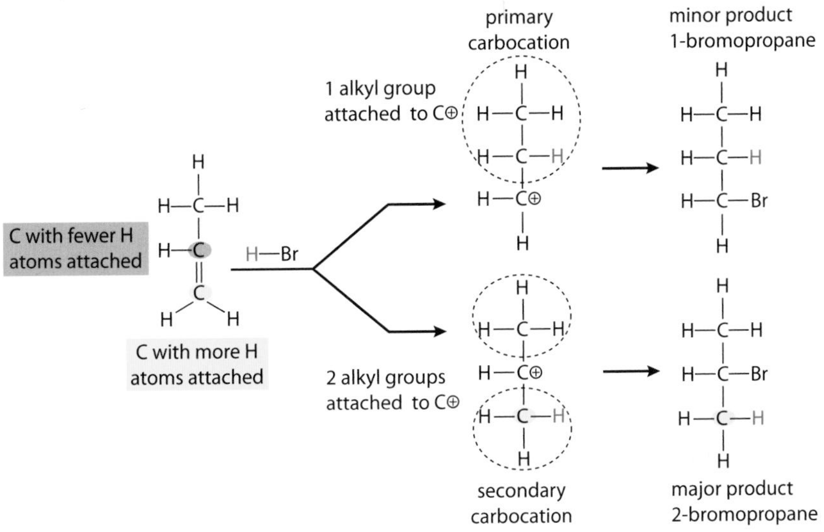

Figure 10.28

Markovnikov's rule can be used to predict the major product – when H–X adds to the double bond of an alkene, the H atom becomes attached to the C atom that has the larger number of H atoms already attached.

This can be explained in terms of the stability of the carbocation intermediates formed. In the scheme above, the secondary carbocation is more stable than the primary carbocation because there are **more electron-releasing alkyl groups** attached to the positively charged C in the secondary carbocation.

For interhalogen compounds, the less electronegative halogen atom behaves as an H atom (it is δ+) and adds to the C atom with more Hs already attached.

> **hint**
> You must explain the major/minor product in terms of the stability of the carbocation – it is never enough to just say 'because of Markovnikov's rule'.

TEST YOURSELF 10.12

Predict the major products of the reactions between:

1 but-1-ene and hydrogen bromide, 2 bromo butane

2 2-methylpent-2-ene + Cl–Br.

Reactions of benzene

Benzene undergoes electrophilic substitution reactions. The ring of six delocalised electrons is a region of high electron density and attracts electrophiles. Substitution is more favourable than addition as the stable delocalised aromatic system is preserved.

Benzene undergoes a nitration reaction in the presence of a mixture of concentrated nitric and sulfuric acids.

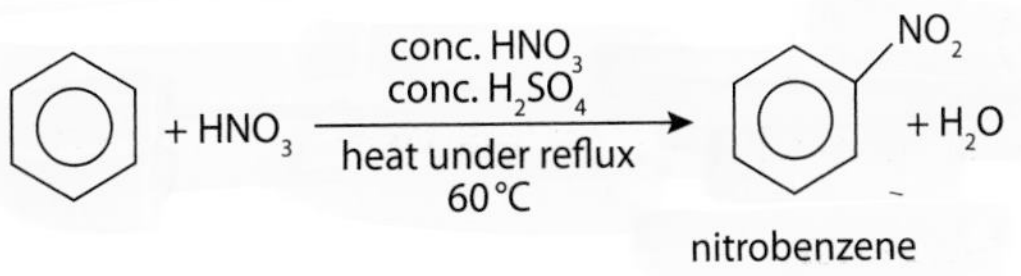

Figure 10.29

You need to remember the mechanism for this reaction.

Formation of electrophile: $HNO_3 + H_2SO_4 \rightarrow NO_2^+ + H_2O + HSO_4^-$

Mechanism:

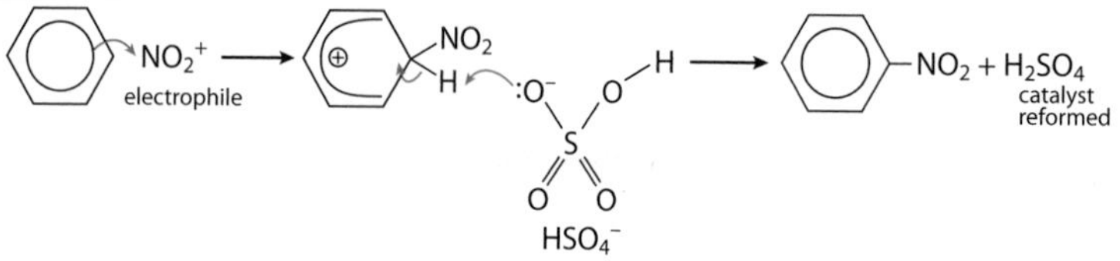

Annotated exemplar answer 10.1

Describe the mechanism for the nitration of benzene. **[4]**

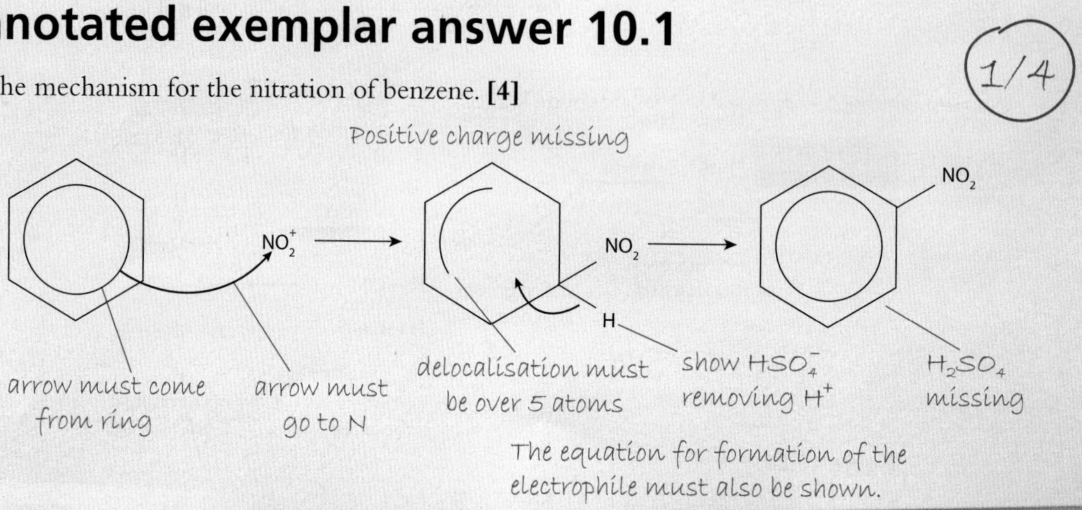

Reduction reactions

Aldehydes, ketones and carboxylic acids can be reduced to alcohols. These reactions are essentially the reverse of the oxidation reactions of primary and secondary alcohols.

Reducing agents:

- aldehydes and ketones – $NaBH_4$ (sodium borohydride)
- carboxylic acids – $LiAlH_4$ (lithium aluminium hydride)

Aldehydes are reduced to primary alcohols, see Figure **10.30**.

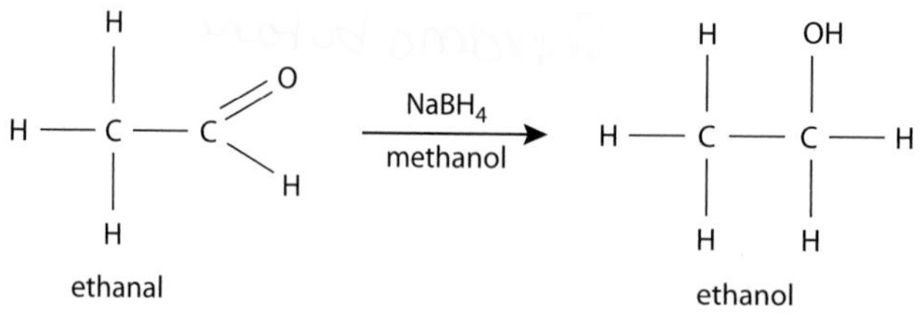

Figure 10.30

Ketones are reduced to secondary alcohols, see Figure **10.31**.

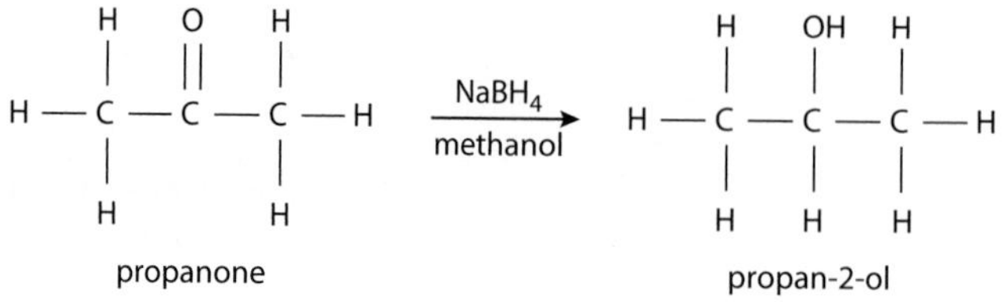

Figure 10.31

A stronger reducing agent ($LiAlH_4$) is required to reduce carboxylic acids. Carboxylic acids are reduced to primary alcohols, e.g.

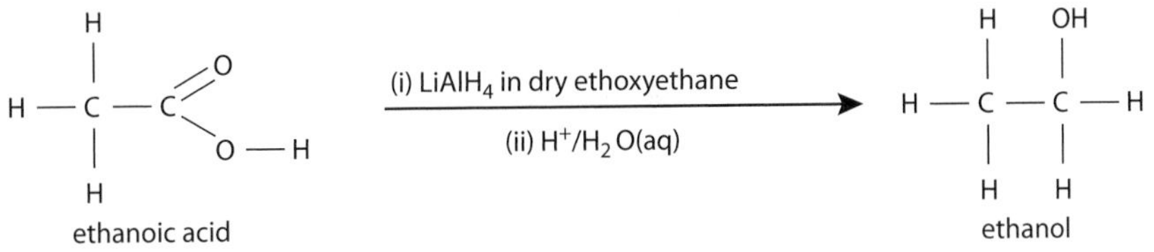

Figure 10.32

The reduction cannot be stopped at the aldehyde stage.

Nitrobenzene can be reduced to phenylamine in a two-stage reaction using a reducing agent such as tin and concentrated HCl.

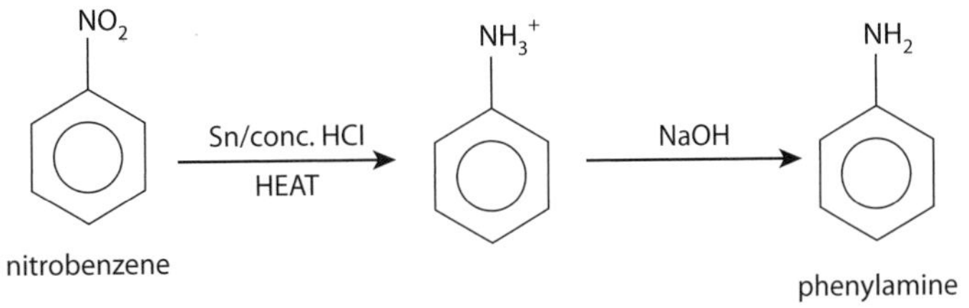

Figure 10.33

First stage: $C_6H_5NO_2 + H^+ + 6[H] \rightarrow C_6H_5NH_3^+ + 2H_2O$

Second stage: $C_6H_5NH_3^+ + OH^- \rightarrow C_6H_5NH_2 + H_2O$

TEST YOURSELF 10.13

Name the products of the following reactions:

1 $CH_3CH_2CH_2CHO + NaBH_4$ **2** $(CH_3)_2CHCOCH_3 + LiAlH_4$

butan-1-ol 3 methyl butanol

10.4 Synthetic routes (HL only)

Organic chemists often adopt a retrosynthetic approach to designing syntheses for organic molecules – they start with the target molecule and work back through known reactions to an available starting material.

Worked example 10.3

Design a synthetic pathway to make propyl ethanoate using only suitable halogenoalkanes as organic starting materials.

Firstly, draw the structure of propyl ethanoate:

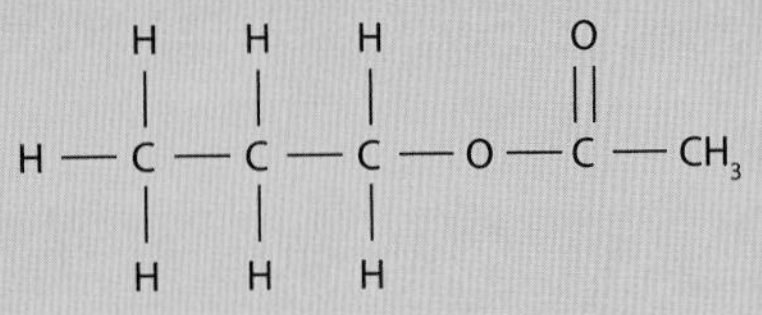

Figure 10.34

We will adopt a retrosynthetic approach, working out what molecules can be made from.

Propyl ethanoate is an ester – it can be made from an alcohol and a carboxylic acid:

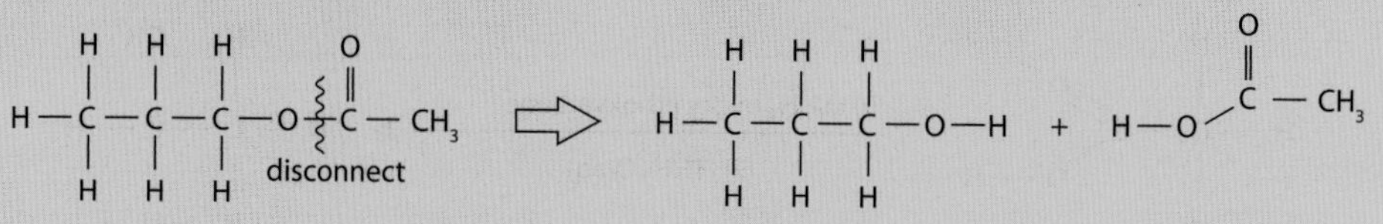

propan-1-ol ethanoic acid

Ethanoic acid can be made from ethanol by complete oxidation:

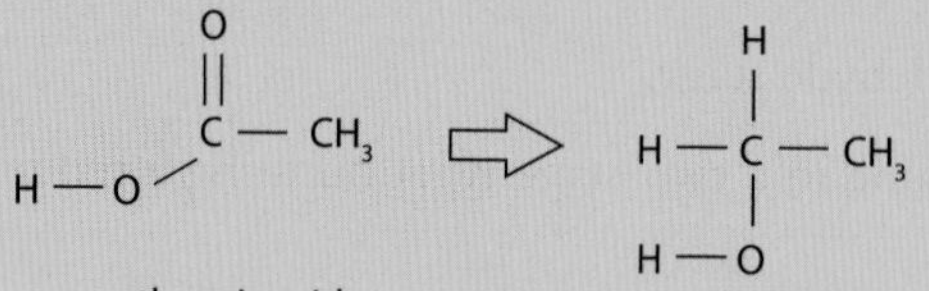

ethanoic acid ethanol

Ethanol and propan-1-ol can be made from halogenoalkanes in nucleophilic substitution reactions:

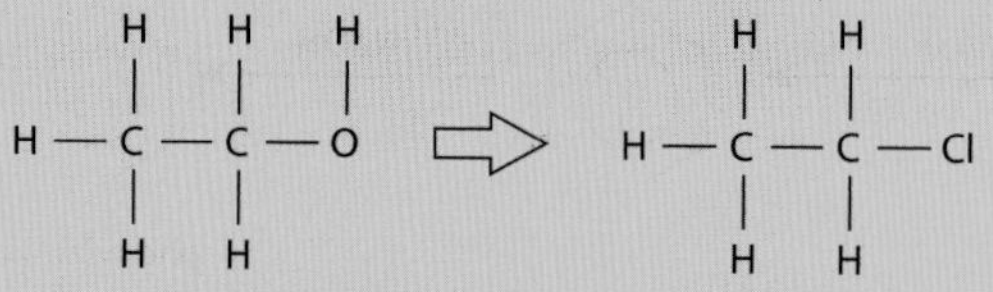

chloroethane

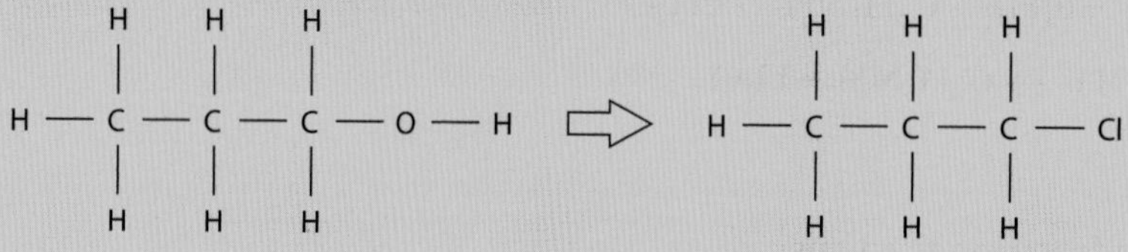

1-chloropropane

Overall, the synthetic route is:

$$CH_3CH_2Cl \xrightarrow[\text{heat}]{\text{NaOH(aq)}} CH_3CH_2OH$$

$$CH_3CH_2CH_2Cl \xrightarrow[\text{heat}]{\text{NaOH(aq)}} CH_3CH_2CH_2OH$$

$$CH_3CH_2OH \xrightarrow[\text{heat under reflux}]{Cr_2O_7^{2-}/H^+} CH_3C(=O){-}O{-}H$$

$$CH_3C(=O)OH + CH_3CH_2CH_2OH \underset{\text{heat}}{\overset{\text{conc } H_2SO_4}{\rightleftharpoons}} CH_3{-}C(=O){-}O{-}CH_2CH_2CH_3 + H_2O$$

This can also be written as a series of balanced equations:

$CH_3CH_2Cl + NaOH \rightarrow CH_3CH_2OH + NaCl$

$CH_3CH_2CH_2Cl + NaOH \rightarrow CH_3CH_2CH_2OH + NaCl$

$CH_3CH_2OH + 2[O] \rightarrow CH_3COOH + H_2O$

$CH_3COOH + CH_3CH_2CH_2OH \rightleftharpoons CH_3COOCH_2CH_2CH_3 + H_2O$

hint

You can be asked about any of the reactions covered in Sections 10 and 20 of the IB guide and so you must learn all the reactions with essential conditions.

TEST YOURSELF 10.14

Deduce a reaction sequence for the synthesis of methyl propanoate using alkanes as the only organic starting materials.

Nature of Science. Scientists should consider ethical and environmental issues when designing syntheses for organic compounds such as new drugs.

10.5 Stereoisomerism (HL only)

DEFINITION

STEREOISOMERS molecules that have same structural formula (i.e. the atoms are joined together in the same way – same connectivity) but the atoms are arranged differently in space.

Conformational isomerism

Conformational isomers are the **same molecule** in different conformations due to rotation about a σ bond, e.g. 1,2-dichloroethane can exist in different conformations according to how the Cl atoms are arranged relative to each other. Two of the conformers are shown in Figure **10.35**.

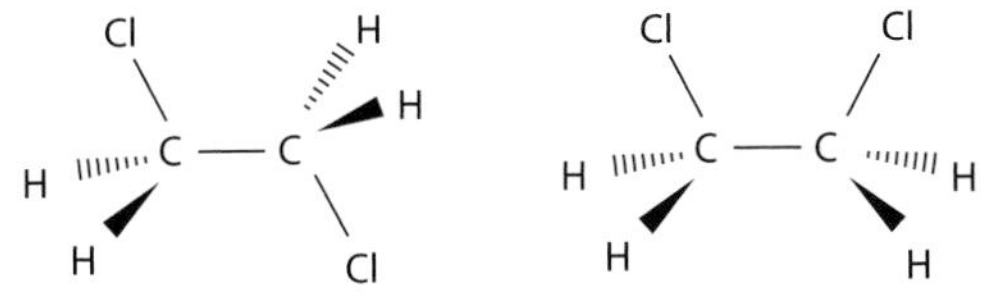

Figure 10.35

These are both the same molecule and interconvert rapidly – individual conformers cannot be isolated. No bonds are broken when conformers interconvert.

Configurational isomers

These can be divided into *cis–trans* and optical isomers and the different isomers can only be interconverted by breaking and re-forming (a) covalent bond(s).

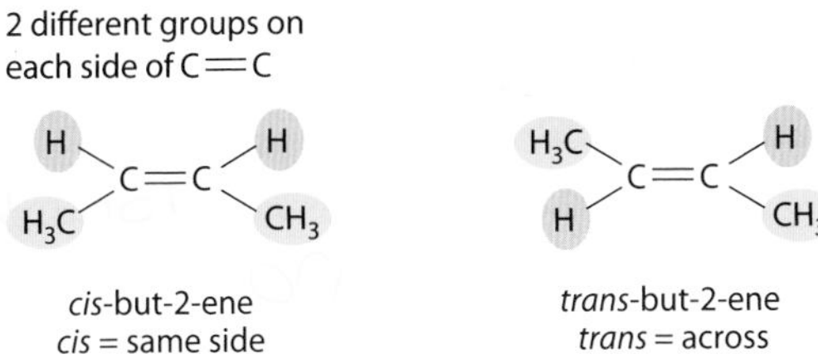

Figure 10.36

Alkenes and cycloalkanes exhibit *cis–trans* isomerism

In order for a molecule to exhibit *cis–trans* isomerism there must be **two different groups** on **both sides** of the double bond. Thus *cis–trans* isomers are possible for but-2-ene (Figure **10.36**) but not for but-1-ene (Figure **10.37**).

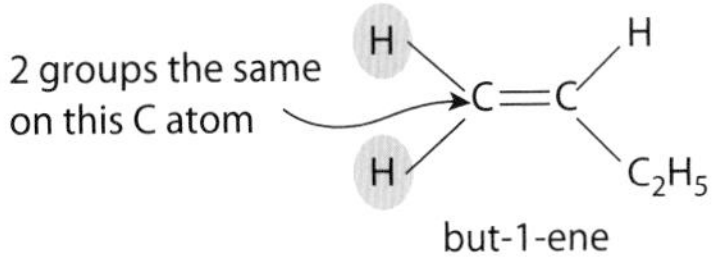

Figure 10.37

Cis–trans isomerism occurs because the π component of the C=C **prevents rotation of groups about the bond** – the groups cannot be rotated around the bond without breaking the π component.

Cis–trans isomerism is also possible for substituted cycloalkanes - the ring structure prevents rotation of a group from the top of the ring to the bottom.

The *E/Z* naming system

Cis–trans isomers can be named using the *E*/*Z* system. Each group attached to the C=C is given a priority according to the Cahn–Ingold–Prelog priority rules (CIP) – a higher priority is given to atoms attached to the C=C that have a higher **atomic number**. We look at each C in the C=C separately and assign a priority (1 or 2) to each

hint

You might want to remember that 'transatlantic' means 'across the Atlantic'.

> **hint**
> Remember that *cis–trans* isomers have the same names except for the *cis–trans* or *E/Z* part, so but-1-ene is a structural isomer of but-2-ene – the name is different.

of the groups attached. If the two groups with highest priority (labelled 1) are on the same side of the C=C (dashed line in Figure **10.38**), the isomer is labelled *Z* and if they are on opposite sides of the C=C, the isomer is labelled *E*. For example, the isomer shown here is *E* because the groups labelled 1 are on opposite sides of the C=C (diagonally opposite).

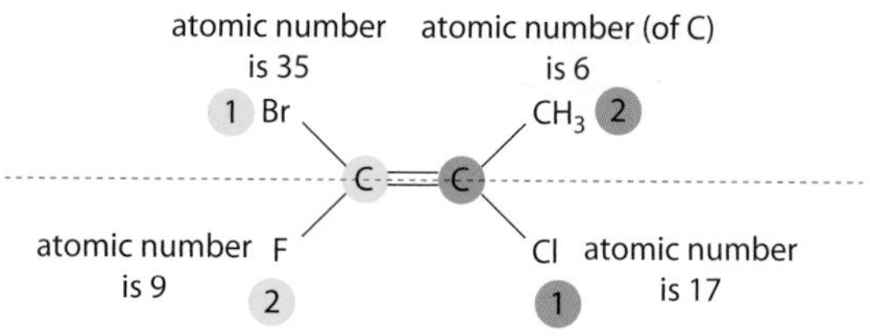

(*E*)-1-bromo-2-chloro-1-fluoroprop-1-ene

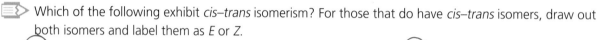

Figure 10.38

TEST YOURSELF 10.15

Which of the following exhibit *cis–trans* isomerism? For those that do have *cis–trans* isomers, draw out both isomers and label them as *E* or *Z*.

1 1-chlorobut-1-ene **2** 2-methylbut-2-ene **3** 3-methylpent-2-ene

Optical isomerism

> **hint**
> Remember, to show these enantiomers clearly you must draw the diagrams using wedges and dashed lines to show the 3-D structure.

To exhibit optical isomerism, there must be **four different groups attached to a C atom.** A carbon atom with four different atoms or groups attached to it is called a **chiral centre** and molecules that exhibit optical isomerism are called **chiral.** The individual optical isomers of a compound are called **enantiomers.**

Butan-2-ol exhibits optical isomerism. There are two forms of this compound that are mirror images of each other – the mirror images are not superimposable – Figure **10.39**.

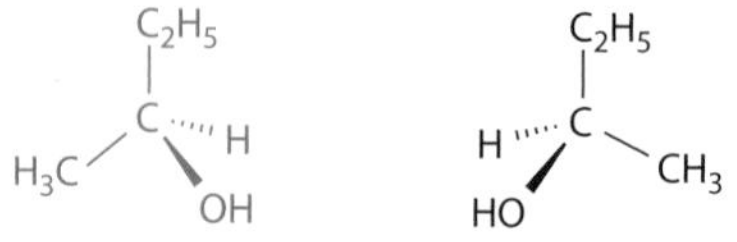

Figure 10.39

TEST YOURSELF 10.16

State which of the following will exhibit optical isomerism:

1 1-bromobutane no **2** 2-bromobutane yes **3** $CH_3CH_2CH(OH)COOH$ yes

DEFINITION

PLANE-POLARISED LIGHT light that vibrates in one plane only.

Plane-polarised light is passed through the enantiomers in a **polarimeter.** The two enantiomers of an optically active compound **rotate the plane of polarisation of plane-polarised light in opposite directions by equal amounts.**

DEFINITION

RACEMIC MIXTURE an equimolar mixture of the two enantiomers of a chiral compound.

A racemic mixture has no effect on plane-polarised light (optically inactive) since the rotations of the two enantiomers cancel each other out.

Properties of enantiomers

Physical properties of enantiomers are identical – the only difference is in the direction of rotation of the plane of polarisation of plane-polarised light.

The chemical properties of enantiomers are identical for reactions with compounds which are not optically active. Enantiomers may, however, react differently with other optically active compounds.

Diastereomers

DEFINITION

DIASTEREOMERS stereoisomers which are not mirror images of each other.

Cis–trans isomers are diastereomers. Diastereomers can also arise when more than one chiral centre is present in a molecule. For example, the molecules shown have two chiral centres.

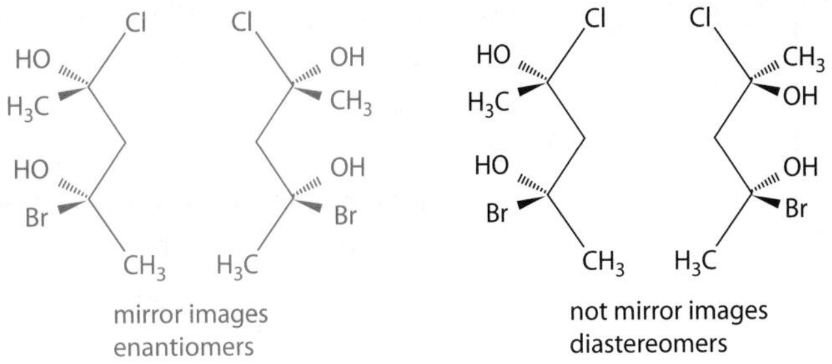

Figure 10.40

Checklist

At the end of this chapter you should be able to:

- ☐ Deduce the names of functional groups present in a molecule.
- ☐ Name organic molecules according to the IUPAC system.
- ☐ Draw isomers for molecules.
- ☐ Write equations for combustion reactions of hydrocarbons and alcohols.
- ☐ Work out the products of reactions of alkanes with halogens, alkenes with hydrogen, halogens, water and hydrogen halides.
- ☐ Write out the mechanism for the reaction of alkanes with halogens.
- ☐ Deduce the products of oxidation of primary, secondary and tertiary alcohols.

Higher Level only

- ☐ Describe S_N1 and S_N2 mechanisms and explain the factors that affect the rates of these reactions.
- ☐ Describe the electrophilic addition mechanism of alkenes and explain Markovnikov's rule for the reactions of unsymmetrical alkenes.
- ☐ Write equations for reduction reactions of aldehydes, ketones, carboxylic acids and nitrobenzene.
- ☐ Design synthetic pathways using the reactions in this topic.
- ☐ Distinguish between configurational and conformational isomers.
- ☐ Recognise *cis–trans* isomers and use the *E/Z* system to name them.
- ☐ Draw out optical isomers and explain their properties.

11 MEASUREMENT, DATA PROCESSING AND ANALYSIS

Analytical Chemistry is a branch of Chemistry that deals with the separation, determination of composition and determination of the structures of substances. Scientists working in this field need to be aware of the limitations of their data.

This chapter covers the following topics:

- ☐ Qualitative and quantitative data
- ☐ Uncertainties and errors in measurements and results
- ☐ Graphical techniques
- ☐ Spectroscopic Identification of organic compounds

11.1 Qualitative and quantitative data

DEFINITION

QUALITATIVE DATA non-numerical data obtained from an experiment.

These will be observations made during an experiment.

DEFINITION

QUANTITATIVE DATA numerical data.

11.2 Uncertainties and errors in measurement and results

Random errors

Random errors arise due to the limitations of the measuring apparatus. All measurements must be quoted with an uncertainty (e.g. $69.0 \pm 0.3\ cm^3$). The uncertainty indicates the size of the random error.

Analogue apparatus – the uncertainty on a measurement is half the smallest division that you take a reading to.

Digital instruments – the uncertainty of a measurement is quoted as ± the smallest division, e.g. a reading taken with a 2 decimal place electronic balance should be quoted as: 2.46 ± 0.01 g.

The effects of random uncertainties should mean that the measurements taken will be distributed either side of the mean.

The random uncertainties can never be completely eliminated but the effect of random uncertainties can be reduced by repeating the measurements more often.

Systematic errors

A systematic error can be introduced into an experiment due to the apparatus used or the procedure. Systematic errors are always in the same direction. For example, heat loss to the surroundings will always cause a measured value for an enthalpy change of combustion to be less exothermic than expected.

The presence of systematic errors can be identified by comparison with accepted literature values for quantities.

We can use the **percentage error** to compare the experimental value with the accepted literature value:

$$\text{Percentage error} = \left|\frac{\text{experimental value} - \text{accepted value}}{\text{accepted value}}\right| \times 100$$

hint

| value | indicates the modulus of the value – ignore the overall sign.

If the percentage error is greater than the percentage uncertainty due to random errors the experiment involves some systematic errors, but if the percentage error is smaller than the percentage uncertainty, any deviation from the literature value can be explained in terms of random errors, that is, the limitations of the measuring apparatus.

☆ Model answer 11.1

A student carried out an experiment to measure the enthalpy change of neutralisation. 50 cm^3 of each solution was measured using measuring cylinders and the temperature change upon mixing measured using a temperature probe. The student obtained a value for the enthalpy of neutralisation of –54±1 kJ mol^{-1}. The accepted literature value is –57.3 kJ mol^{-1}. Calculate the percentage error for this experiement and discuss whether there are systematic and/or random errors in the experiment. Suggest possible sources of random and systematic errors and ways to reduce them.

$$\text{percentage error} = \left|\frac{54 - 57.3}{57.3}\right| \times 100 = 5.8\%$$

The ±1 indicates the uncertainties due to random errors.

The percentage uncertainty due to random uncertainties: $\frac{1}{54} \times 100 = 1.9\%$

Since the percentage error is larger than the percentage uncertainty there are systematic as well as random errors in the experiment.

Possible sources of systematic errors:

- heat loss to the surroundings
- the solutions were not of the stated concentrations – if they were less concentrated less heat would be given out.

Improvements to reduce systematic errors:

- insulate the reaction vessel to reduce heat loss to the surroundings,
- check the concentrations of the solutions by titration against standard solutions.

Possible sources of random error:

- measuring the volumes using a measuring cylinder,
- there will also be an uncertainty associated with the temperature measurement but as a temperature probe was used this is likely to be quite small.

Random errors can be reduced by:

- repeating the experiment,
- using more precise measuring apparatus (e.g. pipettes to measure the volumes),
- using more concentrated solutions to give a larger temperature change.

hint

With systematic errors you should always consider the direction in which it occurs – there would have been no point in suggesting heat loss to the surroundings as a systematic error if the experimental value had been more exothermic than the literature value.

hint

Systematic errors cannot be reduced by repeating an experiment without changing anything else.

hint

You may have to use your judgement to decide whether a particular error has a large or a small effect on your experimental value – look at the experimental value and the literature value and try to judge the effect of a particular flaw. As a general rule, heat loss/gain is a major source of error in calorimetry experiments.

Quoting values with uncertainties

The uncertainty is usually quoted to 1 significant figure – values should be stated so that the uncertainty is in the last significant figure, i.e. no figures quoted after the uncertainty.

Calculated value±uncertainty	Value to quote
157.47±0.1	157.5±0.1
363.2±8	363±8

TEST YOURSELF 11.1

Quote each of the following values to the appropriate number of significant figures:

1 23.78367291±0.005 23·784 ± 0·005
2 678.96±10 680 ± 10
3 $1.3429\times10^{-3}\pm2\times10^{-5}$ $1\cdot34\times10^{-3}\pm2\times10^{-5}$

Precision

Precision relates to the reproducibility of results. High precision means that repeat values are close together and close to the mean.

Accuracy

Accuracy refers to how close a measurement is to the actual value of a particular quantity.

Low accuracy is due to **systematic errors** within the experimental procedure.

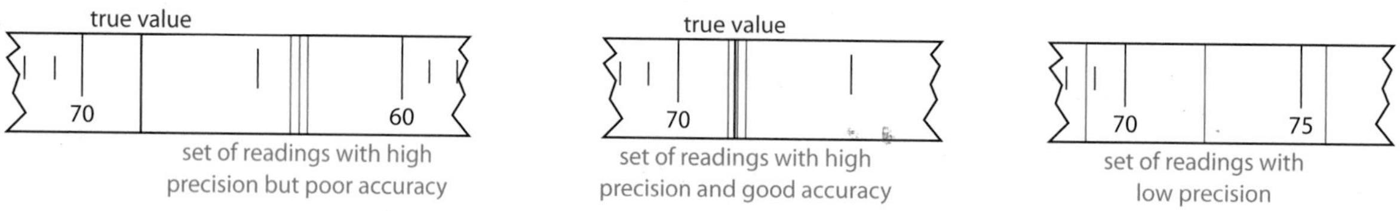

Figure 11.1

Significant figures and decimal places

Adding or subtracting numbers

The final answer should be quoted to the same number of **decimal places** as the piece of original data that has the fewest decimal places.

e.g. $12.4-8.2=4.2$ or $4.77+9.33395=14.11$

Multiplying or dividing numbers

The final answer should be quoted to the same number of **significant figures** as the piece of data with the fewest significant figures.

e.g. $37.82\times3.68=139$ or $\frac{1.47\times10^{-5}}{6.538\times10^{-3}}=2.25\times10^{-3}$

TEST YOURSELF 11.2

In an experiment to determine the enthalpy change of solution of magnesium chloride, a student added 1.00 g of $MgCl_2$ (M_r 95.21) to 50.00 cm^3 of water originally at 20.10 °C. The temperature rose to 27.64 °C. The specific heat capacity of water is 4.18 J g^{-1} °C^{-1}. The value of the enthalpy change of solution of magnesium chloride from this experiment is:

A -1.5×10^2 kJ mol^{-1} **C** -1.500×10^2 kJ mol^{-1}

B -1.50×10^2 kJ mol^{-1} **D** -150.037 kJ mol^{-1}

Absolute and percentage uncertainties

An uncertainty may be reported either as an absolute value, e.g. 1.23±0.02 m, or as a percentage value, e.g. 1.23 m±2%.

$$\text{Percentage uncertainty} = \frac{\text{absolute uncertainty}}{\text{value}} \times 100$$

For example, for 1.27±0.01 g, the percentage uncertainty $= \frac{0.01}{1.27} \times 100$, i.e. 0.8%

$$\text{Absolute uncertainty} = \frac{\text{percentage uncertainty}}{100} \times \text{value}$$

If the final value of a calculation is 23.27±1% the absolute uncertainty is given by:

absolute uncertainty $= \frac{1}{100} \times 23.27$, i.e. 0.2 (to 1 significant figure), therefore, the final answer is 23.3±0.2.

hint

Percentage uncertainty has no units.

Adding and/or subtracting quantities with uncertainties

When quantities with uncertainties are added or subtracted then the **absolute uncertainties** are **added**, e.g.

0.342±0.001+0.278±0.002 = 0.620±0.003

Multiplying or dividing

When multiplying or dividing quantities with uncertainties the **percentage uncertainties** are **added** to give the percentage uncertainty of the final value.

Worked example 11.1

Calculate the absolute uncertainty when 9.78±0.04 is divided by 3.349±0.005 and quote the final answer to an appropriate number of significant figures.

9.78/3.349 = 2.920274709

Percentage uncertainties: 0.04/9.78×100 = 0.41% 0.005/3.349×100 = 0.15%

Total percentage uncertainty: 0.41+0.15 = 0.56%

Absolute uncertainty = 0.56/100×2.920274709 i.e. 0.02 to 1 significant figure

The absolute uncertainty is in the second decimal place and therefore no figures should be quoted beyond that. The final answer should be quoted as 2.92±0.02.

When multiplying or dividing a quantity with an uncertainty by a pure number (e.g. 2) the **percentage uncertainty stays the same**.

TEST YOURSELF 11.3

In a particular experiment 45.6±0.2 cm^3 of gas was produced in 20±1 s. Calculate the rate of reaction in cm^3 s^{-1} and give your answer to an appropriate number of significant figures, with an uncertainty.

2·3 ± 0·1

11.3 Graphical techniques

Graphs are used extensively in science to show the relationship between two variables. Graphs may be either:

- Sketched graphs – the axes are labelled but there is no scale.
- Drawn graphs – the axes are labelled and have scales.

DEFINITIONS

INDEPENDENT VARIABLE what is changed in an experiment – what is being investigated.

DEPENDENT VARIABLE what is measured in an experiment.

In a set of experiments to investigate the effect of concentration on reaction time, concentration is the independent variable and reaction time is the dependent variable.

Tips for drawing graphs

- Make the graph as large as possible – choose your scales and axes to, as far as possible, retain the precision of the data.
- The independent variable should be plotted along the horizontal (x) axis and the dependent variable (or a quantity derived from it) should be plotted along the vertical (y) axis.
- Label the axes with the quantity and units, e.g. volume/cm^3.
- Plot the points clearly, so they can be seen.
- Draw a line of best fit – a straight line or a curve – do not join points (except for discrete data).

Graphs

The line of best fit must be a straight line and pass through the origin for a propotional relationship.

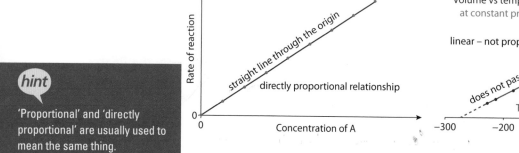

hint

'Proportional' and 'directly proportional' are usually used to mean the same thing.

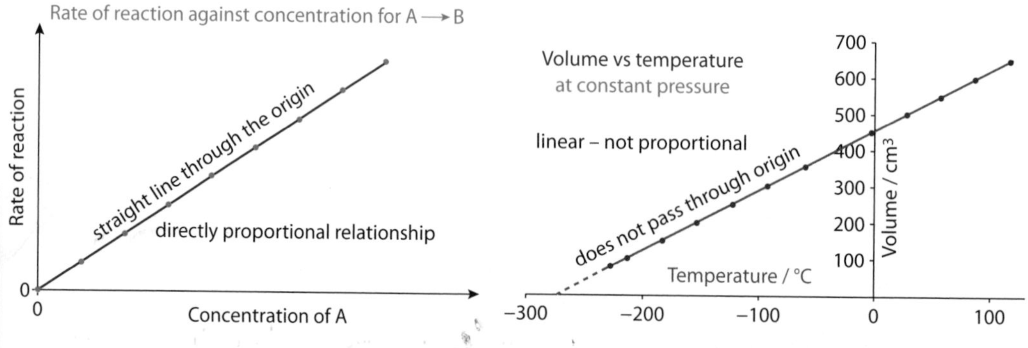

Figure 11.2

Deriving quantities from graphs

The gradient (slope) gives us an idea how much one quantity (the dependent variable) is affected by another quantity (the independent variable). If the gradient is large then a small change in the independent variable has a large effect on the dependent variable.

$$\text{Gradient} = \frac{\text{change in } y}{\text{change in } x}$$

The units of the gradient are obtained by dividing the units of the quantity on the y-axis by the units of the quantity on the x-axis.

For a curve draw a tangent and determine the gradient of the tangent:

$$\text{gradient} = \frac{\text{change in volume}}{\text{change in time}} = \frac{70}{32}$$

gradient $= 2.2\ \text{cm}^3\ \text{s}^{-1}$

The intercept of a line on the y-axis may also be required in a question. The units of the intercept are the same as the units of the quantity on the y-axis.

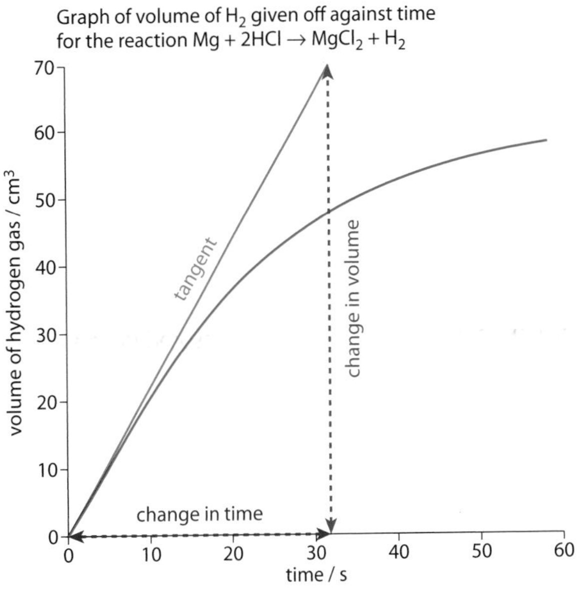

Figure 11.3

11.4 Spectroscopic identification of organic compounds

Index of hydrogen deficiency

The index of hydrogen deficiency (IHD) indicates the number of double bond equivalents (the number of double bonds or rings) in a compound. A triple bond counts as two double bond equivalents.

Molecule	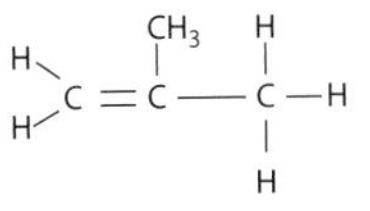	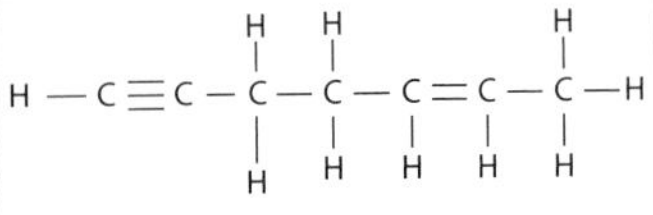	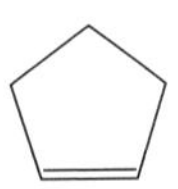	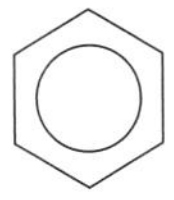
IHD	1	3	2	4

Figure 11.4

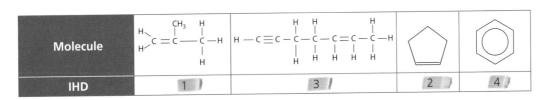

For a molecule of formula $C_cH_hN_nO_oX_x$ where X is a halogen atom, the IHD is given by:

$\text{IHD} = \frac{1}{2}[2c + 2 - h - x + n]$

Annotated exemplar answer 11.1

Calculate the IHD of C_4H_7NO and suggest two possible structures. **[3]**

IHD = $\frac{1}{2}[2\times4+2-7-0+1] = 2$

Possible structures are:

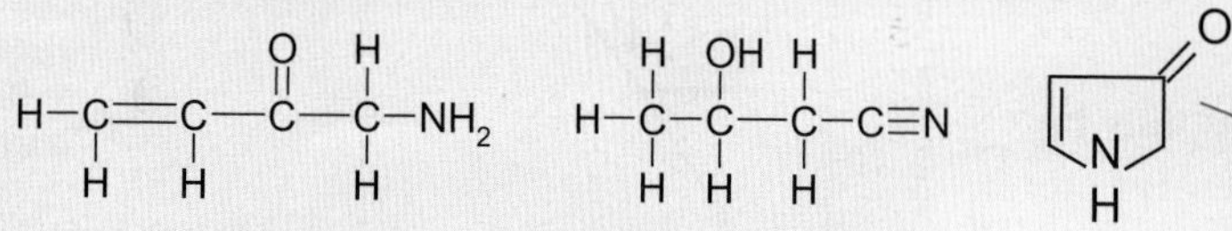

This answer is not correct as with two double bonds and a ring it has an IHD of 3. Never give more answers than required in the question – if you get one wrong it will cancel out marks for correct answers. This answer would have scored 3/3 without this last structure.

2/3

TEST YOURSELF 11.4

Calculate the IHD for each of the following:

1 C_4H_7Cl 1 **2** $C_6H_8N_2O$ 4 **3** $C_8H_{12}NO_2Br$ 3

11.5 Spectroscopic techniques

hint

In questions on this section you may also have to work out empirical and molecular formulae using percentage composition data – this is covered in Chapter 1.

Organic compounds can be identified using spectroscopic techniques.

Infrared (IR) spectroscopy

Infrared spectroscopy can be used to determine the bonds (functional groups) present in a molecule. The bonds responsible for absorption bands in the spectrum can be determined by comparison to the frequencies in the table in the *IB Data Booklet*.

The IR spectrum of ethanoic acid shows absorption bands in the region:

2500–3300 cm^{-1} O–H hydrogen bonded in carboxylic acids

1700–1750 cm^{-1} C=O in aldehydes, ketones, carboxylic acids and esters

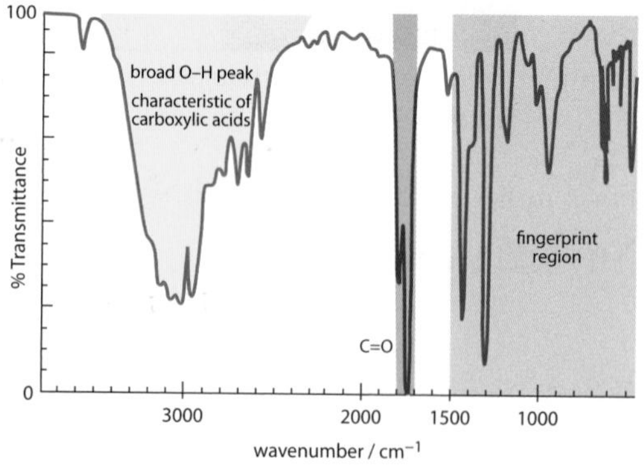

Figure 11.5

All organic molecules also show an absorption in the region 2850–3090 cm^{-1} due to the C–H bonds present. This is swamped by the O–H absorption in the spectrum of ethanoic acid.

The region below 1500 cm^{-1} is called the 'fingerprint region' and is characteristic of the molecule as a whole. The region contains many absorptions and is difficult to interpret. We only look at the fingerprint region to confirm the presence of a particular vibration. For example, a band in the 1050–1410 cm^{-1} region does not confirm the presence of a C–O bond in a molecule, but the absence of a band in this region means that C–O is not present. There is a band in the region 1050–1410 cm^{-1} in the IR spectrum of ethanoic acid (contains C–O) – if there was no absorption there we would have to conclude that we had made a mistake and the spectrum is not that of ethanoic acid.

TEST YOURSELF 11.5

Match these IR spectra to the following molecules:

$CH_3CH_2CH_2COOH$ $\quad$ $CH_3CH_2CH_2OH$ $\quad$ $CH_3COOCH_2CH_3$

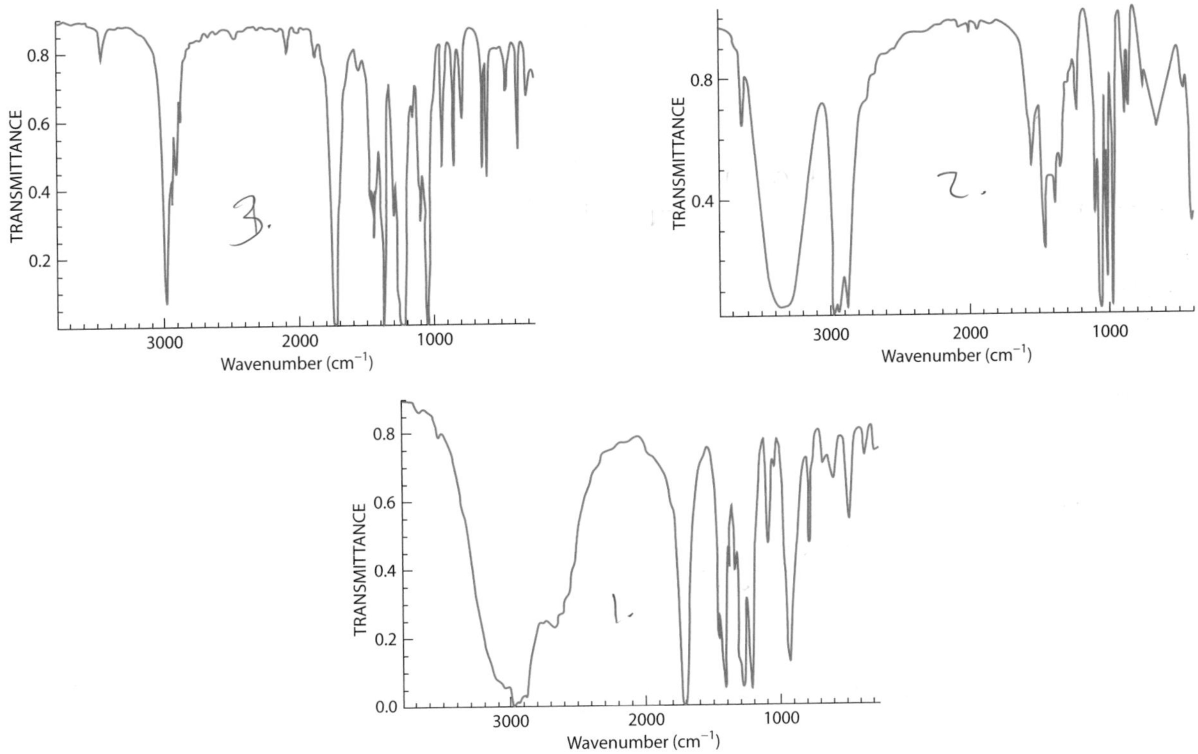

Figure 11.6

Mass spectrometry

In a mass spectrometer an organic compound is bombarded with high-energy electrons to produce positive ions. Positive ions pass through the mass spectrometer where they are separated according to mass and detected.

When a sample of propanoic acid (CH_3CH_2COOH) is introduced into a mass spectrometer some of the molecules are converted into the $CH_3CH_2COOH^+$ ion – the molecular ion, M^+. Other molecules will be split up into smaller fragments and any positive ions formed will produce a signal in the spectrum – this is the fragmentation pattern.

The molecular ion peak is at the highest mass value (not necessarily the largest peak) in the spectrum and indicates the relative molecular mass of the compound. In the spectrum of propanoic acid the molecular ion peak is at $m/z = 74$, which is the relative molecular mass.

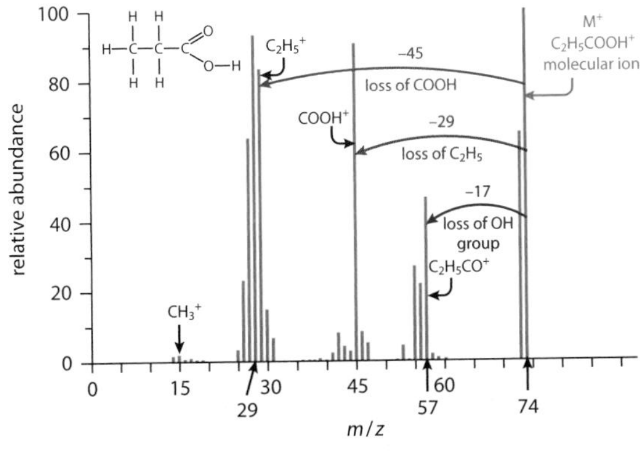

Figure 11.7

A mass spectrum can be interpreted either in terms of the fragments lost from the whole molecule or in terms of the masses of the fragment ions formed. For example, in the spectrum of propanoic acid, there is a peak at 57, which corresponds to the loss of OH (mass 17) from the molecular ion. The fragment responsible for the peak at 57 is thus ($C_2H_5COOH^+ - OH$), that is, $C_2H_5CO^+$.

hint

The scale on the *x*-axis is mass:charge ratio (*m/z* or *m/e*) and essentially corresponds to mass.

Worked example 11.2

There are peaks at *m/z* = 45 and *m/z* = 77 in the mass spectrum of benzoic acid (molecular formula $C_7H_6O_2$). Suggest the identity of the peaks.

Benzoic acid contains C (A_r = 12), H (A_r = 1) and O (A_r = 16). Possible combinations that make 45 are:

$C_2H_5O^+$ and $COOH^+$

We know, however, that benzoic acid contains a COOH group but it does not contain C_2H_5O, therefore the peak at 45 is $COOH^+$ (this is characteristic of carboxylic acids).

The only possible fragments for *m/z* = 77 are: $C_6H_5^+$ and C_5HO^+. If we notice that the relative molecular mass of benzoic acid is 122 then 122−45 = 77. We know that 45 corresponds to COOH and if we take COOH from $C_7H_6O_2$ we get C_6H_5. Therefore, the peak at 77 is $C_6H_5^+$ – this is characteristic of aromatic compounds containing a phenyl ring.

hint

Remember that only positive ions form a peak in the spectrum – don't forget the positive charge if asked to give the formulae of species responsible for peaks in the mass spectrum.

TEST YOURSELF 11.6

1 Suggest the identity of the peaks corresponding to the following *m/z* values in the mass spectrum of a hydrocarbon:

15 29 26 77

2 Suggest the identity of the peaks corresponding to the following *m/z* values in the mass spectrum of a compound containing C, H and O:

29 31 45

Nuclear magnetic resonance spectroscopy (NMR)

This technique looks at the hydrogen atoms (specifically, hydrogen nuclei-protons) in a molecule and requires electromagnetic radiation in the radio-frequency range.

- The number of signals in the spectrum corresponds to the number of different chemical environments for H.
- The area under the peaks depends on the ratio between the numbers of H atoms in each environment. This is also represented by the vertical height of the steps in the integration trace.

In the NMR spectrum of propanal (Figure **11.8**) there are three peaks as there are three different environments for H atoms (shown in different colours). The areas under the peaks (and the vertical heights of the steps in the integration trace) are in the ratio 1:2:3.

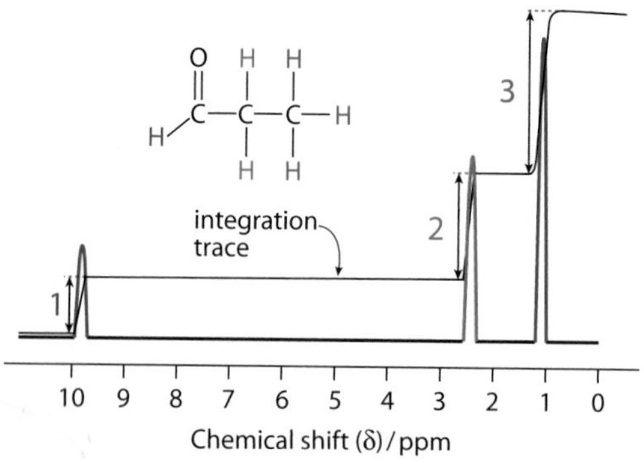

Figure 11.8

When determining the number of hydrogen atoms in different chemical environments remember that H atoms attached to the same C are all in the same environment and take care to notice whether the molecule is symmetrical. Some examples are given in the table:

	Number of different chemical environments for H	Ratio of number of H atoms in each environment
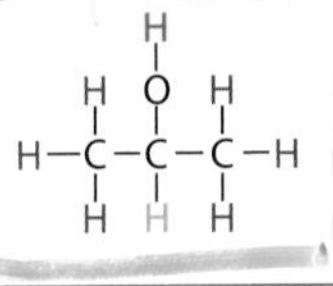	3	6:1:1
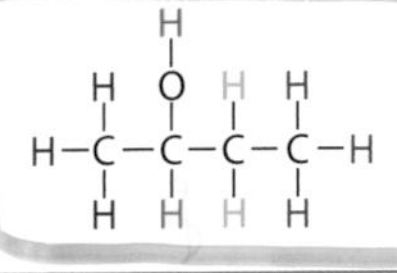	5	3:1:1:2:3

Nature of Science. Advances in technology and instrumentation have enabled scientists to work out the structures of complex organic molecules.

TEST YOURSELF 11.7

State the number of signals in the NMR spectrum of each of the following and the ratio of the areas under the peaks.

1	2	3
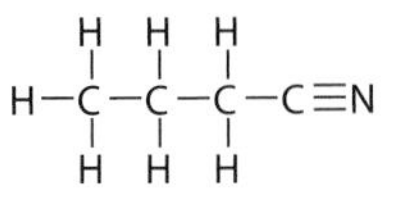	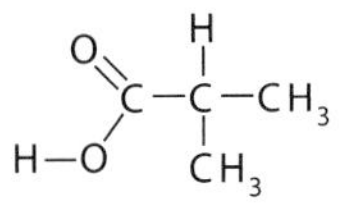	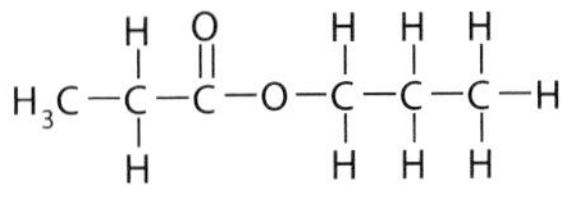

3 3 5

3:2:2 1:1:6 3:2:2:2:3

High resolution NMR spectra (HL only)

Splitting patterns

In high resolution NMR spectra it can be seen that the signals from low resolution NMR spectra are split into clusters of smaller peaks – the splitting is due to the presence of H atoms on adjacent atoms. If a particular signal is split into n smaller peaks (n is called the multiplicity) this indicates that the number of H atoms on the adjacent atom is $n-1$.

Signal	Multiplicity	H atoms on adjacent C
singlet	1	0
doublet	2	1
triplet	3	2
quartet	4	3

Figure **11.9** is a high resolution NMR spectrum of chloroethane.

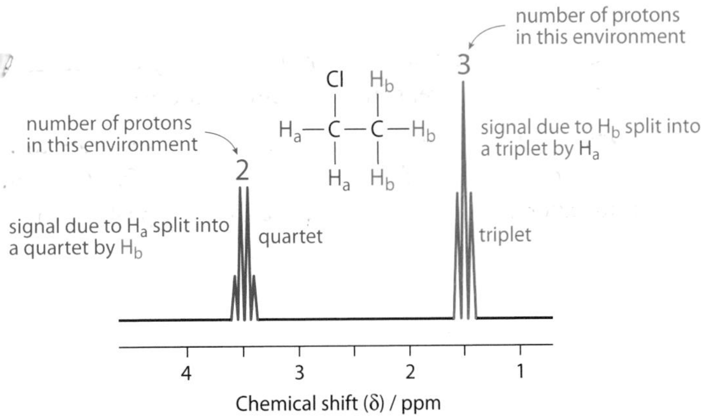

Figure 11.9

TEST YOURSELF 11.8

Suggest the splitting pattern for each of the following:

1	2	3
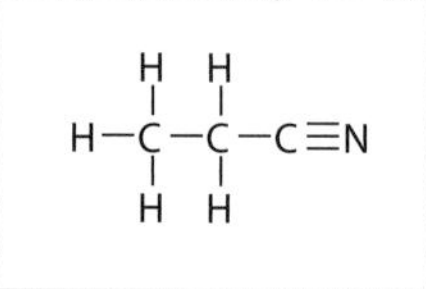	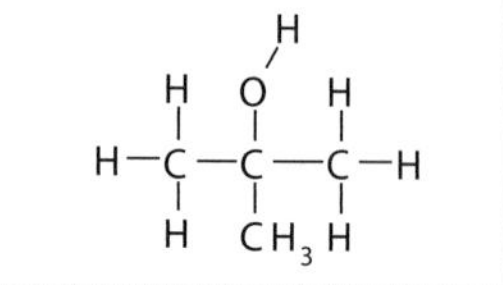	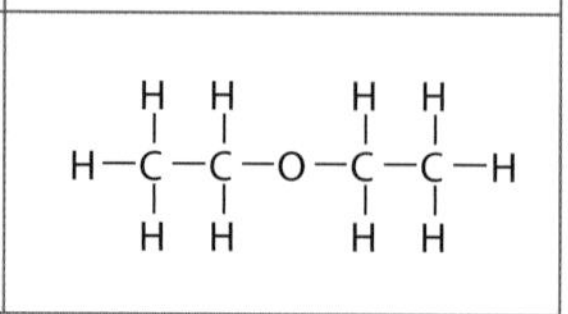

triplet quartet · 2x singlets · triplet quartet

Chemical shift

The chemical shift scale provides information about the environment that the H atoms are in – the chemical shifts in a spectrum can be matched up to the H atoms in a molecule by reference to the *IB Data Booklet*, e.g. for propanal:

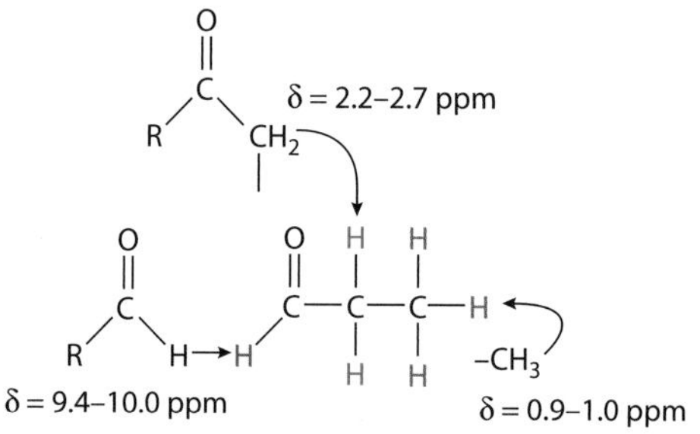

Figure 11.10

hint

When using the *IB Data Booklet* to work out the chemical shifts of protons (except for alkyl groups), ignore the number of protons in the environment e.g. if there were only one proton next to the CHO of an aldehyde group the chemical shift would also be in the range 2.2–2.7 ppm.

TMS

TMS (tetramethylsilane) is used as a reference standard for the chemical shift scale – the protons in TMS are assigned a chemical shift of 0.0.

TMS (Figure **11.11**) was chosen as the standard because:

- it has 12 protons all in the same environment
- the chemical shift of the protons in TMS is well away from (lower than) the chemical shifts of protons in most organic molecules
- it is non-toxic and inert.

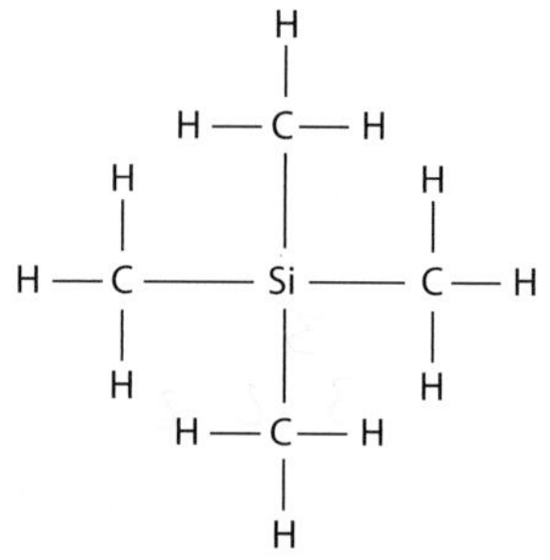

Figure 11.11

TEST YOURSELF 11.9

Suggest the chemical shift range for each group of protons in the following structures:

1	2	3
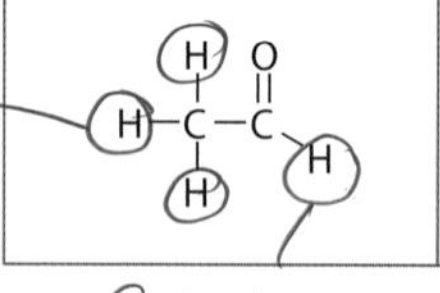	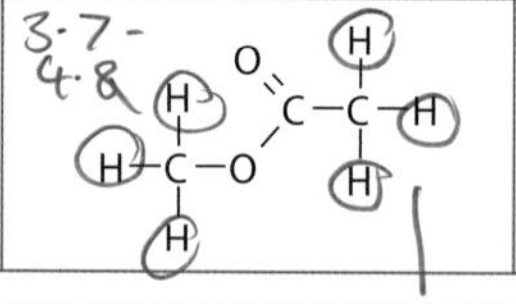	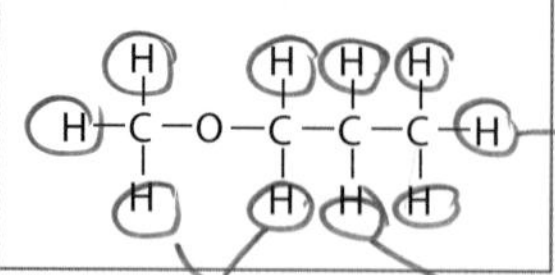

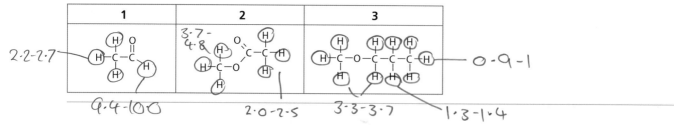

2.2-2.7 · 9.4-10.0 · 3.7-4.8 · 2.0-2.5 · 3.3-3.7 · 1.3-1.4 · 0.9-1

X-ray crystallography (HL only)

This technique uses X-rays to probe the structure of molecules. X-ray crystallography provides a detailed 3-D structure for a molecule including bond lengths and bond angles.

Determination of structure (HL only)

Information from a range of spectroscopic techniques can be used to work out the structure of a molecule.

Worked example 11.3

Compound X contains carbon, hydrogen and oxygen only. Information about X from various spectroscopic techniques is given in the table. Deduce the structure of X and explain your reasoning.

IHD	**1**
IR spectrum	**Major absorptions at 2980 cm^{-1} and 1738 cm^{-1}. No significant absorptions above 3007 cm^{-1}. Three strong bands between 1172 and 1367 cm^{-1}.**
Mass spectrum	**Molecular ion peak at 116 and other significant peaks include 43 and 57.**
NMR spectrum	**the spectrum consists of two singlets in the ratio 3:1.**

The molecular ion peak allows us to deduce that the relative molecular mass is 116. Combinations of C, H and O that add up to 116 are:

$C_4H_4O_4$ $C_6H_{12}O_2$

$C_5H_8O_3$ $C_7H_{14}O$

Using IHD = ½[2c+2−h−x+n] we can work out that the only one with an IHD of 1 is $C_6H_{12}O_2$

	IHD
$C_4H_4O_4$	3
$C_5H_8O_3$	2
$C_6H_{12}O_2$	1
$C_7H_{14}O$	0

A major absorption in the range 1700–1750 cm^{-1} suggests the presence of C=O. With two oxygen atoms there is a good chance that this could be a carboxylic acid or an ester. Absorption bands in the region 1050–1410 cm^{-1} confirm the presence of C–O. No absorption above 3007 cm^{-1} suggests that no O–H is present. The absorption at 2980 cm^{-1} is due to C–H.

With 12 H atoms, two singlets in the ratio 3:1 indicate that there are 9H atoms in one environment and 3H in the other. A carboxylic acid would have 1H (from the COOH) in a different environment, therefore this can be ruled out. 9H in the same environment suggests 3 CH_3 groups attached to the same C. The signals are all singlets, therefore, none of the H atoms have any H atoms on the adjacent C.

A peak in the mass spectrum at 57 would correspond to $C(CH_3)^+$ and one at 43 could be CH_3CO^+. The structure is thus (Figure **11.12**):

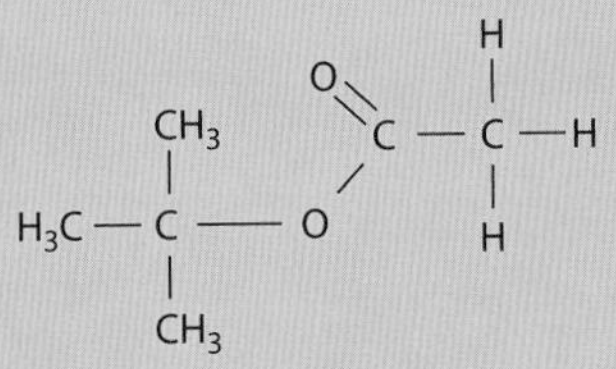

Figure 11.12

hint

When answering a question like this you should try to use all the data given.

TEST YOURSELF 11.10

1 Compound Y contains carbon, hydrogen and oxygen only. Deduce the structure of Y from the information in the table:

IHD	1
IR spectrum	Major absorption at 1749 cm^{-1}.
Mass spectrum	Molecular ion peak at 58 and major peak at 43
NMR spectrum	Two singlets in the ratio 1:1

CH_3COCH_3

2 Compound Z contains carbon, hydrogen and oxygen only. Deduce the structure of Z from the information in the table:

IHD	1
IR spectrum	Major absorption around 2900 cm^{-1}. No other major absorbtion above 1500 cm^{-1}. Several absorptions in the range 1050–1410 cm^{-1}
Mass spectrum	Molecular ion peak at 60 and major peaks at 45, 29 and 15
NMR spectrum	A singlet, a triplet and a quartet in the ratio 3:3:2

$CH_3OCH_2CH_3$

Checklist

At the end of this chapter you should be able to:

- ☐ **Record all measurements with uncertainties and suggest sources of random and systematic errors in experimental work.**
- ☐ **Work out percentage errors.**
- ☐ **Distinguish between accuracy and precision.**
- ☐ **Calculate absolute and percentage uncertainties and record data to an appropriate number of significant figures/decimal places.**
- ☐ **Draw and interpret graphs of experimental data.**
- ☐ **Calculate gradients and uncertainties from graphs.**
- ☐ **Calculate the IHD for various molecules.**
- ☐ **Deduce information relevant to the structures of molecules from infrared, mass and NMR spectra.**

Higher Level only

- ☐ **Deduce information from splitting patterns and chemical shift data in NMR spectra.**
- ☐ **Deduce the structure of organic molecules from spectroscopic data.**

MATERIALS

A

Materials are the substances that the things around us are made from. A knowledge of the properties of materials is important to people from many walks of life . . . scientists, engineers, architects, doctors etc.

This chapter covers the following topics:

- ☐ An introduction to materials science
- ☐ Metals and inductively-coupled plasma (ICP) spectroscopy
- ☐ Catalysts
- ☐ Liquid crystals
- ☐ Polymers
- ☐ Nanotechnology
- ☐ The environmental impact of plastics
- ☐ Superconduction and X-ray crystallography (HL only)
- ☐ Condensation polymers (HL only)
- ☐ The environmental impact of heavy metals (HL only)

A.1 Introduction to materials science

Classifying materials

Materials can be classified based on their properties, uses, bonding or structure. Typically materials are classified as **metals**, **ceramics**, **polymers** or **composites**.

- Metals (including alloys) conduct electricity and heat well, and are malleable and ductile. Examples include iron and steel.
- Ceramics are typically inorganic compounds comprising a metal and a non-metal. They are brittle, heat and chemical resistant, hard and strong. Examples include aluminium oxide and concrete.
- Polymers are usually long carbon chain molecules (see separate subtopic).
- Composites are mixtures in which materials are composed of two distinct phases; a reinforcing phase that is embedded in a matrix phase. Examples include fibreglass.

Relationship between physical properties and bonding and structure

Melting point

Typically, metals and ceramics have high melting points due to:

- The strong electrostatic attraction between positively charged ions and the delocalised electrons in a metal.
- The strong electrostatic attraction between oppositely charged ions in an ionic compound.
- The presence of strong covalent bonds in giant covalent structures, e.g. silicon dioxide.

These all require large amounts of heat energy to break.

In polymers, only weak intermolecular forces between polymer chains need to be overcome and so they tend to have low melting points.

Permeability

In general, the more closely packed the particles in a material, the less permeable to water it will be. Metals and most ceramics have tightly packed structures and are impermeable to water. Some, e.g. concrete, have a porous structure and water is able to penetrate them. Hydrocarbon polymers are generally impermeable to water due to the non-polar nature of the molecules. Increasing the crystallinity of a polymer (i.e. how closely packed the chains are) decreases water-permeability.

Conductivity

Metals are generally good conductors of both heat and electricity due to the presence of mobile delocalised electrons which can move throughout the structure. Ceramics and polymers do not contain delocalised electrons and act as electrical and thermal insulators.

Elasticity

Elasticity occurs when a material is stretched due to an applied force but returns to its original shape when the force is removed. Metals can be stretched slightly, forcing the ions apart a little; however the electrostatic forces are capable of pulling the structure back in-to place. If metals are stretched too far the layers of ions slide over one another and the structure is deformed. Polymers can exhibit similar properties (although explanations here must be in terms of molecules and intermolecular forces).

Brittleness

Ceramics tend to be brittle due to their ionic nature. Displacement of a layer of ions brings like charges together – the repulsion created causes the layers to separate and the solid breaks apart. Metals are malleable because the layers of metal ions can slide over one another without changing the bonding – the ions are still attracted to the delocalised electrons.

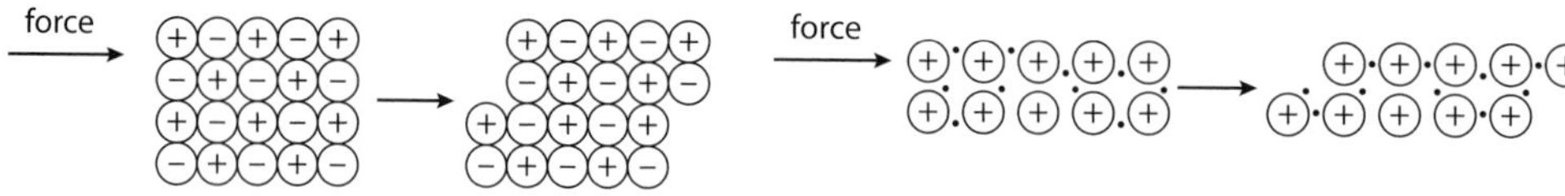

Figure A.1

Bonding triangles

The properties of a material are based on the degree of covalent, ionic or metallic character in the compound. The type of bonding can be deduced from its position in a bonding triangle.

hint

The *IB Data Booklet* contains a bonding triangle so you do not need to memorise one.

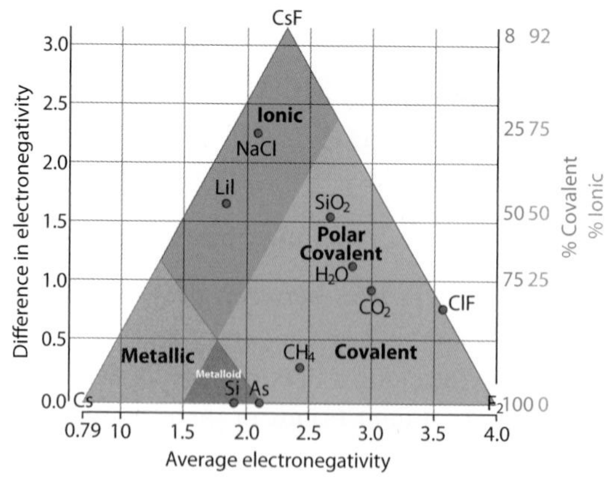

Figure A.2

You may be asked to use a bonding triangle to classify binary compounds (containing just two elements) from electronegativity data.

Worked example A.1

Determine the percentage ionic character in strontium sulfide (SrS).

Find the electronegativities of both elements using the table in the *IB Data Booklet*:

Strontium = 1.0 Sulfur = 2.6

Calculate the difference in electronegativity and the average electronegativity:

Difference $= 2.6 - 1.0 = 1.6$ Average $= \frac{2.6+1.0}{2} = 1.8$

Plot these values on the bonding triangle where average electronegativity is on the *x*-axis and the difference on the *y*-axis.

Read off the corresponding value for ionic nature based on the position of the compound = **~45%**

TEST YOURSELF A.1

1 Determine the percentage ionic character of the following compounds:
a MgO **b** PH_3 **c** $AlCl_3$ **d** OsO_4

2 Deduce the nature of the bonding in:
a $SnCl_4$ **b** SiH_4 **c** CaO **d** H_2O

A.2 Metals

Extraction of metals

The method used to extract a metal from its **ore** depends on its position on the activity series.

You should be able to write balanced equations for the extraction of various metals.

hint

An activity series is given in the *IB Data Booklet*.

Reduction with carbon or a metal

Metals below carbon in the series can be extracted by reduction with coke (carbon). This is typically used to extract iron, zinc or lead from their oxides at high temperatures. The product of the reaction will be the metal and either carbon monoxide or carbon dioxide.

Annotated exemplar answer A.1

Write an equation, including state symbols, for the extraction of molten iron from iron(III) oxide using carbon at high temperature. **[2]**

$Fe_2O_3 + C \rightarrow Fe + CO$

The chemical equation is not balanced, losing a mark – all chemical equations in answers must be balanced.

0/2

The question specifically asks for state symbols – they must be included.

Equations showing the formation of CO or CO_2 should both be marked correct, e.g. in the question: $2Fe_2O_3(s) + 3C(s) \rightarrow 4Fe(l) + 3CO_2(g)$ or $Fe_2O_3(s) + 3C(s) \rightarrow 2Fe(l) + 3CO(g)$ would both get full marks.

hint

If the question does not ask for state symbols it is generally better to leave them out than get them wrong.

TEST YOURSELF A.2

Construct balanced equations for the extraction of:

1 zinc from zinc oxide

2 lead from lead(IV) oxide

For all metals, it is possible to extract them from their compounds by reduction with a metal higher in the activity series.

For most metals above carbon in the activity series, electrolysis is used to extract the metal from its compounds. The production of aluminium is carried out in this manner (Figure **A.3**).

Production of aluminium

1 Aluminium oxide, Al_2O_3, is dissolved in molten cryolite, Na_3AlF_6, at about 900 °C.

2 A very large current is passed through the electrolyte.

3 Aluminium ions are reduced at the cathode to produce molten aluminium: $Al^{3+}(l) + 3e^- \rightarrow Al(l)$.

4 Oxide ions are oxidised at the anodes to produce oxygen gas: $2O^{2-}(l) \rightarrow O_2(g) + 4e^-$.

5 Some oxygen gas reacts with the graphite to produce carbon dioxide.

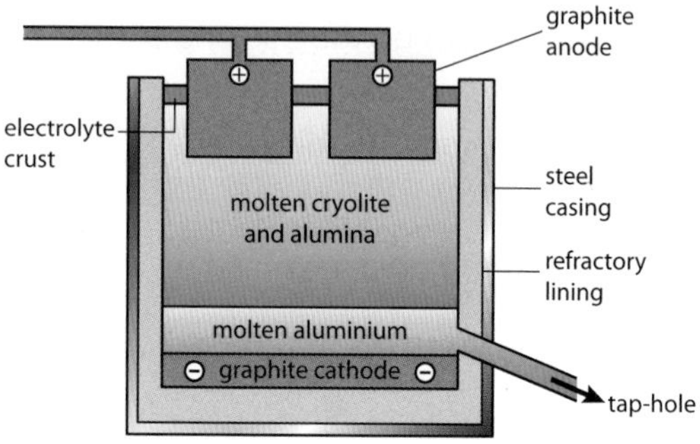

Figure A.3

Quantitative electrolysis

The amount of product formed in electrolysis can be calculated using the following procedure:

1 Calculate the amount of charge that flows using $Q = I \times t$ (remember that current must be in amps and the time in seconds).

2 Divide the charge by the Faraday constant (96 500 C mol^{-1}) to give the number of moles of electrons.

3 Write the half-equation for the reaction at the electrode to produce one mole of product.

4 Divide the number of moles of electrons by the coefficient of the electrons in the half-equation. This gives the number of moles of product formed.

5 Convert the number of moles of product formed to a mass.

☆ Model answer A.1

Calculate the mass of aluminium (in kg) produced if a current of 100 kA is passed for 1.00 hour through an electrolysis cell containing aluminium oxide dissolved in molten cryolite.

$Q = I \times t$

$I = 100\,000$ A $\quad t = 1 \times 60 \times 60 = 3600$ s

$Q = 100\,000 \times 3600 = 3.6 \times 10^8$ C

$$\text{Moles of } e^- = \frac{\text{charge}}{96\,500} = \frac{3.6 \times 10^8}{96\,500} = 3.73 \times 10^3 \text{ mol}$$

$Al^{3+} + 3e^- \rightarrow Al$

$$\text{Moles of Al} = \frac{\text{moles of } e^-}{3} = 1244 \text{ mol}$$

Mass of aluminium = moles of aluminium $\times A_r$ = 1244×26.98 = 33 550 g = **33.5 kg**

TEST YOURSELF A.3

Calculate the mass of metal produced during the electrolysis of:

1. molten NaCl for 30 minutes at 3.0 A
2. molten $PbBr_2$ for 5 minutes at 0.50 A
3. molten CaO for 2 hours at 10 A

hint

You need to remember the equation $Q = I \times t$, however, Faraday's constant is given in the *IB Data Booklet*.

Alloys

DEFINITION

ALLOYS are homogeneous mixtures of metals with other metals or non-metals.

Alloying metals can change the properties of the metal significantly. Desirable changes may include increased melting point, stiffness, strength and chemical-resistance. Most changes of properties occur due to the disruption of the regular **lattice structure** by incorporating atoms/ions of different radii. This prevents the layers from sliding over one another effectively.

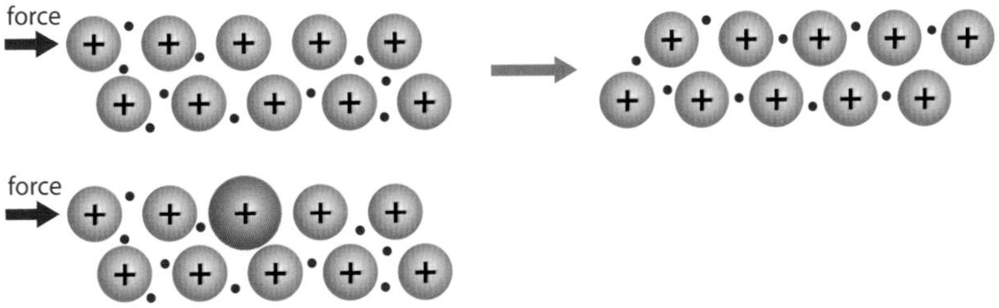

Figure A.4

Paramagnetism and diamagnetism

DEFINITIONS

PARAMAGNETIC SUBSTANCES these are attracted by a magnetic field.

PARAMAGNETISM is caused by the presence of unpaired electrons.

DIAMAGNETIC SUBSTANCES these are repelled slightly by a magnetic field.

DIAMAGNETISM is caused by the presence of paired electrons.

To determine if a compound of a metal will exhibit paramagnetism or diamagnetism you will need to draw orbital diagrams for the atoms/ions in question. If the ion formed by the metal contains any unpaired electrons then the compound will be paramagnetic – if all the electrons are paired then the metal will be diamagnetic. The more unpaired electrons that are present the stronger the paramagnetic effect.

Ti^{2+}	[Ar] $3d^2$ [↑][↑][ ][ ][ ]	paramagnetic
Mg^{2+}	$1s^2$ [↑↓] $2s^2$ [↑↓] $2p^6$ [↑↓][↑↓][↑↓]	diamagnetic

Figure A.5

TEST YOURSELF A.4

Determine whether compounds containing the following ions would be diamagnetic or paramagnetic.

1 Mn^{2+} **2** Ca^{2+} **3** Pb^{2+}

Inductively-coupled plasma (ICP) spectroscopy

Inductively-coupled plasma (ICP) spectroscopy can be used in conjunction with mass spectrometry (ICP-MS) and optical emission spectroscopy (ICP-OES) to identify and quantify trace amounts of a metal by ionising them with an argon gas plasma. Typical uses of ICP spectroscopy include analysis of biological samples, geological samples, water, food and medicines.

DEFINITION

PLASMA is a fully or partially ionised gas consisting of positive ions and electrons (the plasma is usually electrically neutral overall).

The plasma is created by initially ionising argon gas with a spark from a Tesla coil. A strong electromagnetic field created by an induction coil causes argon ions to collide with argon atoms and create more ionised argon. When a sample is injected into this plasma the atoms are immediately ionised.

In ICP-OES, excited metal ions in the sample emit spectral lines of specific frequencies. Both the exact frequencies and intensities of this emitted light are detected. The frequencies emitted are used to identify the elements present and the intensities are related to the concentrations present.

It is necessary to calibrate the machinery beforehand by creating a calibration curve using known quantities of the metals being detected. The sample intensity can be read directly from this calibration curve.

In ICP-MS, the sample ionised by the ICP is injected directly into a mass spectrometer. The mass spectrometer separates the metal ions based on their mass:charge ratio. The exact mass : charge ratios detected along with their intensities can be used to both identify and quantify the elements present.

You may be asked to work out the concentration of metal ions present in a sample from a calibration curve.

Worked example A.2

A 250 cm^3 sample of water was collected for analysis of its calcium content. A 25 cm^3 portion of this sample was subjected to ICP-OES. An arbitrary intensity value of 54 was recorded. Use this information and the calibration curve to determine the mass of calcium present in the original sample.

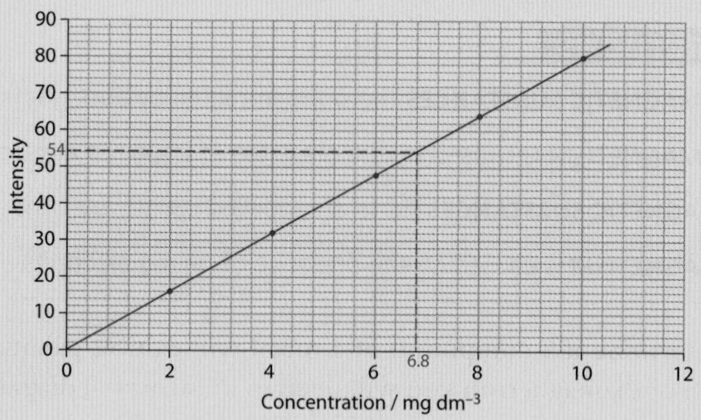

Figure A.6

Determine concentration of calcium that corresponds to an intensity of 54 using the calibration curve.

Concentration of calcium = 6.8 mg dm^{-3}

Calculate the mass present in the 250 cm^3 sample using: mass = concentration × volume in dm^3

Mass of calcium present = 6.8 × 0.25 = 1.7 mg.

A.3 Catalysts

DEFINITIONS

HETEROGENEOUS CATALYSTS are those in a different phase of matter to the reactants.

HOMOGENEOUS CATALYSTS are those in an identical phase of matter to the reactants.

> **hint**
> Do not confuse adsorption with absorption: adsorption means that gas molecules bind to a surface.

Heterogeneous catalysts

Heterogeneous catalysts tend to be solids; transition metals are typical examples (Figure **A.7**). They are able to **adsorb** gaseous molecules on to **active sites** on their surface. The covalent bonds within the reactants are weakened and the activation energy is lowered as a result. Once the reaction has occurred, the products **desorb**.

As heterogeneous catalysis can only occur at the surface, these catalysts are typically used in the form of **nanoparticles**, i.e. particles <100 nm diameter. This gives them a very large surface area to mass ratio to increase the rate of reaction.

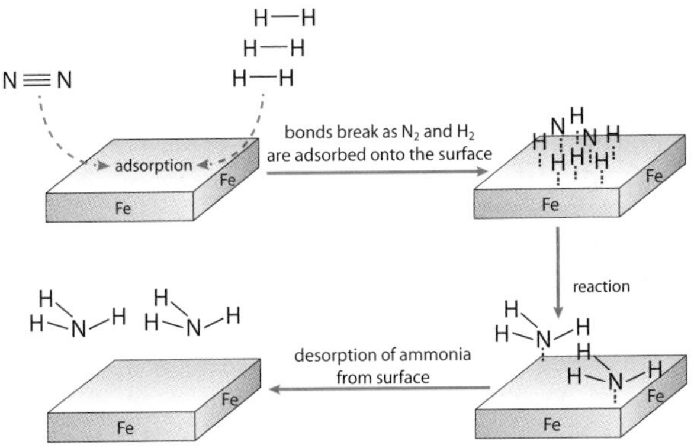

Figure A.7

Carbon nanocatalysts

Carbon nanotubes are capable of acting as catalysts themselves or as scaffold for other catalysts – they have an extremely high surface area to mass ratio and are able to coordinate other atoms.

Zeolites

Zeolites are aluminosilicate catalysts with a cage-like structure. The presence of pores produces a large surface area and makes the catalysis very selective (Figure **A.8**).

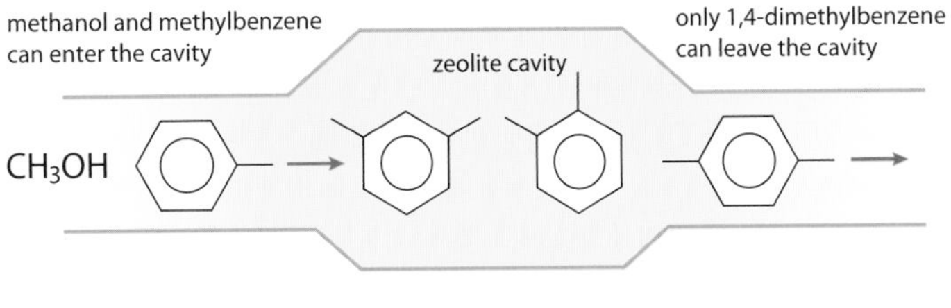

Figure A.8

Homogeneous catalysts

Homogeneous catalysts chemically combine with the reactants to form a temporary activated complex (in which the catalyst stabilises the transition state) or a reaction intermediate. Transition metal compounds may act as homogeneous catalysts due to their ability to exhibit variable oxidation states and also to coordinate other molecules and ions.

Choice of catalyst

Selectivity – homogeneous catalysts tend to be chosen when high selectivity is required, e.g. using supported enzymes for biochemical reactions.

Efficiency – homogeneous catalysts are more efficient than heterogeneous ones as **all** the particles are available rather than just those at the surface.

Conditions – if the reaction requires harsh conditions (e.g. high temperature and pressure) then heterogeneous catalysts are better suited to the process.

Impurities – both types of catalyst can be 'poisoned' by impurities. Poisoned heterogeneous catalysts can be cleaned of deposits but homogeneous catalysts will need replacing.

Environmental considerations – the disposal of heavy metal catalysts can lead to significant environmental problems due to their toxicity.

Historically, the choice of catalyst was a matter of guesswork and trial-and-error. This drove the development of models and theories to explain their mechanism of action.

A.4 Liquid crystals

DEFINITION

LIQUID CRYSTALS are fluids whose physical properties are dependent on their molecular orientation to some fixed axis in the material. (Such properties include electrical, optical and elastic ones.)

In a liquid crystal, the molecules exhibit uniform orientation (point in the same direction) but are able to move randomly and flow past one another.

Lyotropic liquid crystals

DEFINITION

LYOTROPIC liquid crystals are solutions that show the liquid-crystal state over a certain range of concentrations.

A mixture of soap/detergent and water is an example of a lyotropic liquid crystal. Typical soaps contain stearate ions ($C_{17}H_{35}COO^-$) that have a polar carboxylate 'head' and a non-polar alkyl 'tail'. At low concentrations the stearate ions are separated by relatively large distances and are orientated randomly. As the concentration increases and the stearate ions are pushed together, they tend to orientate themselves in the same direction, creating a liquid-crystal phase.

Thermotropic liquid crystals

DEFINITION

THERMOTROPIC liquid crystals are pure substances that show liquid-crystal behaviour over a specific temperature range.

Biphenyl nitriles are molecules that exhibit thermotropic liquid-crystal properties, e.g.

These rod-like molecules possess a polar nitrile 'head' which means that in the liquid crystal phase they are orientated, on average, in the same direction. They are not fixed in position so they can flow like a liquid. This arrangement of molecules is called the **nematic phase**.

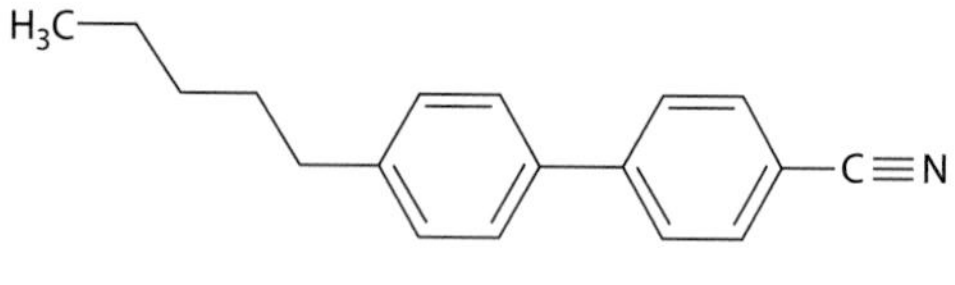

Figure A.9

DEFINITION

The NEMATIC LIQUID CRYSTAL PHASE is characterised by rod-shaped molecules that are randomly distributed but on average align in the same direction.

Liquid crystal displays (LCDs)

Thermotropic liquid crystals are ideal for use in liquid crystal displays (LCDs).

Desirable properties of a liquid crystal for use in LCDs include:

- chemical stability
- having a phase that is stable over a suitable temperature range
- polarity – they can change orientation when an electric field is applied (to permit or prevent the passage of light)
- rapid switching speed between the different orientations (light/dark).

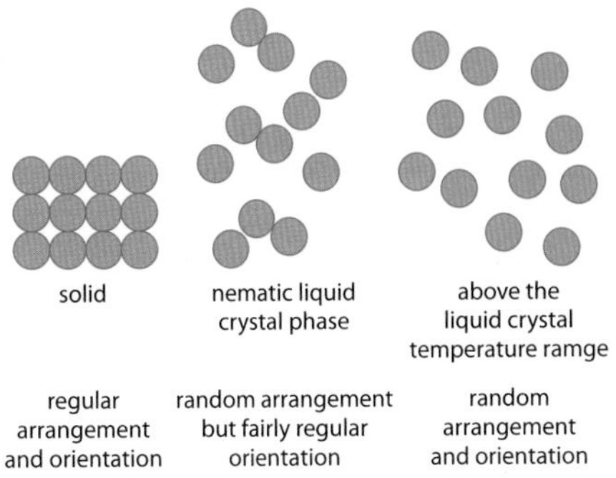

Figure A.10

The discovery of liquid crystals is another example of serendipity. Friedrich Reinitzer accidently discovered flowing liquid crystals in 1888 while experimenting on cholesterol.

A.5 Polymers

DEFINITIONS

THERMOPLASTICS are those that soften when heated and harden when cooled.

THERMOSETTING POLYMERS are pre-polymers (soft solids or viscous liquids) that change irreversibly into a hardened thermoset by curing.

ELASTOMERS are flexible polymers that can be deformed under force but will nearly return to their original shape once the stress is released.

Thermoplastics

Thermoplastics like poly(ethene) are made of long polymer chains held together by weak London forces. When heated the London forces are weakened or broken, allowing the polymer chains to slide past one another and the plastic becomes soft.

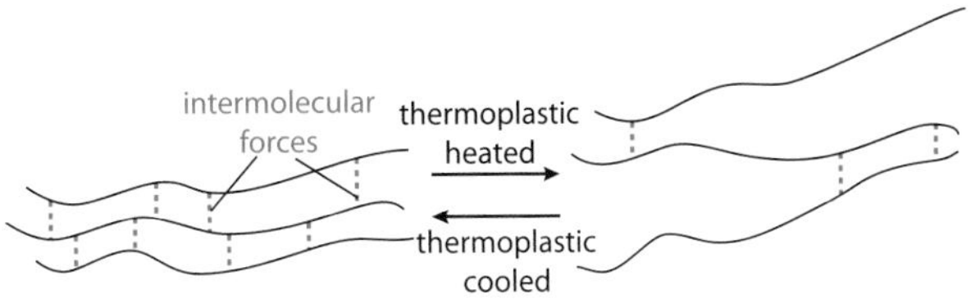

Figure A.11

Poly(ethene)

Poly(ethene) can be manufactured in low-density and high-density forms. The main differences between these two polymers are:

Low-density poly(ethene) (LDPE)	High-density poly(ethene) (HDPE)
Contains many branched chains	Mainly straight chains – minimal branching
Low percentage crystallinity	High percentage crystallinity
Softer and more flexible	Harder and more rigid
Chains pack less efficiently	Chains pack together efficiently

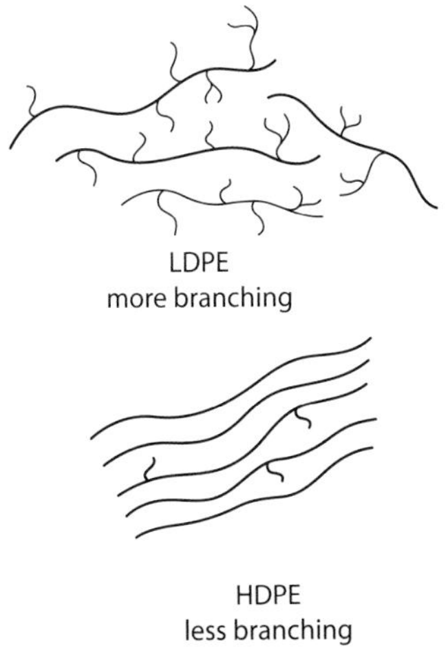

Figure A.12

Poly(propene)

DEFINITIONS

An ISOTACTIC addition polymer is one in which all of the substituent groups are on the same side of the chain.

An ATACTIC addition polymer is one in which the substituent groups are positioned randomly.

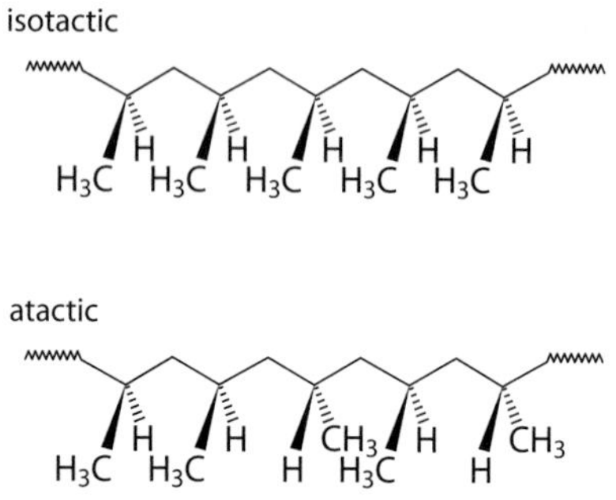

Figure A.13

In isotactic poly(propene) all of the methyl groups are on the same side of the polymer chain – this regular arrangement means the chains can pack together well, creating a strong, rigid polymer. In atactic poly(propene) the methyl groups are placed randomly and the chains are unable to pack as effectively – atactic poly(propene) is soft and rubbery.

Plasticisers

Rigid polymers such as poly(chloroethene) (polyvinyl chloride) can be made more flexible by the addition of a plasticiser. The plasticiser is typically a small molecule that inserts itself between the polymer chains, forcing them apart in places. This weakens the force of attraction between the chains and allows them to move more freely.

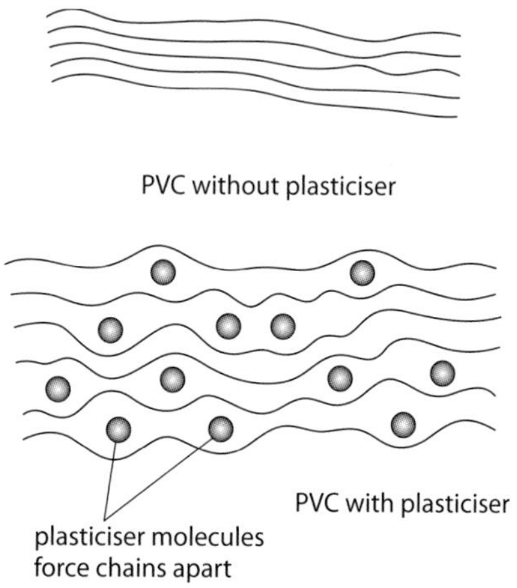

Figure A.14

Expanded polymers

Poly(phenylethene) [poly(styrene)] is a rigid, glassy polymer. However, a volatile hydrocarbon such as pentane can be dissolved in the liquid polymer during manufacture. When the polymer is heated in steam, the pentane vaporises and the gas bubbles created cause the polymer to expand to 60–70 times its original volume. The result is the white, low-density packaging material – Styrofoam™.

Thermosetting polymers

Thermosetting polymers contain long polymer chains linked to each other through covalent cross-links. This increases the rigidity of the polymer. When the polymer is heated, the strong covalent cross-links prevent the chains sliding past each other and so the polymer remains stiff at high temperatures.

Elastomers

Elastomers such as rubber contain polymer chains that are coiled like a spring. When the polymer is stretched, the chains straighten. When the stretching force is removed, the polymer chains spring back into their original coiled forms.

Poly(2-methylpropene)

2-methylpropene forms an elastomeric polymer. Typically, monomers join head-to-head and the two methyl side groups are placed on every other carbon atom on the chain. Occasionally, a monomer adds on to the chain the other way around and two consecutive carbons have two methyl side groups.

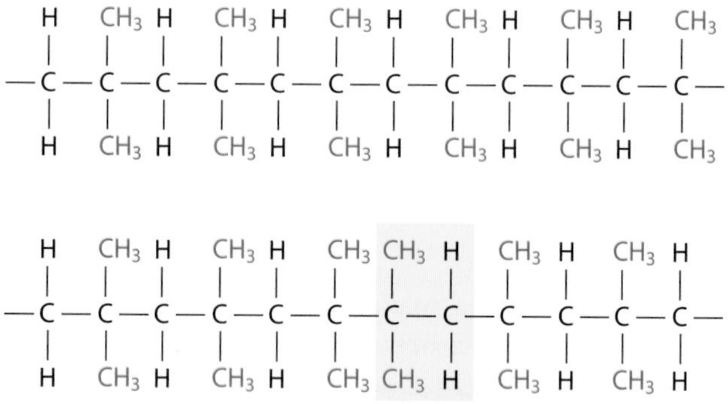

Figure A.15

Atom economy

DEFINITION

ATOM ECONOMY is a measure of efficiency applied in green chemistry:

$$\text{Atom economy} = \frac{\text{molar mass of desired product(s)}}{\text{total molar mass of all reactants}} \times 100$$

The production of ethanoic acid provides a good illustration of atom economy. Previously, ethanoic acid was produced by the direct oxidation of butane:

$$2C_4H_{10}(g) + 5O_2(g) \rightarrow 4CH_3COOH(l) + 2H_2O(l)$$

$$\text{Atom economy} = \frac{4 \times 60}{(2 \times 58) + (5 \times 32)} \times 100 = 87.0\%$$

In a new process methanol is reacted with carbon monoxide using a catalyst.

$$CH_3OH(g) + CO(g) \rightarrow CH_3COOH(l)$$

As only one product is made the atom economy is **100%**. This represents the application of green chemistry to the synthesis. The conditions required are less harsh (lower pressure and temperature) and atom economy is greater.

hint

Do not confuse atom economy and the yield of a reaction. Atom economy is theoretical and provides an idea of how much waste is produced. Yield is an experimentally derived quantity. Reactions with 100% atom economy may have low yield and *vice versa*.

TEST YOURSELF A.5

Calculate the atom economy of the following processes:

1. Formation of poly(ethene) from ethene.
2. Production of ethanol via the fermentation of glucose.
3. The synthesis of aspirin from 2-hydroxybenzoic acid and ethanoic anhydride.

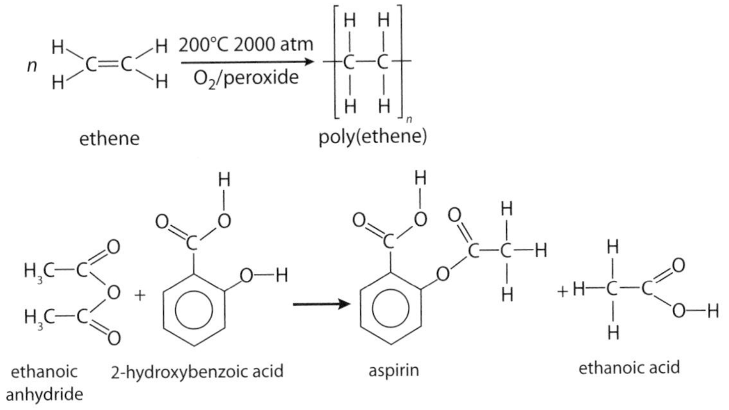

Figure A.16

A.6 Nanotechnology

Nanotechnology involves the production and use of particles or structures that have at least one dimension of less than 100 nm.

Molecular self-assembly

Molecular self-assembly is the bottom-up assembly of nanoparticles (using physical or chemical techniques) and can occur by selectively attaching molecules to specific surfaces. Self-assembly also may occur spontaneously in solution; for example, the formation of micelles by detergent molecules in solution.

Physical and chemical techniques

Physical techniques involve the manipulation and positioning of atoms to specific requirements. The **scanning tunnelling microscope** can be used to pick up atoms and move them into specific positions.

Chemical techniques involve carrying out chemical reactions to position atoms within molecules and structures. The formation of enzymes through covalent bonding of amino acids is an example.

Carbon nanotubes

Carbon nanotubes are single sheets of graphite (graphene) rolled up into a tube and capped at the ends. They contain carbon atoms arranged in hexagons (and hexagons and pentagons at the capped ends). They are typically 1–2 nm in diameter but may be several centimetres in length.

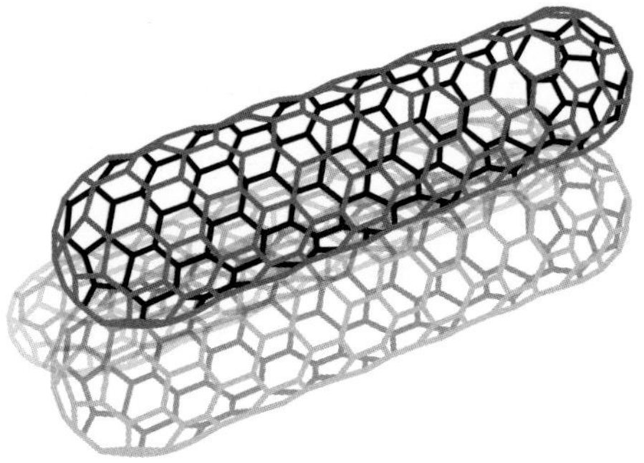

Figure A.17

Carbon nanotubes can be synthesised in a number of ways:

Arc discharge

Either:

1. Two carbon electrodes are placed 1–2 mm apart. A large current (about 100 A) creates an electric arc between the electrodes. The high temperature vaporises carbon at the anode which is then deposited as carbon nanotubes at the cathode.
2. Two metal electrodes are placed 1–2 mm apart in a liquid hydrocarbon solvent. A large current creates an arc between the electrodes causing the hydrocarbon to decompose – small rod-shaped structures are deposited at the anode. The oxidation state of carbon in a hydrocarbon is always negative but in carbon nanotubes it is zero, therefore this process involves oxidation of the carbon.

Chemical vapour deposition (CVD)

A carbon-containing gas, e.g. methane, is subjected to very high temperatures in the presence of a metallic nanocatalyst. The covalent bonds in the gas are broken at the surface of the catalyst and the carbon atoms are able to bond to form nanotubes. This technique must be carried out in an inert atmosphere, e.g. argon gas, to avoid any reaction of the gas with oxygen forming carbon dioxide.

High-pressure carbon monoxide deposition (HIPCO)

Carbon monoxide and iron pentacarbonyl, $Fe(CO)_5$, enter a reaction vessel at very high pressure and temperature. $Fe(CO)_5$ decomposes to form catalytic iron nanoparticles and more carbon monoxide. The carbon monoxide disproportionates (same species oxidised and reduced) on the surface of the iron catalyst forming carbon nanotubes.

$$2CO(g) \rightarrow CO_2(g) + C(s)$$

Properties of carbon nanotubes

Carbon nanotubes:

- are strong – due to the presence of strong covalent bonds only
- are conductors of electricity – C only forms three bonds so there are delocalised electrons that can move throughout the structure carrying charge
- can act as catalysts.

Implications and applications of nanotechnology

Potential applications of nanoparticles include:

- the creation of strong, low-density composite materials
- heterogeneous catalysis
- antibacterial agents in clothing
- improving the effectiveness of sunscreen.

Concerns regarding the use of nanotechnology include:

- lack of knowledge regarding the toxicity of nanoscale substances
- concern that the human defence system may not be effective against nano-sized particles
- ensuring that industries and governments are responsible and regulated.

A.7 Environmental impact of plastics

Incineration

The presence of strong covalent bonds in plastics means that they do not degrade easily – this makes disposal challenging. One method of disposal is incineration.

You should be able to construct a balanced equation for the combustion of a given plastic or monomer.

e.g. combustion of poly(ethene): $\text{-}(CH_2\text{—}CH_2)\text{-}_n + 3nO_2 \rightarrow 2nCO_2 + 2nH_2O$

Combustion of a plastic can be represented by an equation based on the monomer/repeating unit

e.g. combustion of chloroethene: $CH_2CHCl + O_2 \rightarrow 2CO + HCl + H_2O$

Burning polyvinyl chloride (the polymer of chloroethene) produces carbon monoxide, hydrogen chloride and dioxins, all of which are toxic.

Domestic fires may release toxic products due to plastic household items. To minimise the potential release of the toxins, low-smoke, zero-halogen cabling is often used in wiring in public buildings.

hint

The monomer and repeating unit of a plastic have the same molecular formula.

Dioxins

DEFINITIONS

DIOXINS are molecules that contain unsaturated six-membered heterocyclic rings with two oxygen atoms, usually in positions 1 and 4.

HETEROCYCLIC ring in which there is more than one type of atom.

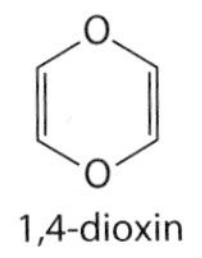

Figure A.18

Dioxins may also be chlorinated and form polychlorinated dibenzodioxins.

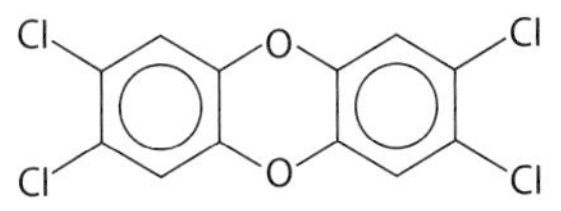

Figure A.19

These chlorinated dioxins are highly toxic, causing disruption to hormone signalling that may lead to cellular and genetic damage. They pose a particular threat because they do not decompose in the environment and can be passed on in the food chain.

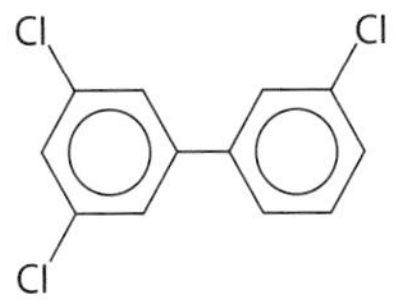

Figure A.20

The **polychlorinated biphenyls** (PCBs) are a group of dioxin-like compounds that are believed to have undesirable environmental and health effects; they are thought to be carcinogenic.

You must be able to compare the structures of PCBs and dioxins – PCBs differ in structure from the dioxins as they do not contain heterocyclic rings nor do they contain oxygen atoms.

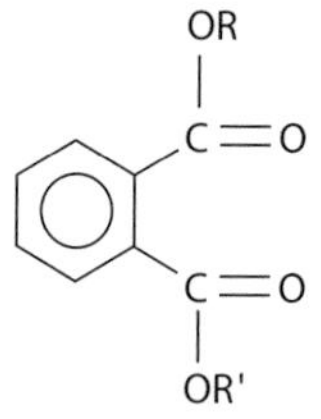

Figure A.21

Plasticisers

One of the most common types of plasticiser used in polymer production is the **phthalate ester**.

There are health concerns about these volatile, fat-soluble molecules including their potential to cause cancer as well as disrupt hormone signalling and sexual development.

The environmental impact of the use of plastics

Negative effects:

- plastics are produced from a non-renewable resource – crude oil
- plastics persist in the environment for a long time – litter and a danger to wildlife
- disposal of plastic in landfill sites takes up space and adds to pollution
- incineration of plastics produces toxic compounds.

Positive effects:

- plastic bottles are lighter than glass bottles – less energy needed to transport them
- production of plastic is more energy efficient than that of glass or aluminium
- plastic piping needs replacing less often than metal/ceramic piping.

> **hint**
> RICs for different plastics are given in the *IB Data Booklet*.

Recycling of plastics

Due to the number of different types of polymer used to make plastics and the presence of additives, recycling them is a more energy-intensive process than for many other materials. The plastic has to be sorted into resin type, cleaned, melted, extruded and reformed.

Plastics can be identified by a Resin Identification Code (RIC) and this is used to separate them during the recycling process.

Infrared (IR) spectroscopy can be used to distinguish the different Resin Identification Codes of plastics.

It is difficult to distinguish the hydrocarbon-based plastics like poly(ethene) and poly(propene) as they contain the same bonds. The C–Cl bond vibration in poly(chloroethene) is in the fingerprint region and is often masked by other peaks. Polystyrene has a characteristic series of peaks in the range 1500–1600 cm^{-1} due to the presence of the phenyl group.

Worked example A.3

The infrared spectrum of a plastic is shown here. Using information from the *IB Data Booklet*, deduce the possible Resin Identification Code(s) of this plastic.

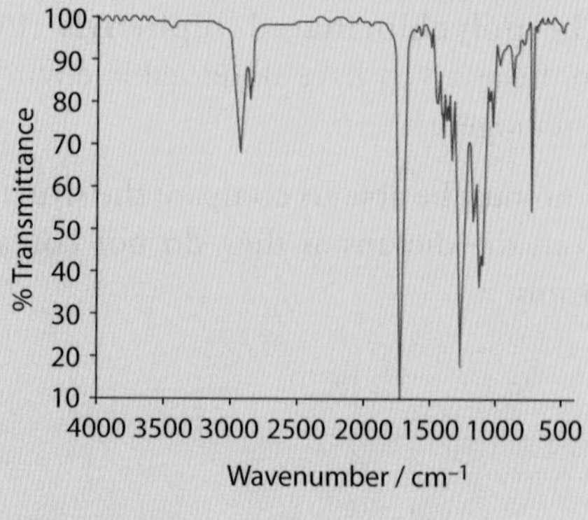

Figure A.22

Using the table of wavenumbers from the *IB Data Booklet*, it is possible to identify major peaks at 2900 cm^{-1} and 1700 cm^{-1}. These correspond to C–H and C=O bonds.

Only two RICs are assigned to compounds that have (or may have) a C=O; they are **Code 1** (PET) and **Code 7** (Other).

A.8 Superconducting metals and X-ray crystallography (HL only)

Superconductors

DEFINITION

SUPERCONDUCTORS are materials that have zero electrical resistance below a critical temperature.

Resistance in a metal occurs when delocalised electrons collide with the vibrating positive metal ions. As the temperature falls, the metal ions vibrate less (decrease in amplitude) and resistance drops.

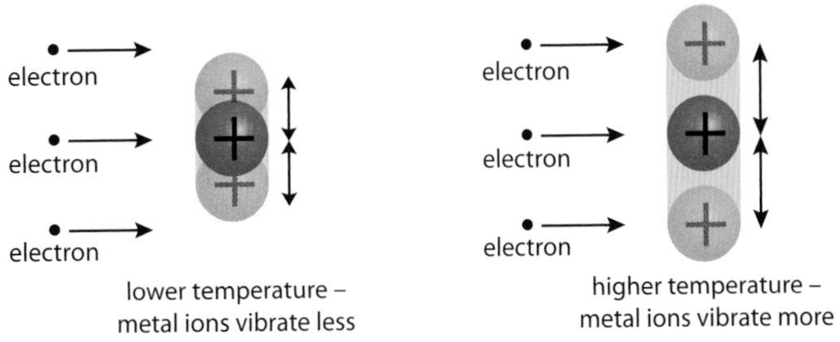

Figure A.23

The Bardeen–Cooper–Schrieffer theory

Superconductivity can be explained using the **Bardeen–Cooper–Schrieffer** theory of superconductors. Below the critical temperature, as a delocalised electron travels through the lattice it distorts the lattice (by attracting positive ions to it) creating a region with very slightly more positive charge. This region attracts another electron, which is then coupled with the first electron to form a **Cooper pair**. The Cooper pair is able to travel freely through the lattice – there is zero resistance.

The Meissner effect

Superconductors are able to expel an external magnetic field by creating a **mirror image** of the field. This is termed the **Meissner effect**. This explains why superconductors levitate in a magnetic field.

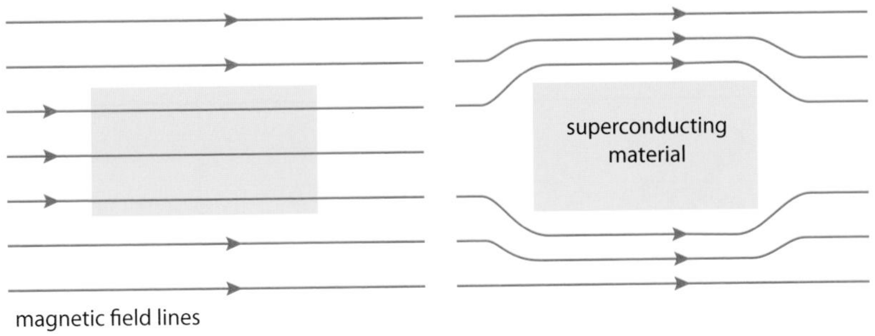

Figure A.24

Type 1 and Type 2 superconductors

Superconductivity can be destroyed by a strong enough magnetic field even below the critical temperature. Type 1 superconductors (Figure **A.25**) exhibit **sharp** transitions between non-superconductivity and superconductivity, whereas Type 2 superconductors (Figure **A.26**) have a more **gradual** transition.

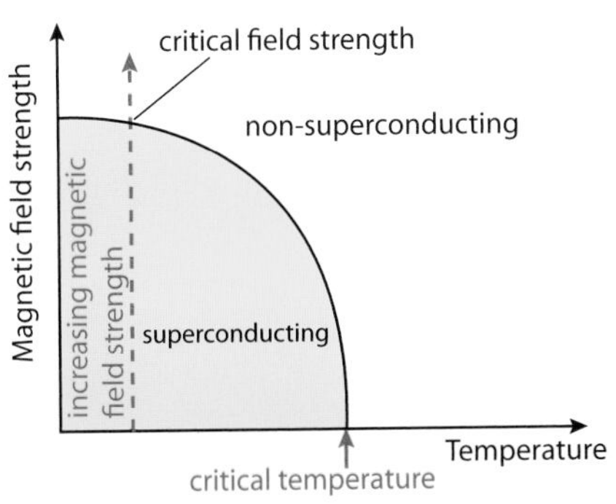

Figure A.25

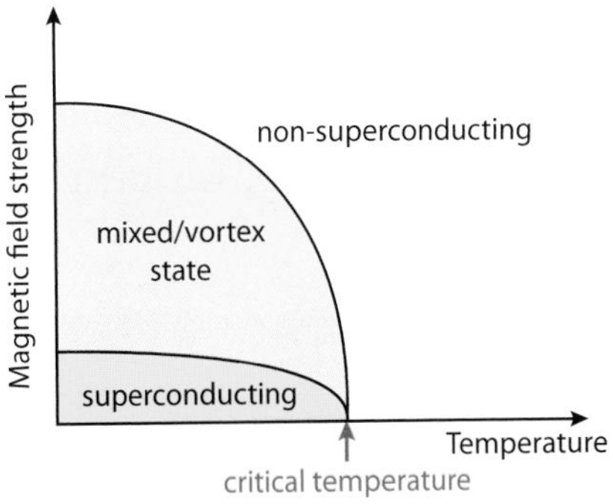

Figure A.26

A Materials

X-ray crystallography

X-ray crystallography is a technique that uses X-ray diffraction to analyse the structure of metallic and ionic lattices.

Lattice structures

DEFINITIONS

A **UNIT CELL** is the simplest repeating unit from which a whole crystal can be constructed.

The **COORDINATION NUMBER** of an atom/ion is the number of nearest neighbours it has.

In a unit cell, the atoms/ions on the faces and vertices are shared with neighbouring unit cells. Atoms/ions on faces are shared between two unit cells and atoms/ions on vertices between eight.

	contribution to unit cell
atom/ion at a vertex	$^1/_8$
atom/ion on a face	$^1/_2$

Given information about a crystal, you may be expected to deduce the type of lattice it has from a choice of three.

Lattice	Unit cell		Number of ions/ atoms in unit cell	Coordination number
Simple cubic	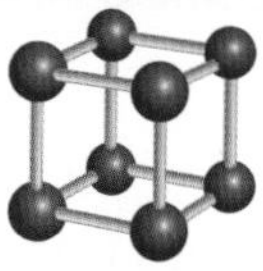		$8 \times {}^1/_8 = 1$	6
Body-centred cubic	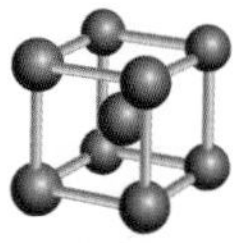		$8 \times {}^1/_8 + 1 = 2$	8
Face-centred cubic	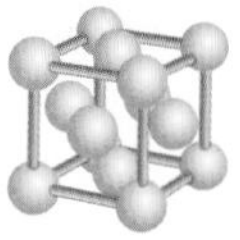	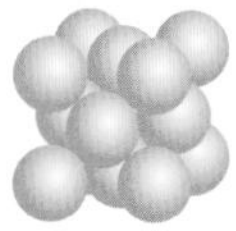	$8 \times {}^1/_8 + 6 \times \frac{1}{2} = 4$	12

X-ray diffraction

X-rays fired at a metallic or ionic crystal will undergo diffraction by the planes of ions. Diffracted X-rays may combine constructively or destructively depending on the distance between the planes – only constructive interference can be detected.

For X-rays to combine constructively the **Bragg equation** must be obeyed:

$$n\lambda = 2d\sin\theta$$

λ is the wavelength of the X-ray

d is the distance between the planes of ions

θ is the angle at which the X-rays strike the planes of ions

n is an integer

When $n = 1$ it is termed **first-order reflection**, when $n = 2$ it is termed **second-order reflection** and so on.

hint

You will only be asked to apply the Bragg equation to simple cubic structures in which the value of *d* is equal to the length of the unit cell.

TEST YOURSELF A.6

1 Calculate the length of the unit cell in a simple cubic crystal if X-rays with a wavelength of 1.54×10^{-10} m have a first order reflection of 16.4°.
2 Calculate the angle of the second order reflection of the crystal in Question **1**.

Model answer A.2

A metal with a simple cubic structure is bombarded with X-rays with a wavelength of 1.36×10^{-10} m. First order reflection of the X-rays occurs when the angle of incidence is 14.2°. Calculate the length of the unit cell from these data.

$$n = 1 \qquad \lambda = 1.36\times10^{-10} \qquad \theta = 14.2$$

$n\lambda = 2d\sin\theta$

Therefore: $d = \dfrac{n\lambda}{2\sin\theta}$

$$d = \frac{1\times1.36\times10^{-10}}{2\times\sin 14.2°} = \mathbf{2.77\times10^{-10}\ m}$$

d is the length of the unit cell in a simple cubic lattice, therefore the length of the unit cell is **2.77×10^{-10} m.**

It is possible to calculate the density of a metal from information about its crystal packing structure and its atomic radius.

Worked example A.4

The metal polonium ($A_r = 209$) has an atomic radius of 1.68×10^{-10} m and adopts a simple cubic crystal structure. Use this information to calculate the density of polonium, giving your answer in g cm^{-3}.

Calculate the mass of one atom of polonium:

$$\text{Mass of one atom} = \frac{\text{mass of 1 mole}}{\text{Avogadro's number}} = \frac{209}{6.02\times10^{23}} = 3.47\times10^{-22}\ \text{g}$$

Work out the number of atoms per unit cell from the structure and determine the mass of the unit cell.

A simple cubic unit cell contains **1 atom**, so a unit cell has a mass of 3.47×10^{-22} g.

Calculate the volume of the cell using the atomic radius and information about the structure.

In a simple cubic unit cell, the length of the cell is equal to $2r$, where r is the atomic radius.
Therefore, the length of the unit cell is $2\times1.68\times10^{-10} = 3.36\times10^{-10}$ m
The volume of the unit cell is $(3.36\times10^{-10})^3 = 3.79\times10^{-29}$ m^3

Dividing the mass of the unit cell by the volume of the unit cell will give the density of the metal.

$$\text{Density of the metal} = \frac{3.47\times10^{-22}}{3.79\times10^{-29}} = 9150\,000\ \text{g m}^{-3}$$

Convert to g cm^{-3} by diving by 10^6, therefore **density = 9.15 g cm^{-3}.**

hint

Do a common sense check. Does your density look sensible? If you are getting values outside the range 0.5–25.0 g cm^{-3}, you have probably made an error somewhere.

This calculation is made more challenging if the metal has face or body-centred **lattice structure**. In these cases you will need to remember that the number of atoms per unit cell is different and that the volume of the unit cell has to be calculated differently. You can draw the unit cell and use Pythagoras' Theorem to do this. The relationship between the atomic radius of an atom and the volume of its unit cell are summarised here and it might be easier just to learn the data in the table rather than trying to work it out.

Crystal structure	Volume of unit cell
Simple cubic	$8r^3$
Face-centred cubic	$16\sqrt{2}r^3$
Body-centred cubic	$\frac{64}{3\sqrt{3}}r^3$

TEST YOURSELF A.7

1. Calcium adopts a face-centred cubic structure. Given the atomic radius of calcium is 1.97×10^{-10} m, calculate the density of calcium.
2. Iron adopts a body-centred cubic structure. Given the atomic radius of iron is 1.26×10^{-10} m, calculate the density of iron.

A.9 Condensation polymers (HL only)

Condensation polymerisation occurs when monomers, each with two functional groups, react with elimination of a small molecule such as water, hydrogen chloride or, occasionally, ammonia.

Comparison of addition and condensation polymers

Addition	Condensation
Monomers are alkenes and must contain C=C bonds.	Monomers are not alkenes.
Monomers join together to form a long chain without the loss of anything.	Each time two monomers join together a small molecule is eliminated.
Empirical formula of the polymer is the same as that of the monomer (ignoring end groups).	Empirical formula of the polymer is not the same as those of the monomers.
Typically only one monomer is used.	Typically two different monomers are used.
Monomer needs only one functional group.	Monomers must contain two functional groups.
Polymer contains mostly non-polar groups and strong bonds and is therefore chemically inert – not biodegradable.	Polymer contains polar groups (ester and amide groups) – can be hydrolysed to reform the monomers and may be biodegradable.

Polyesters

Polyesters are made from the reaction of a dihydric alcohol (two OH groups) with a dicarboxylic acid (Figure **A.27**). One hydroxyl group on each end of the alcohol reacts with one carboxyl group in the dicarboxylic acid to create an ester linkage – the other carboxyl group is free to react with other alcohol molecules.

You should be able to write equations for the formation of a polyester from its monomers as well as draw the repeating unit.

To draw the repeating unit (in brackets; see Figure **A.28**) – join the two monomers together with the elimination of water to form an ester linkage, then remove the OH from the other COOH and H from the other hydroxyl group:

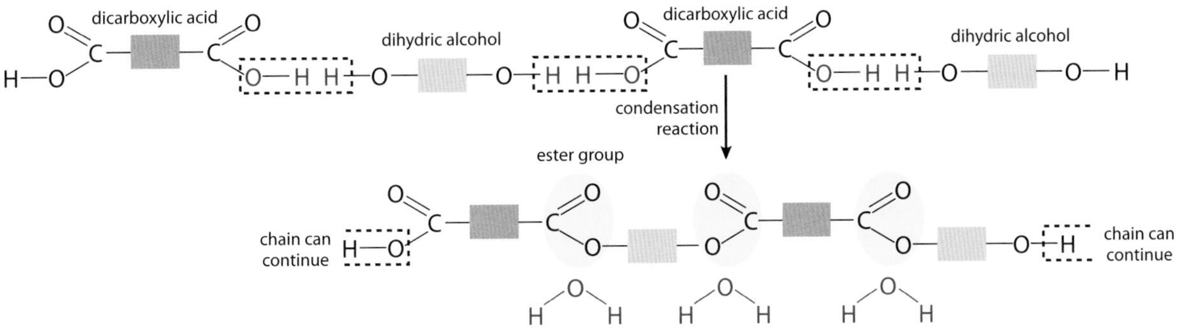

Figure A.27

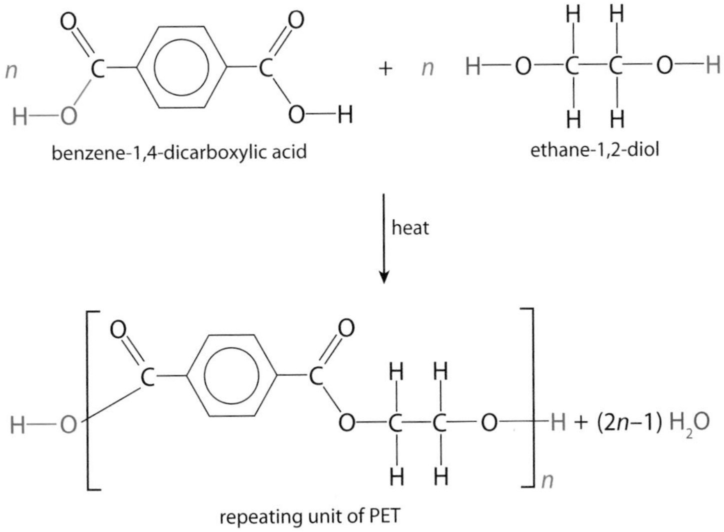

Figure A.28

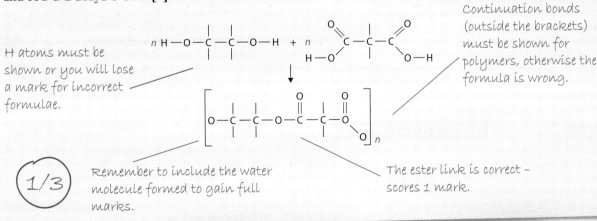

Polyesters can also be produced from diacyl chlorides (instead of dicarboxylic acids) in which the carboxyl hydroxyl groups are replaced by chlorine atoms. In this case, HCl is eliminated rather than water.

Annotated exemplar answer A.2

Write an equation, using structural formulae, for the formation of a condensation polymer from $HOCH_2CH_2OH$ and $HOOCCH_2COOH$. **[3]**

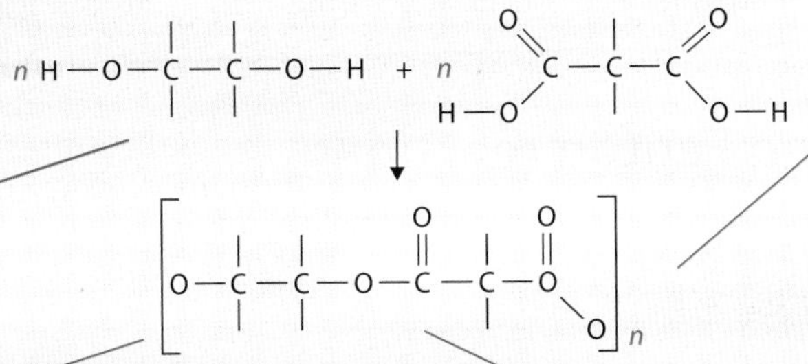

H atoms must be shown or you will lose a mark for incorrect formulae.

Continuation bonds (outside the brackets) must be shown for polymers, otherwise the formula is wrong.

1/3

Remember to include the water molecule formed to gain full marks.

The ester link is correct – scores 1 mark.

Polyamides

Polyamides form in a similar way to polyesters except that the diol monomer is replaced with a diamine. In this case, the NH_2 group reacts with the carboxyl group to form an amide linkage with the elimination of water. Diacyl chlorides can also be used in the production of polyamides.

The polyamide produced in this reaction is called Kevlar®. It forms long polymers chains in which there is extensive hydrogen bonding between the N–H hydrogen atoms and the C=O oxygen atoms of another strand. This gives Kevlar a highly ordered structure.

The hydrogen bonding means that the polymer chains in Kevlar require large amounts of energy to separate; this makes Kevlar incredibly strong, hence its use in protective clothing, ropes and sporting equipment.

However, in the presence of concentrated sulfuric acid, C=O and N–H groups in the polymer form hydrogen bonds to sulfuric acid molecules rather than to other polymer strands. This separates the strands and causes the polymer to dissolve in the acid solvent.

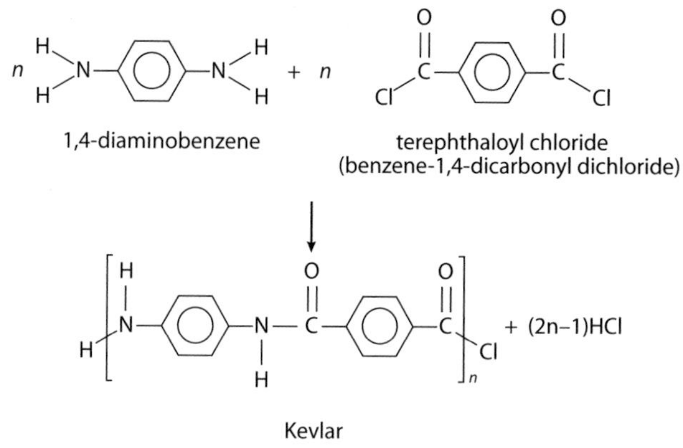

Figure A.29

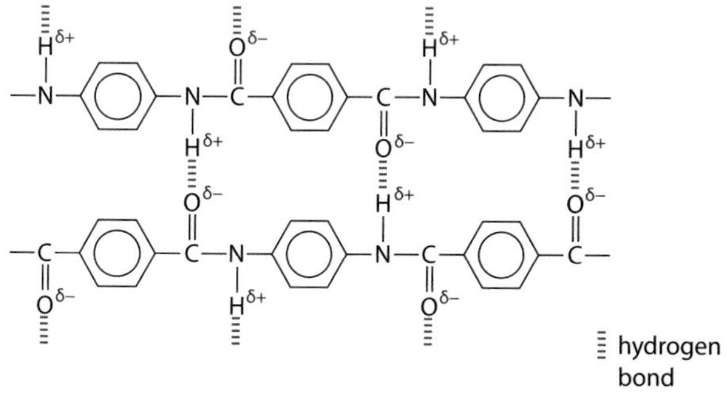

Figure A.30

Green polymers

Part of the problem with addition polymers is their lack of biodegradability due to strong C–C single bonds which are hard to hydrolyse. Even in polyamides and polyesters, the linkages between monomers only break down slowly. Green chemistry principles can be used to make polymers such as poly(lactic acid) and thermoplastic starch. These are made from renewable resources such as corn starch. The linkages between the monomers can be broken by the action of microbial organisms.

A.10 Environmental impact – heavy metals (HL only)

The presence of heavy (transition) metals in the environment can have disastrous consequences. At toxic doses, transition metals interfere with redox processes occurring within cells – this can lead to cell damage and cell death.

The Haber–Weiss and Fenton reactions

The formation of radical species, in particular the hydroxyl radical (HO•), inside cells can lead to the damage of DNA, proteins etc. Hydroxyl radicals can be produced via the **Haber–Weiss reaction**, in which superoxide ions ($\bullet O_2^-$) react with hydrogen peroxide:

$$\bullet O_2^- + H_2O_2 \rightarrow HO\bullet + OH^- + O_2$$

This is a very slow reaction inside cells and unlikely to contribute much to cellular damage.

The **Fenton reaction** is more likely to be involved in the production of hydroxyl radicals in cells:

$Fe^{2+} + H_2O_2 \rightarrow Fe^{3+} + HO\bullet + OH^-$

Superoxide ions present in cells can then react with Fe^{3+} ions to convert them back to Fe^{2+} ions:

$Fe^{3+} + \bullet O_2^- \rightarrow Fe^{2+} + O_2$

which can react with more H_2O_2 in the Fenton reaction.

Overall the process is equivalent to the Haber–Weiss reaction catalysed by Fe^{2+}

$Fe^{2+} + H_2O_2 \rightarrow Fe^{3+} + HO\bullet + OH^-$

$Fe^{3+} + \bullet O_2^- \rightarrow Fe^{2+} + O_2$

$\bullet O_2^- + H_2O_2 \rightarrow HO\bullet + OH^- + O_2$

hint

Fe^{2+} is a catalyst here – it is used up in the first step but produced again in the second.

A.11 Methods for removing heavy metals (HL only)

It is important to reduce the amount of heavy metal present in the environment and **adsorption**, **chelation** and **precipitation** are used to do this.

Adsorption

Various solid substances such as zeolites, carbon nanotubes, clay minerals and metal oxides can be added to solution containing heavy metal ions. The metal ions are **ad**sorbed to the surface of the solid and removed from solution. The adsorbed particles can be separated from the solvent by filtration.

Precipitation

Transition metals tend to form insoluble hydroxides, so the addition of substances like calcium hydroxide, ($Ca(OH)_2$), cause the heavy metal hydroxides to form and precipitate from the solution.

e.g. $Cd^{2+}(aq) + 2OH^-(aq) \rightarrow Cd(OH)_2(s)$

It is also possible to precipitate certain heavy metals like mercury by bubbling hydrogen sulfide through the solution to form insoluble sulfides.

e.g. $Hg^{2+}(aq) + H_2S(g) \rightarrow HgS(s) + 2H^+(aq)$

These precipitates are dense and can be separated easily by sedimentation and filtration.

Solubility product constant

Many substances that we consider to be insoluble in water do dissolve to a very small extent. In a saturated solution of these substances, an equilibrium exists:

e.g. $Cd(OH)_2(s) \rightleftharpoons Cd^{2+}(aq) + 2OH^-(aq)$

The solubility product constant, K_{sp}, of this system at equilibrium is given by:

$K_{sp} = [Cd^{2+}(aq)][OH^-(aq)]^2$

The value of K_{sp} at a certain temperature can be calculated if the solubility of the substance is known.

hint

The concentration of the solid is constant and is not included in the expression.

hint

Remember that values of K_{sp} have no units and are temperature dependent.

☆ Model answer A.3

The solubility of calcium fluoride at 298 K is 2.14×10^{-4} mol dm^{-3}. Calculate the value of K_{sp} for calcium fluoride at 298 K.

$$CaF_2(s) \rightleftharpoons Ca^{2+}(aq) + 2F^-(aq)$$

$K_{sp} = [Ca^{2+}(aq)][F^-(aq)]^2$

For every mole of CaF_2 that dissolves, one mole of Ca^{2+} and two moles of F^- ions are produced. So,

$[Ca^{2+}(aq)] = 2.14\times10^{-4}$ mol dm^{-3} and $[F^-(aq)] = 2\times2.14\times10^{-4} = 4.28\times10^{-4}$ mol dm^{-3}

$K_{sp} = 2.14\times10^{-4}\times(4.28\times10^{-4})^2 = \mathbf{3.92\times10^{-11}}$

It is also possible to work backwards and use the solubility product of a compound to predict its maximum solubility at a given temperature.

☆ Model answer A.4

The value of K_{sp} for barium carbonate, $BaCO_3$, at 298 K is 2.58×10^{-9}. Calculate the maximum solubility of barium carbonate at 298 K.

$$BaCO_3(s) \rightleftharpoons Ba^{2+}(aq) + CO_3^{2-}(aq)$$

Let the solubility of $BaCO_3$ be s

For every $BaCO_3$ unit that dissolves, 1 Ba^{2+} and 1 CO_3^{2-} ion are produced:

i.e. $[Ba^{2+}(aq)] = [CO_3^{2-}(aq)] = s$

$K_{sp} = [Ba^{2+}(aq)][CO_3^{2-}(aq)]$

Therefore: $K_{sp} = s^2$

$2.58\times10^{-9} = s^2$

$s = 5.08\times10^{-5}$ mol dm^{-3}

The maximum solubility of $BaCO_3$ at 298 K is **5.08×10^{-5} mol dm^{-3}**

hint

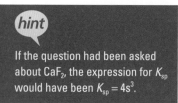

If the question had been asked about CaF_2, the expression for K_{sp} would have been $K_{sp} = 4s^3$.

The common ion effect

Adding ions such as hydroxide ions can be used to precipitate heavy metals out of solution.

For example, K_{sp} for $Cd(OH)_2$ is 7.2×10^{-15} at 298 K.

$Cd(OH)_2(s) \rightleftharpoons Cd^{2+}(aq) + 2OH^-(aq)$

If a solution has $[Cd^{2+}(aq)] = 1.0\times10^{-5}$ mol dm^{-3} and $[OH^-(aq)] = 2.0\times10^{-5}$ mol dm^{-3} then

$[Cd^{2+}(aq)][OH^-(aq)]^2 = 1.0\times10^{-5}\times(2.0\times10^{-5})^2 = 4.0\times10^{-15}$

This value is less than K_{sp} and therefore all the Cd^{2+} ions remain in solution. If, however, enough NaOH is added to this solution to increase $[OH^-(aq)]$ to 0.10 mol dm^{-3} then:

$[Cd^{2+}(aq)][OH^-(aq)]^2 = 1.0\times10^{-5}\times0.10^2 = 1.0\times10^{-7}$

This value is larger than K_{sp} therefore $Cd(OH)_2$ must precipitate out of the solution to reduce the concentration of ions to that required for $[Cd^{2+}(aq)][OH^-(aq)]^2$ to equal K_{sp}.

Worked example A.5

The concentration of lead(II) ions in a sample of river water is 4.5×10^{-10} mol dm^{-3}. Sodium hydroxide is added to the water until the concentration of hydroxide ions is 2.3×10^{-5} mol dm^{-3}. Deduce, showing your reasoning, whether or not lead ions are precipitated as lead(II) hydroxide at 298 K.

$Pb(OH)_2(s) \rightleftharpoons Pb^{2+}(aq) + 2OH^-(aq)$

$K_{sp} = [Pb^{2+}(aq)][OH^-(aq)]^2$

$[Pb^{2+}(aq)][OH^-(aq)]^2 = 4.5 \times 10^{-10} \times (2.3 \times 10^{-5})^2 = 2.4 \times 10^{-19}$

The value of K_{sp} for lead(II) hydroxide from the *IB Data Booklet* is $K_{sp} = 1.43 \times 10^{-20}$

The calculated value is larger than the value of K_{sp} so lead(II) hydroxide **will be precipitated**.

Adding hydroxide ions works well as most transition metal hydroxides have low values of K_{sp}. The same is true of heavy metal sulfides – the sulfide ion concentration is increased by bubbling H_2S through the solution.

hint

Carbonate and sulfate ions can also be added to precipitate heavy metal salts.

TEST YOURSELF A.8

1 Use the data in the table to calculate K_{sp} values for $BaSO_4$ and $Ni(OH)_2$ at 298 K.

	solubility at 298 K/mol dm^{-3}
$BaSO_4$	1.0×10^{-5}
$Ni(OH)_2$	5.2×10^{-6}

2 Use the values in the table to calculate the solubility of each substance at 298 K.

	K_{sp} at 298 K
$PbCO_3$	7.40×10^{-14}
Ag_2SO_4	1.20×10^{-5}

3 A sample of water contains silver ions at a concentration of 2.0×10^{-3} mol dm^{-3}. Sulfuric acid is added to the water until the concentration of sulfate ions is 1.0×10^{-2} mol dm^{-3}. Explain whether silver sulfate will precipitate from the solution.

hint

The concentration of hydroxide ions due to the dissociation of water (1×10^{-7} mol dm^{-3}) is ignored as it is significantly smaller than the concentration of hydroxide ions from sodium hydroxide.

Chelation

DEFINITIONS

MONO-/BI-/POLY DENTATE LIGANDS are ligands that possess one, two or more lone pairs capable of forming coordinate bonds.

A CHELATE structure is a complex ion in which at least one ligand is bonded to a central metal ion through more than one donor atom to form a ring including the metal ion.

Types of ligand

Monodentate ligands, e.g. water, ammonia, chloride and hydroxide ions, form a single coordinate bond per ligand. It is possible to have up to six ligands coordinated to the central metal ion, e.g. $[Cr(H_2O)_6]^{3+}$.

Bidentate ligands, such as ethane-1,2-diamine (en), possess two N atoms far enough apart that they are capable of forming two coordinate bonds to the metal ion. Typically, up to three of these ligands may bond to a single metal ion, e.g. $[Co(en)_3]^{2+}$.

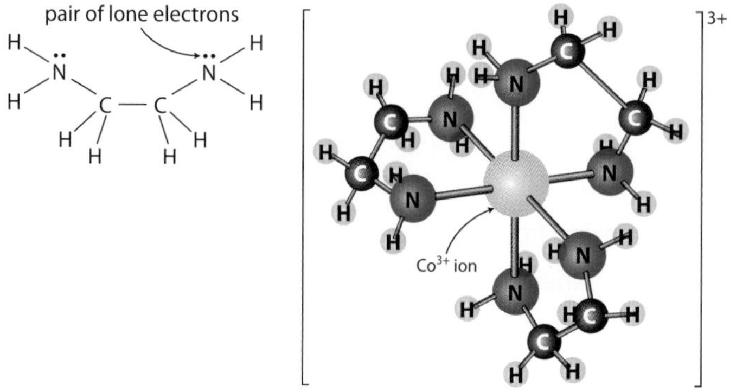

Figure A.31

Hexadentate ligands, such as $EDTA^{4-}$, have six donor atoms (with lone pairs) and can form six coordinate bonds to the central metal ion.

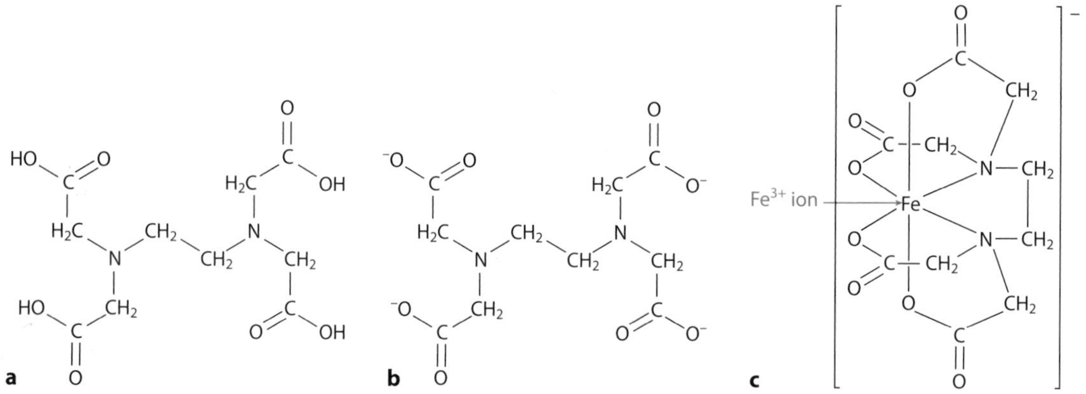

Figure A.32

Chelation of a heavy metal ion with a ligand such as $EDTA^{4-}$ has the effect of reducing the available concentration of metal ions present in a solution – surrounding the metal ion with a polydentate ligand prevents it from interacting as it would if it were free.

> **hint**
> The type of ligand can be identified by looking at the number of atoms that have a formal negative charge or an available lone pair of electrons. In addition, these atoms should normally be separated by at least three bonds.

The chelate effect

Chelate structures formed by polydentate ligands, such as ethane-1,2-diamine (en) or $EDTA^{4-}$, are usually more stable than complex ions formed by monodentate ligands with the same donor atoms.

This can be explained in terms of entropy.

Consider the reaction where six NH_3 ligands (6 N donor atoms) are replaced with three en-ligands (6 N donor atoms).

$[Ni(NH_3)_6]^{2+} + 3H_2NCH_2CH_2NH_2 \rightleftharpoons [Ni(H_2NCH_2CH_2NH_2)_3]^{2+} + 6NH_3$

The enthalpy change will be very small because there are the same number and type of bonds on both sides (ignoring interactions with solvent molecules). However, the reaction in the forward direction involves an increase in entropy (4 particles going to 7) and is favourable.

Checklist

At the end of this chapter you should be able to:

- ☐ Explain how materials are classified based on their properties and determine the bonding in a binary compound.
- ☐ Describe the extraction processes used in the production of metals and perform quantitative electrolysis calculations.
- ☐ Deduce the magnetic properties of a metal.
- ☐ Analyse data from ICP-OES/MS experiments.
- ☐ Describe and explain the action of heterogeneous and homogeneous catalysts.
- ☐ Identify liquid crystal molecules and describe their properties.
- ☐ Distinguish between thermoplastic and thermosetting polymers.
- ☐ Describe the production of HDPE and LDPE as well as the use of plasticisers and polymer modification.
- ☐ Describe the different ways in which carbon nanotubes are created.
- ☐ Explain and describe the environmental impact of producing, using and disposing of plastics.

Higher Level only

- ☐ Describe superconductivity and X-ray crystallography.
- ☐ Apply the Bragg equation to simple cubic structures and determine the density of a metal.
- ☐ Write equations for the production of different condensation polymers.
- ☐ Describe the environmental impact of heavy metals.

B BIOCHEMISTRY

Biochemistry is concerned with the interactions of the molecules found within living organisms. These biological molecules fall into categories such as proteins, carbohydrates and lipids displaying diverse functions dependent on their shape and structure.

This chapter covers the following topics:

- ☐ An introduction to biochemistry
- ☐ Proteins and enzymes
- ☐ Lipids
- ☐ Carbohydrates
- ☐ Vitamins
- ☐ Biochemistry and the environment
- ☐ More on proteins and enzymes (HL only)
- ☐ Nucleic acids (HL only)
- ☐ Biological pigments (HL only)
- ☐ Stereochemistry in biomolecules (HL only)

B.1 Introduction to biochemistry

Metabolic reactions occur in a highly-controlled aqueous environment – factors such as pH, ion concentration and temperature must be kept in very narrow ranges.

Catabolism and anabolism

DEFINITION

CATABOLISM reactions involving the breakdown of larger molecules into smaller ones releasing energy, e.g. the hydrolysis of polymers like starch and proteins.

Respiration involves catabolism – glucose, $C_6H_{12}O_6$, is converted to carbon dioxide, water and energy in a series of steps.

$$C_6H_{12}O_6 + 6O_2 \rightarrow 6CO_2 + 6H_2O$$

DEFINITION

ANABOLISM reactions involving the synthesis of, generally larger, molecules from precursors requiring an input of energy, e.g. formation of a protein from amino acids.

Photosynthesis is an anabolic process in which carbon dioxide from the atmosphere is taken in by plants and combined with water to form energy-rich molecules such as glucose.

$$6CO_2 + 6H_2O \rightarrow C_6H_{12}O_6 + 6O_2$$

Respiration and photosynthesis help to balance atmospheric levels of both CO_2 and O_2. These equations show how the numbers of moles of gas produced and reacted are equal.

TEST YOURSELF B.1

Distinguish between catabolism and anabolism using an example of each. Include chemical equations to justify your distinction.

Condensation and hydrolysis

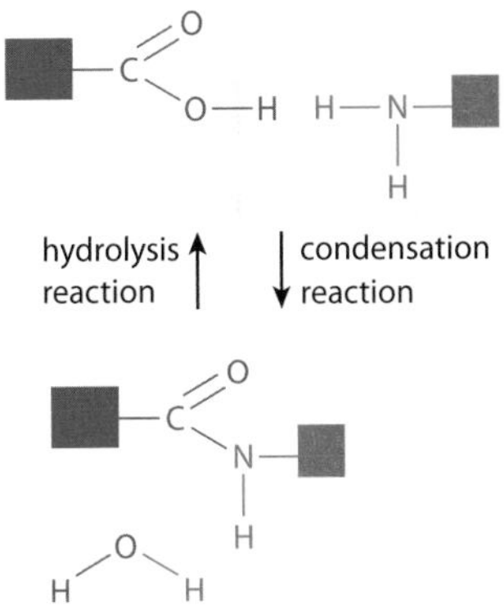

Figure B.1

DEFINITIONS

CONDENSATION the joining together of two molecules with the formation of a covalent bond and the elimination of water.

HYDROLYSIS breaking a covalent bond by the addition of water – essentially the reverse of condensation.

B.2 Proteins and enzymes

Amino acids

2-amino acids undergo **condensation polymerisation** to form proteins. The link between the monomers is termed an **amide link**.

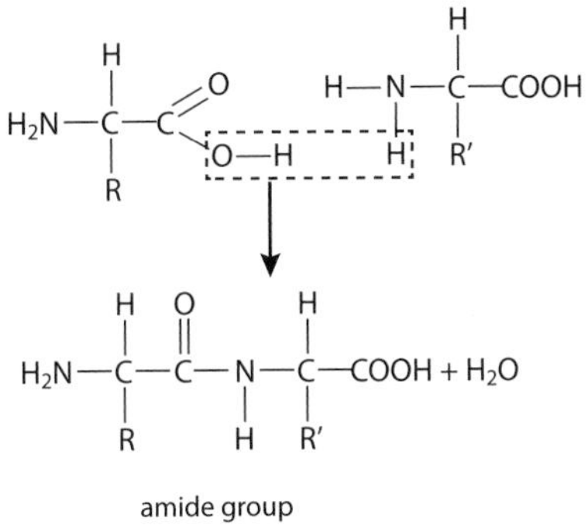

Figure B.2

You must be able to draw the structure of a peptide (two or more amino acids) given the structures of the monomers (amino acids). Figure **B.2** shows the general condensation reaction between two amino acids.

Proteins may be hydrolysed (by heating with acid) to form amino acids.

To work out the structure of the amino acids formed:

- break apart the peptide between the C=O and N–H in each peptide bond
- add an O–H group to each C=O group
- add a hydrogen atom to each N–H group.

hint

In IB Biology, amide links are called **peptide bonds**.

hint

Do not forget the molecule of water released when writing the equation.

TEST YOURSELF B.2

1 Draw the structures of the two different dipeptides that can be formed when alanine combines with serine (structural formulae are in the *IB Data Booklet*).

2 Draw the structures of the amino acids formed during the hydrolysis of:

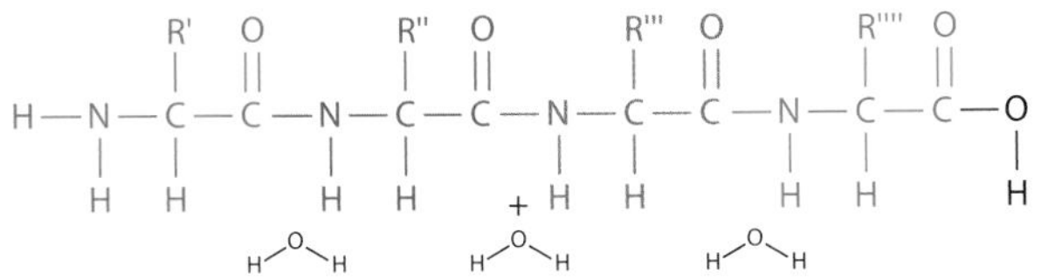
Figure B.3

hint

Two different dipeptides can be formed when any two different amino acids combine depending on which way around the COOH and NH_2 groups are joined together.

hint

The structures of amino acids are shown in the *IB Data Booklet*

Zwitterions

In aqueous solution, 2-amino acids can exist in one of three forms:

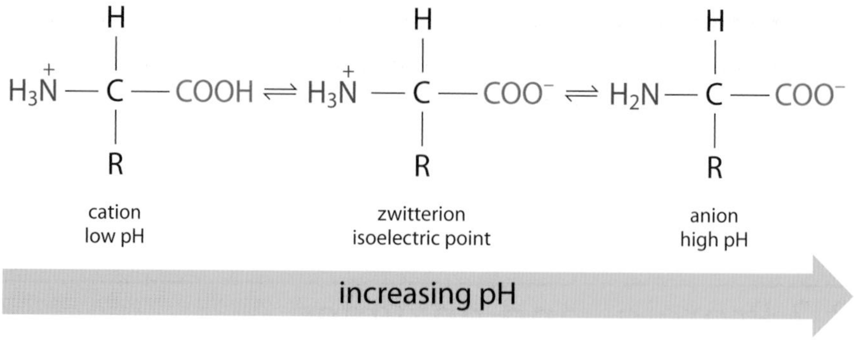

Figure B.4

The amine group acts as a base and the carboxyl group as an acid, thus making the molecule **amphoteric**.

In general, the cation form (ammonium) predominates at low pH (high H^+ concentration to protonate the NH_2 group) and the anionic form (carboxylate) at high pH (low H^+ concentration). The zwitterion is the main form present at intermediate values of pH; it has an overall neutral charge as it has one full negative and one full positive charge.

> **hint**
> Remember that **amphoteric** describes a compound that can act as both a base and an acid.

Amino acids have much higher melting points than carboxylic acids of similar molar mass:

- Amino acids exist as zwitterions in the solid state.
- The electrostatic attractions between the NH_3^+ groups and COO^- groups on adjacent amino acids are stronger than hydrogen bonds between carboxylic acid molecules.

Amino acids exhibit high solubility in water and low solubility in non-polar solvents due to the charges present (NH_3^+ and COO^-). The zwitterion has lower water solubility than the cationic or anionic forms as it is neutrally charged.

Isoelectric point

DEFINITION

The **ISOELECTRIC POINT** of an amino acid or protein is the pH at which the population of molecules exhibits no overall charge, i.e. all amino acids exist as zwitterions.

The isoelectric points for each amino acid are given in the *IB Data Booklet*.

- If the pH is lower than the isoelectric point the protein/amino acid will have a positive charge (more $-NH_3^+$ present).
- If the pH is higher than the isoelectric point the protein/amino acid will have a negative charge (more $-COO^-$ present).

> **hint**
> Don't forget that the side groups of some amino acids can also be charged depending on the pH of the solution.

TEST YOURSELF B.3

The amino acid phenylalanine ($R = CH_2C_6H_5$) has an isoelectric point of 5.5. Draw the structure of the species present at:

1 pH = 5.5 **2** pH = 2.0 **3** pH = 9.4

Protein structure

- **Primary structure** – the linear sequence of amino acids in the polypeptide chain.
- **Secondary structure** – regions of alpha-helix or beta-pleated sheet formed as the primary structure coils or folds back on itself.
- **Tertiary structure** – the three-dimensional structure of a protein chain.
- **Quaternary structure** – two or more protein units associated with each other – hemoglobin has four subunits.

Protein structure and shape are determined by a number of interactions between amino acids that you will need to learn. These include:

- **Hydrogen bonds:** between N–H and C=O groups as well as O–H groups on some side groups.
- **London forces:** between non-polar side groups, e.g. alkyl or phenyl groups.
- **Electrostatic interactions:** between NH_3^+ and COO^- groups on side groups.
- **Disulfide bridges:** two sulfhydryl groups (R–S–H) on cysteine residues react to form a covalent crosslink, R–S–S–R.

Hydrogen bonding between amide N–H and C=O groups exclusively is responsible for the maintenance of protein secondary structure. Tertiary and quaternary structure is maintained by a combination of the interactions listed here.

The role a protein plays is, in part, dependent on its structure. Fibrous proteins such as keratin and collagen act as structural molecules. Globular proteins have various roles including acting as enzymes (e.g. amylase), hormones (e.g. insulin), transporters (e.g. hemoglobin) and receptors.

Enzymes

Enzymes (generally globular proteins) act as biological catalysts – each has a unique three-dimensional shape containing an **active site**. Substrate molecules interact (through hydrogen bonds, London forces, etc.) with the amino acids in the active site. Only the correct substrate will be the right shape to fit into the active site. The binding of the enzyme and the substrate means that the reaction occurs by a different reaction pathway of lower activation energy.

Annotated exemplar answer B.1

Outline the bonding interactions responsible for the secondary and tertiary structure of proteins. **[4]**

This is wrongly discussing the primary structure – the question asks for secondary and tertiary

There are peptide bonds between the amino acids in the protein to hold the whole structure together. These are strong and difficult to break. There are also alpha-helices and beta-sheets. Intermolecular forces such as hydrogen bonding are responsible for tertiary structure.

Alpha-helices and beta-sheets are mentioned but the bonding interactions are not – to gain a mark, you need to state that hydrogen bonding between N—H and C=O of amide groups on different parts of the protein chain maintains protein secondary structure.

1/4

Mark awarded for mention of hydrogen bonding

To gain the final mark, you need to discuss other important interactions such as: electrostatic forces (between NH_3^+ and COO^- groups), London forces disulfide bridges

The rate of reaction changes with substrate concentration (see Figure **B.5**).

The activity of an enzyme is dependent on its structure otherwise it cannot bind a substrate. Protein structure is highly sensitive to changes in temperature (Figure **B.6**), pH (Figure **B.7**) and the presence of heavy metal ions.

1 **Temperature** – generally, temperatures above 40 °C will cause proteins to **denature**, that is permanently lose their specific three-dimensional shape. This prevents substrate from binding to the enzyme. Low temperatures decrease enzyme activity but do not denature the protein.

2 **pH** – enzymes tend only to function in a small range of pH values around an **optimum** value. Low/high pH may cause amino acid side groups to be protonated/deprotonated – this disrupts the weak interactions between amino acids (causing denaturation) and between substrate and active site.

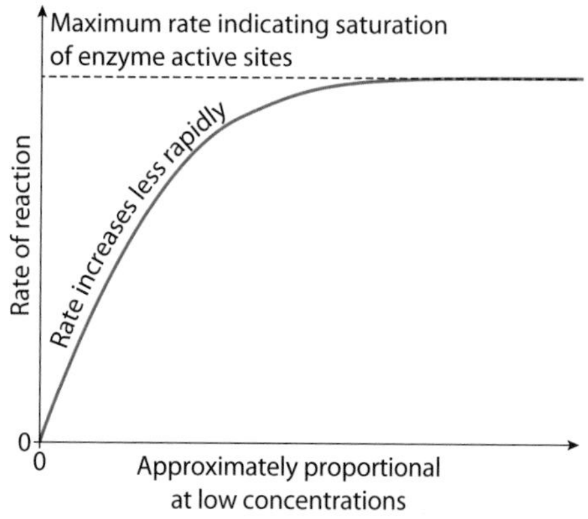

Figure B.5

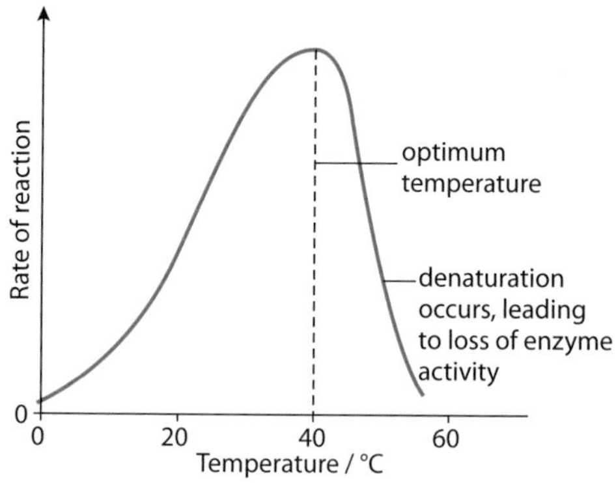

Figure B.6

3 **Heavy metal ions** – compounds of silver(I), mercury(II) and lead(II) react with sulfhydryl groups (R–SH) present on the amino acid cysteine, replacing the hydrogen. This prevents the formation of disulfide bridges essential for maintaining proper shape.

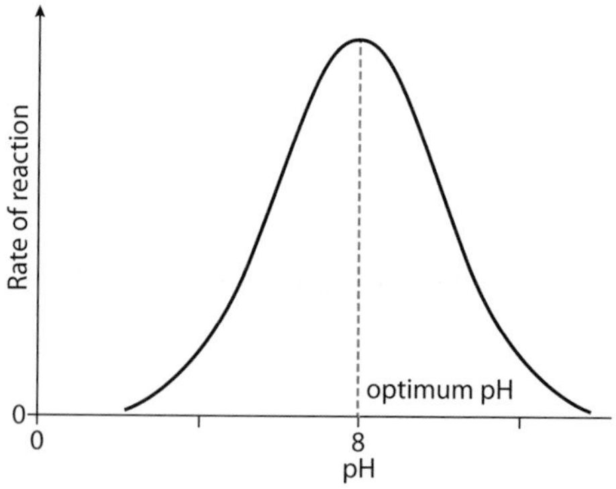

Figure B.7

TEST YOURSELF B.4

Describe and explain the effect of increasing substrate concentration on the rate of enzyme activity.

Chromatography and electrophoresis

Paper chromatography

1 Protein hydrolysed into amino acids by refluxing with 6 mol dm^{-3} hydrochloric acid.

2 Amino acid sample is spotted onto chromatography paper, which is suspended in a suitable solvent (e.g. a water/ethanol mix).

3 Solvent rises up the paper by capillary action carrying amino acids with it.

4 How far an amino acid travels up the paper is dependent on how it partitions between the solvent (mobile phase) and the water adsorbed onto the paper fibres (stationary phase).

hint

Most of the time you are given only dotted lines on which to write. Do not let this put you off drawing a graph if it will make an answer clearer.

5 Amino acids more soluble in the solvent travel further than those that are more soluble in the water on the fibres.

6 A locating agent such as **ninhydrin** is added to turn amino acids purple.

7 **Retardation factor (R_f)** is calculated as:

$$R_f = \frac{\text{distance of amino acid from origin (mm)}}{\text{distance of solvent front from origin (mm)}}$$

8 R_f values can be compared to literature values to identify amino acids present.

Reference samples of known amino acids can also be spotted onto the paper and the migration of the unknown samples with the known references can be compared.

> **hint**
>
> When working out R_f values remember to measure from the origin (base line) and not from the bottom of the paper.

> **hint**
>
> Remember that a protein/amino acid will be positively charged if the pH is below its isoelectric point and negatively charged if the pH is above its isoelectric point.

Gel electrophoresis

1 Samples (amino acids or whole proteins) are injected into a gel matrix, typically poly(acrylamide), immersed in a buffer solution of fixed pH.

2 Electric field applied – molecules with net positive charge migrate towards the cathode; those with net negative charge move towards the anode.

3 Amino acids are located using ninhydrin and proteins using a stain such as Coomassie blue.

4 The distance travelled by a protein depends on its mass : charge ratio; proteins with no net charge will not move.

B.3 Lipids

Uses of lipids

Lipids have many important uses in the body, including:

- structural component of cell membranes
- energy storage (see later)
- thermal insulation
- electrical insulation in nerve cells (myelin)
- the transportation of lipid-soluble vitamins, e.g. vitamins A, E and K
- hormones.

Energy storage

Lipids are more reduced than carbohydrates, meaning that when they are oxidised to CO_2 during metabolic processes they release far more energy per gram. Lipids are insoluble in water and are stored in cells, allowing for slow energy release. Simple carbohydrates are water soluble, allowing them to be transported around the body easily – they are sources of quick-release energy.

TEST YOURSELF B.5

By calculating the average oxidation state of carbon in the triglyceride $C_{55}H_{98}O_6$ and the carbohydrate sucrose, $C_{12}H_{22}O_{11}$, explain why lipids are described as being more reduced than carbohydrates.

Lipid structure and reactions

Lipids are tri-esters formed from one molecule of glycerol (propane-1,2,3-triol) and three fatty acid molecules. The fatty acids may contain no (saturated), one (monounsaturated) or more than one (polyunsaturated) C=C double bond.

Phospholipids, which make cell membranes, are derivatives of triglycerides in which one fatty acid unit is replaced with a substituted phosphate group.

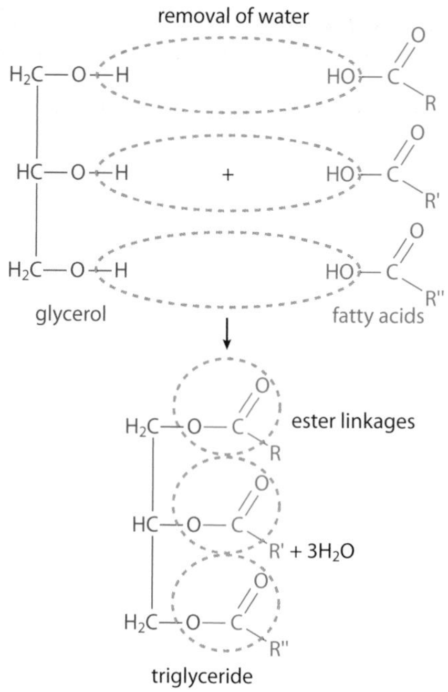

Figure B.8

hint

Where two lipids/fatty acids have similar amounts of unsaturation, the one with the higher molecular mass will tend to have the higher melting point.

Melting points

Mono- and polyunsaturated fatty acids generally have lower melting points than saturated fatty acids of similar relative molecular mass – the more C=C the lower the melting point.

Hydrolysis

Triglycerides and phospholipids can be hydrolysed using acids or alkalis as well as specific enzymes called lipases. Hydrolysis yields glycerol and either the fatty acids (in acidic conditions) or the salt of the fatty acids (in alkaline conditions).

TEST YOURSELF B.6

1 Draw the structural formulae of the products of the complete hydrolysis of this triglyceride using **a** dilute hydrochloric acid and **b** dilute potassium hydroxide.

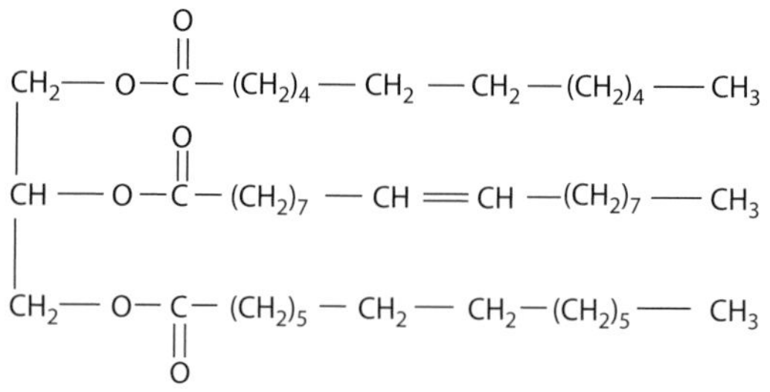

2 Draw the structural formula of the phospholipid formed when one molecule of glycerol reacts with one molecule of phosphoric(V) acid, H_3PO_4, and two molecules of dodecanoic acid, $C_{11}H_{23}COOH$.

Rancidity

Lipids become **rancid** through either hydrolytic or oxidative processes.

Hydrolytic rancidity: ester links in a lipid are broken by water, producing glycerol and unpleasant tasting and smelling fatty acids.

Hydrolytic rancidity is favoured by:

- high water content
- acidic or alkaline conditions
- high temperature
- presence of lipase enzymes.

Oxidative rancidity: fatty acid chains are broken down when oxygen reacts with C=C double bonds in unsaturated lipids. Unpleasant tasting and smelling ketones, alcohols and aldehydes are produced.

Oxidative rancidity is favoured by:

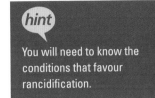

- large number of C=C double bonds
- high temperature
- high availability of oxygen
- high light intensity
- metals such as copper and nickel.

hint

You will need to know the conditions that favour rancidification.

Iodine number

DEFINITION

IODINE NUMBER is defined as the number of grams of iodine that will react with 100 g of lipid.

Iodine reacts with the C=C double bonds in unsaturated lipids. The iodine number of a lipid is an indication of the degree of unsaturation: the greater the iodine number, the greater the degree of unsaturation.

Calculating and using iodine numbers

Model answer B.1

17.0 g of a lipid reacts completely with 23.0 g of iodine. Calculate the iodine number of the lipid.

$$\text{Iodine number} = \frac{\text{mass of iodine reacted}}{\text{mass of lipid reacted}} \times 100$$

$$\frac{23.0}{17.0} \times 100 = 135$$

Model answer B.2

A fatty acid molecule has a relative molecular mass of 302.45 and an iodine number of 420. Calculate the number of C=C double bonds present per fatty acid molecule.

$$\text{mass of iodine per mole of lipid} = \frac{\text{iodine number} \times \text{molar mass of lipid}}{100} = \frac{420 \times 302.45}{100} = 1270 \text{ g}$$

$$I_2 \text{ molecules per lipid molecule} = \frac{\text{mass of } I_2 \text{ per mole of lipid}}{\text{molar mass of } I_2} = \frac{1270}{253.60} = 5.00$$

5.00 molecules of I_2 will react with **five C=C double bonds per lipid molecule.**

TEST YOURSELF B.7

1 The iodine number of a lipid is zero. What can be deduced about the structure of the lipid from this?
2 Determine the iodine number of the lipid if 30 g of iodine reacts completely with 25 g of lipid.
3 Calculate the mass of iodine that reacts with 4.0 g of a lipid with an iodine number of 125.

Impact of lipids on health

> **hint**
> SL students do not need to know the structural details of *trans*-fats.

Over-consumption of lipids (fats) may lead to excess energy intake being stored as fatty deposits leading to obesity.

Lipids are transported around the body in **lipoproteins**. There are two main forms of lipoprotein: low-density lipoprotein (LDL) and high-density lipoprotein (HDL). LDL levels are increased by high intake of saturated lipids and/or *trans*-fats.

LDL particles can release their cholesterol/fat content in the body forming deposits in arteries, increasing the risk of heart attack or stroke. HDL particles can scavenge cholesterol/fat from cells, LDL particles and fatty deposits, returning it to the liver for metabolism. Polyunsaturated and (non-*trans*) monounsaturated lipids are good for health as they increase HDL levels.

Steroids

Steroids contain a fused-ring backbone (Figure **B.9**) which is then modified by the addition of variable side groups.

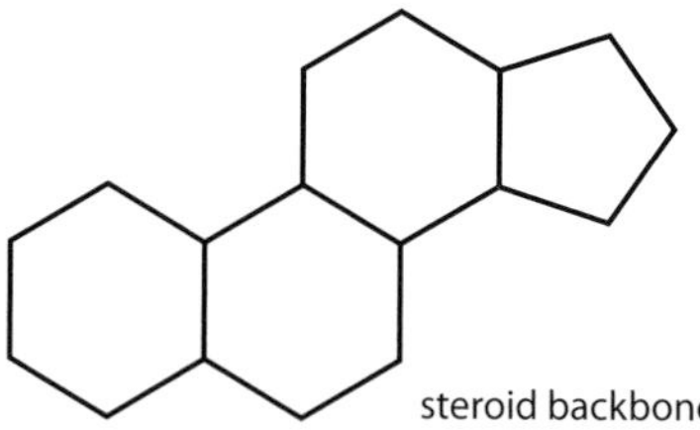

Figure B.9

Cholesterol (Figure **B.10**) is a vital part of the cell membrane and is also used to synthesise steroid hormones such as estrogens, testosterone and many others.

Medical uses of steroids include:

- the contraceptive pill,
- the treatment of skin disorders, such as eczema,
- the treatment of inflammation.

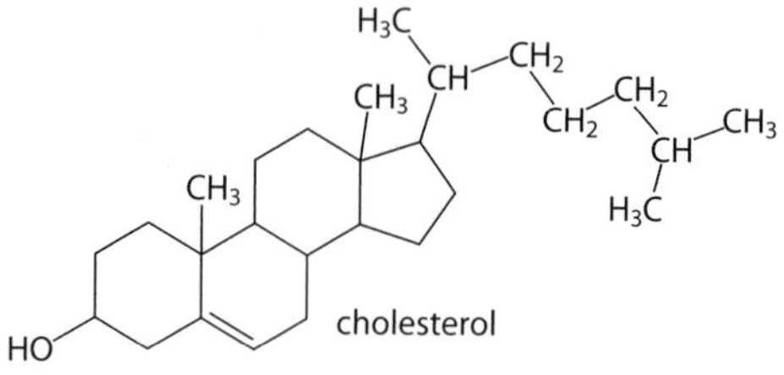

Figure B.10

Steroids such as testosterone and its derivatives are called **anabolic steroids**. Their (over)use leads to increased protein content in cells and extra muscle mass. Anabolic steroids have been subject to abuse by athletes and body-builders to enhance physique and performance. However, anabolic steroids have many serious negative effects such as:

> **hint**
> You need to be able to describe both the use and abuse of steroids.

- increasing levels of aggression and mood disorders
- cardiovascular disease
- feminisation (e.g. breast growth in men)
- masculinisation (e.g. body hair growth in women)
- sterility.

B.4 Carbohydrates

Carbohydrates are vital sources and stores of energy in cells. Simple sugars like glucose are converted into energy easily, whereas the polymers starch and glycogen are energy stores that release glucose when levels are low.

Carbohydrates have the general molecular formula $C_x(H_2O)_y$. For example, fructose, $C_6H_{12}O_6$ can be written $C_6(H_2O)_6$ and sucrose, $C_{12}H_{22}O_{11}$, as $C_{12}(H_2O)_{11}$.

Monosaccharides

These are the simplest carbohydrates comprising 3–6 carbon atoms. When drawn in a linear form, monosaccharides contain either an aldehyde, e.g. glucose, or a ketone group, e.g. fructose, as well as several hydroxyl (–OH) groups.

In aqueous solution, the straight chain form of the monosaccharides is able to cyclise. This forms a ring structure containing an ether link (C–O–C). The rings are typically five- or six-membered.

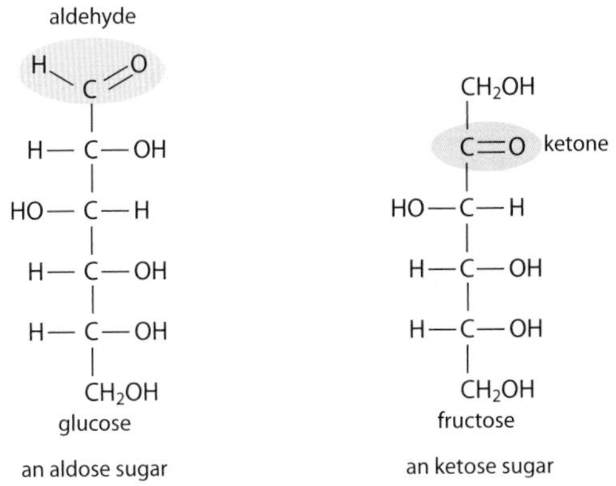

Figure B.11

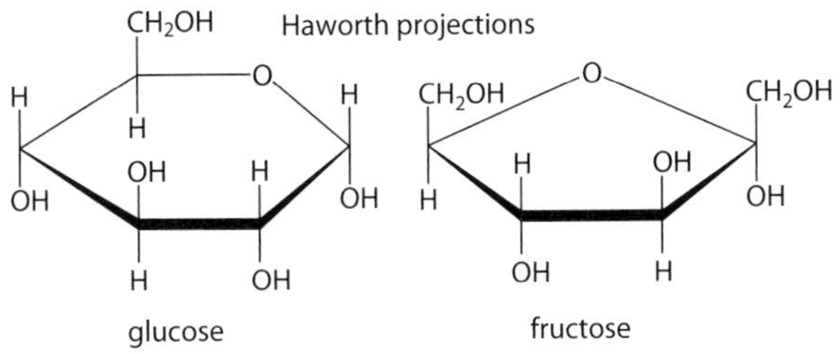

Figure B.12

Haworth projection helps focus on the position of attached groups, improving understanding of a molecule's structural roles.

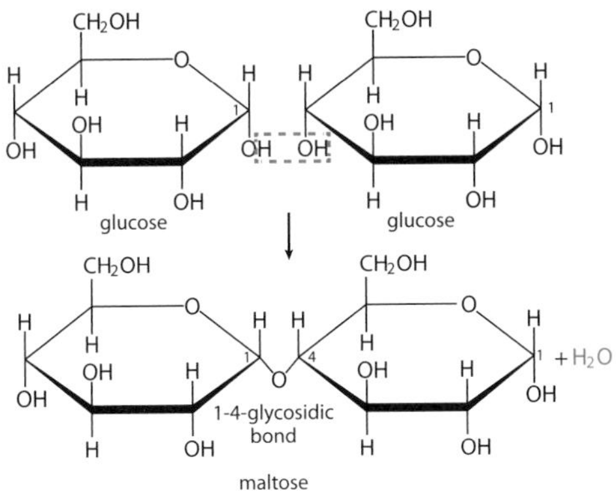

Figure B.13

Disaccharides and polysaccharides

Two monosaccharides may react in a condensation reaction to form a disaccharide, such as sucrose, maltose or lactose. A molecule of water is released in the reaction. The link between the two rings is termed a **glycosidic bond**.

Starch, cellulose and glycogen are **polysaccharides** – polymers made from glucose monomers only. The way the rings are linked gives them different properties.

hint

When drawing the structure of a disaccharide or a polysaccharide, remove the hydrogen from the –OH group from the far right carbon in the first ring (C1) and the –OH group of the far left carbon (C4) of the other ring. Join the two rings by forming a bridge with the remaining oxygen atom.

Structure and function

Carbohydrate	Solubility	Uses	Structural information
Monosaccharides e.g. glucose	Water-soluble	Source of cellular energy. Precursor in biosynthesis	
Starch	Insoluble in water	Energy storage	Combination of amylose and amylopectin. Branched polymer containing 1,4 and 1,6 glycosidic bonds.
Glycogen	Insoluble in water	Energy storage	Branched polymer containing 1,4 and 1,6 glycosidic bonds.
Cellulose	Insoluble in water	Structural – plant cell walls	Linear polymer containing 1,4 glycosidic bonds only.

Hydrogen bonding between the hydroxyl groups on different polysaccharide chains is weakened by branching. The linear nature of cellulose makes the hydrogen bonding stronger between chains making it able to serve a structural purpose.

B.5 Vitamins

DEFINITION

VITAMINS are organic micronutrients that cannot (except Vitamin D) be synthesised in the body and must be obtained from appropriate food sources.

Micronutrients are those required in tiny amounts by the body unlike protein, carbohydrates and lipids, which are termed macronutrients. Some vitamins are highly sensitive to heat; they decompose or are chemically altered at temperatures involved in cooking causing them to lose their biological effect.

Comparison of structures and solubility

Hydrogen-bonding functional groups aid water solubility, e.g. hydroxyl, carboxylate and amine, amide. Non-polar groups like phenyl and alkyl decrease water solubility but increase fat solubility.

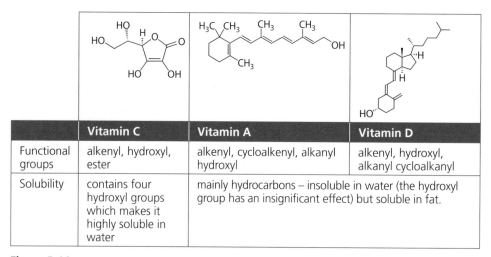

	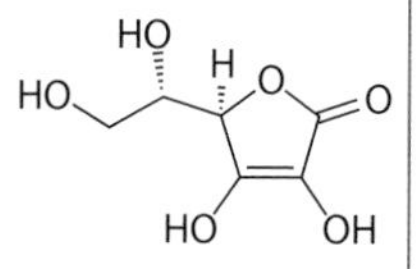	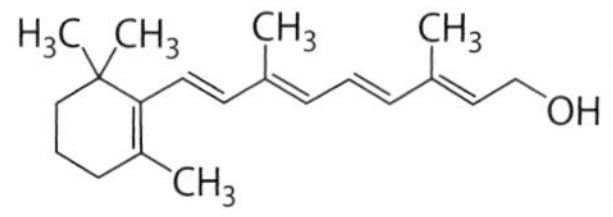	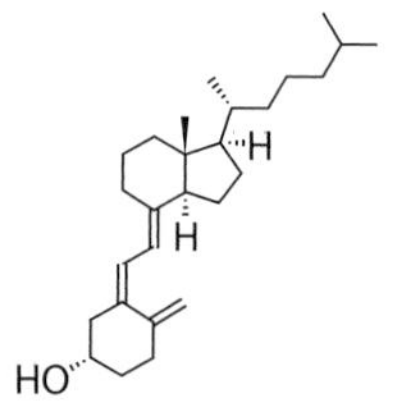
	Vitamin C	**Vitamin A**	**Vitamin D**
Functional groups	alkenyl, hydroxyl, ester	alkenyl, cycloalkenyl, alkanyl hydroxyl	alkenyl, hydroxyl, alkanyl cycloalkanyl
Solubility	contains four hydroxyl groups which makes it highly soluble in water	mainly hydrocarbons – insoluble in water (the hydroxyl group has an insignificant effect) but soluble in fat.	

Figure B.14

TEST YOURSELF B.8

Determine whether each of the following vitamins is likely to be more soluble in water or fat.

hint

You need to be able to identify the key functional groups present in each molecule and explain how the structures affect water and fat solubility.

1

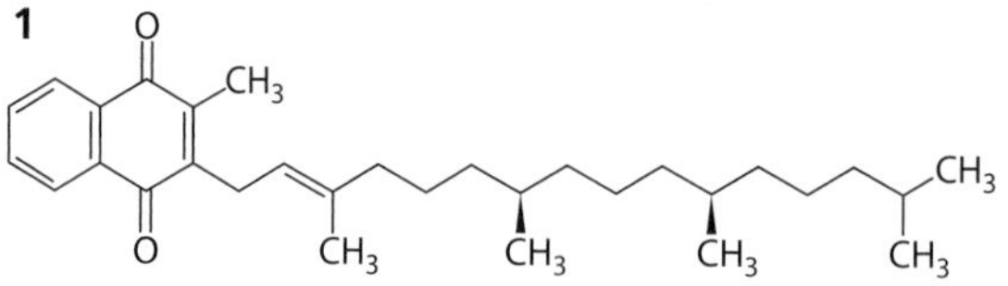

2

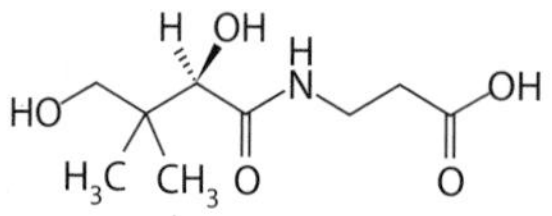

Figure B.15

Vitamin deficiency

Inadequate intake of vitamins is termed **vitamin deficiency**. Severe vitamin deficiencies may lead to diseases or conditions that, if left untreated, can cause significant harm or, in the most severe cases, death, e.g. beri-beri caused by vitamin B_1 deficiency.

The causes of vitamin deficiency vary from country to country. Typical reasons include:

- Famine or harsh agricultural conditions
- Poverty
- Lack of sunlight
- Poor access to or supply of vitamin supplements.

There are several ways to combat these causes:

- Addition of vitamins to food-stuffs
- Genetic modification of crops
- Education
- Medical programmes
- Spend more time outdoors.

TEST YOURSELF B.9

Describe the principal causes of vitamin deficiency and outline two approaches to tackling the problem.

B.6 Biochemistry and the environment

Xenobiotics

DEFINITION

XENOBIOTICS are chemical substances found in an organism that would not normally be present there, e.g. antibiotics in humans.

Xenobiotics such as antibiotics can accumulate in waste water due to overuse and improper disposal. Resulting issues include the development of antibiotic resistance in bacteria and unnecessary exposure resulting in long-term health problems.

Biodegradable and compostable plastics

These are plastics that can be broken down or consumed by bacteria or other living organisms. They offer environmental benefits over petroleum-derived plastics, which are chemically inert and do not biodegrade easily creating pollution hazards and making disposal challenging.

Starch is used to make biodegradable plastics such as:

- **thermoplastic starch** – obtained by mixing starch with plasticisers such as **sorbitol** or **glycerine**.
- **poly(lactic acid)** – corn starch is fermented to make lactic acid and then polymerised.

These polymers decompose into carbon dioxide and water by the action of bacteria and fungi.

Host–guest chemistry

DEFINITION

HOST–GUEST CHEMISTRY describes the **non-covalent** interactions (hydrogen bonding, London forces, dipole-dipole and electrostatic interactions, etc.) between a host molecule (e.g. an enzyme) and a guest molecule (e.g. the substrate).

> **hint**
> You need to know a named example of host–guest chemistry – a good example is that of the calixarenes, which are used to remove radioactive caesium-137 from waste water by forming ion–dipole interactions.

Synthetic molecules can be created to mimic the actions of enzymes, forming host-guest interactions with specific guest molecules. Specially designed host molecules are also able to interact specifically with certain toxic chemicals found in the environment; these host molecules can be used to remove harmful substances from water.

Enzymes and the environment

Enzymes have been developed for use in several 'green' applications. These include:

- Breakdown of oil spillages.
- Treatment of industrial waste.
- Biological detergents that work at low temperatures.

Biomagnification

> **hint**
> You need to be able to describe biomagnification using a named chemical substance. Good examples include heavy metals, e.g. mercury and pesticides, such as DDT.

DEFINITION

BIOMAGNIFICATION the process by which a chemical becomes increasingly concentrated in organisms progressing along a food chain.

Mercury, from gold mining, enters water-courses. The following food chain takes place:

mercury → microscopic plants → microscopic animals → small fish → larger fish → tuna/swordfish

This poses a danger to human consumers of these fish, especially pregnant women, as mercury is a potent neurotoxin.

Green chemistry

> **hint**
> You will not be expected to learn this list but you will need to be able to apply the general principles to a discussion about the challenges and criteria used to assess 'greenness' of a substance.

Green chemistry, also called **sustainable chemistry**, is an approach to chemical research and engineering – the aim is to:

- prevent waste production
- maximise atom economy (see Option A)
- use less hazardous and renewable reactants
- design safer substances
- use safer solvents
- improve energy efficiency
- reduce length (number of steps) of syntheses
- use catalysts as much as possible (lower temperatures can be used – saves energy)
- develop degradable products
- monitor pollution
- improve safety and accident prevention.

B.7 Proteins and enzymes (HL only)

Enzyme kinetics

The rate of a reaction catalysed by an enzyme increases as the substrate concentration increases (to a point). The relationship between these variables can be illustrated by a Michaelis–Menten plot.

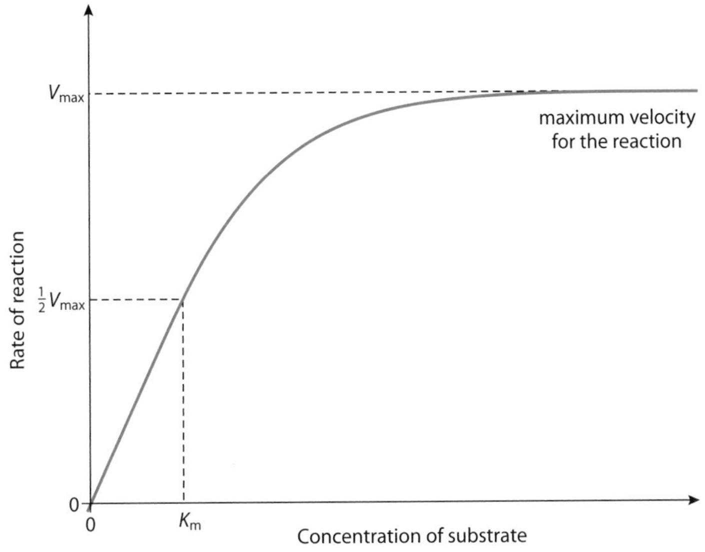

Figure B.16

You will need to be able to describe and explain the shape of a typical Michaelis–Menten plot. You may also be required to plot such a curve given rate and concentration data and/or use the graph to determine two key values, V_{max} and K_m.

DEFINITIONS

V_{max} is the maximum velocity of the enzyme and represents the fastest possible rate of reaction at a given concentration of enzyme. It indicates the point at which an enzyme is saturated with substrate and can work no faster.

K_m is the Michaelis constant and is defined as the concentration of substrate that gives a rate equal to half of V_{max} – the lower the value of K_m, the stronger the affinity of the enzyme for its substrate.

Enzyme inhibitors

Enzyme inhibitors are molecules that are able to bind to an enzyme and either reduce its activity. Inhibition is important for maintaining the correct balance of chemicals within cells.

Product inhibition is a type of negative feedback in which the product of a series of reactions acts as an inhibitor of one of the enzymes involved. This prevents the concentration of product from rising above desired (and safe) levels in the body. As product is used up elsewhere, the amount of inhibition decreases and the rate of synthesis increases again.

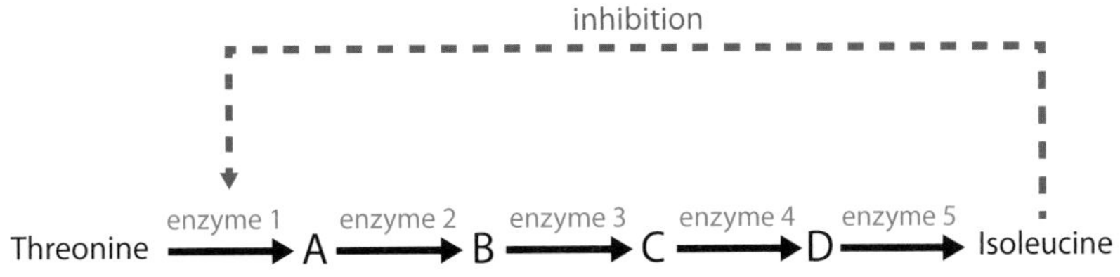

Figure B.17

Competitive inhibition involves a molecule binding in the active site of the enzyme preventing substrate from binding. The inhibitor is likely to have similar structural features to the substrate. Competitive inhibition does **not** affect the value of V_{max} but **increases** the value of K_m.

Non-competitive inhibition involves a molecule binding to an enzyme at a site distinct from the active site called an **allosteric site**. This alters the shape of the enzyme and its active site preventing substrate from binding. Non-competitive inhibition **decreases** the value of V_{max} but does **not** affect the value of K_m.

Buffer solutions, amino acids and protein

You should refer to Chapter 8 before reading this section.

A buffer solution must contain either:

- a weak acid and its conjugate base or,
- a weak base and its conjugate acid.

Consider an amino acid at its isoelectric point – it exists primarily as the zwitterion, $^{-}OOCCHRNH_3^{+}$. If the pH is lowered slightly then some of the zwitterion is protonated to form $HOOCCHRNH_3^{+}$. The solution contains a mixture of weak acid ($HOOCCHRNH_3^{+}$) and its conjugate base ($^{-}OOCCHRNH_3^{+}$) and can act as a buffer.

hint

You do not need to remember the Henderson–Hasselbalch equation. It is provided in the *IB Data Booklet*.

hint

pK_a values can be found in the *IB Data Booklet*. They are a measure of acid strength. Further information can be found in Chapter 8.

If a small amount of acid is added the conjugate base is able to mop it up:

$$^{-}OOCCHRNH_3^{+} + H^{+} \rightleftharpoons HOOCCHRNH_3^{+}$$

If a small amount of alkali is added then the weak acid can neutralise it:

$$HOOCCHRNH_3^{+} + OH^{-} \rightleftharpoons {}^{-}OOCCHRNH_3^{+} + H_2O$$

It is also possible to produce a basic buffer from an amino acid zwitterion by adding alkali.

Proteins in solution may also act as buffers as the side chains of some amino acids in the protein contain a –COOH or –NH_2 group (e.g. aspartic acid, lysine).

Calculating the pH of a buffer solution

You will need to be able to calculate the pH of a buffer using the **Henderson–Hasselbalch equation** (below) if studying this option or option D (Medicinal Chemistry):

$$\text{pH} = \text{p}K_a + \log_{10}\frac{[\text{base}]}{[\text{acid}]}$$

☆ Model answer B.3

Calculate the pH of a buffer solution containing 0.50 mol dm^{-3} HEPES (pK_a = 7.5) and 0.75 mol dm^{-3} of its sodium salt.

$$\text{pH} = 7.5 + \log 10\frac{[0.75]}{[0.5]} = 7.7$$

Worked example B.1

Determine the pH of a buffer solution made by mixing 0.250 dm^3 of 0.150 mol dm^{-3} propanoic acid with 0.400 dm^3 of 0.200 mol dm^{-3} potassium propanoate.

Acid is propanoic acid, C_2H_5COOH ($pK_a = 4.87$); base is the propanoate ion, $C_2H_5COO^-$.

Final concentration of C_2H_5COOH = concentration $\times \frac{V_{HA}}{V_{total}} = 0.150 \times \frac{0.250}{(0.25+0.40)}$

$= 0.0577\ldots$ mol dm^{-3}

Final concentration of $C_2H_5COO^- = 0.200 \times (0.400/(0.25+0.40)) = 0.123\ldots$ mol dm^{-3}

Substitute values into the Henderson–Hasselbalch equation:

$$pH = 4.87 + \log_{10}\frac{0.123...}{0.0577...} = 4.87 + 0.329... = 5.20$$

hint

When confronted with unfamiliar substances, the acid is the species with the extra H.

Protein assay

Ultraviolet-visible (UV-vis) spectrophotometry can be used to determine concentrations of proteins in solution. First, a calibration curve is created – the absorption of several known concentrations of the standard protein solution is recorded at λ_{max} (the wavelength at which the protein absorbs the most light). Typically, this will yield a directly proportional relationship between concentration and absorption.

The absorption of the unknown sample is recorded at the same wavelength and the value for the concentration of protein that produces that absorption can be read directly from the *x*-axis. In the figure, the unknown sample gives an absorption of 0.29 which corresponds to a concentration of 3.6 ppm.

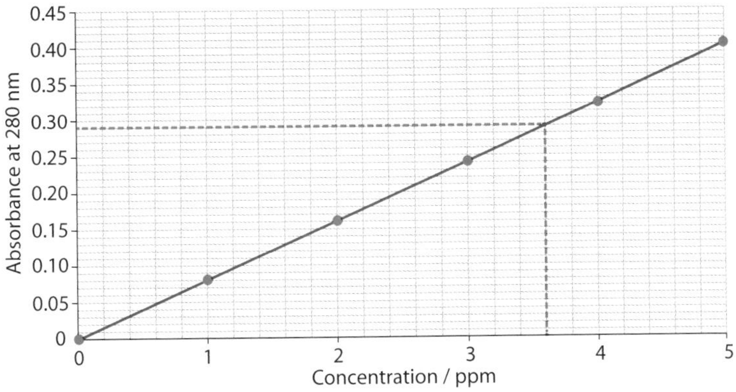

Figure B.18

The **Beer–Lambert Law** can also be used to find the concentration of a protein without the need for a calibration curve. The law is:

$A = \varepsilon cl$

where,

- A is the absorbance reading
- ε is the molar extinction coefficient (measured in $mol^{-1}\ dm^3\ cm^{-1}$)
- c is the concentration of the sample (measured in mol dm^{-3})
- l is the path length – the distance the light travels through the sample (measured in cm).

hint

You may need to determine concentrations from masses of solid dissolved in a given volume of water.

hint

If you are mixing different volumes of different concentrations of acid and salt then you will need to recalculate the final concentrations in the **combined** volume of solution.

hint

If there is a higher concentration of acid than salt then the pH of the buffer will be below that of the pK_a of the acid. If there is a higher concentration of salt than acid then the pH of the buffer will be higher than the pK_a of the acid.

hint

The Beer–Lambert law is given in the *IB Data Booklet*.

Worked example B.2

Calculate the concentration of a protein ($\varepsilon = 130\ mol^{-1}\ dm^3\ cm^{-1}$), if a sample placed in a 1 cm wide cuvette gives an absorbance reading of 0.46.

Rearrange the Beer–Lambert law to make c the subject and substitute in values:

$$c = \frac{0.46}{130 \times 1} = 3.5 \times 10^{-3}\ mol\ dm^{-3}$$

B.8 Nucleic acids (HL only)

Nucleic acids, such as DNA and RNA are comprised of linear chains of subunits called **nucleotides**. Nucleotides are the condensation products of the reactions between a pentose sugar (five carbons), phosphoric acid and a nitrogenous base (one of adenine (A), thymine (T), cytosine (C), guanine (G) or uracil (U); Figure **B.19**).

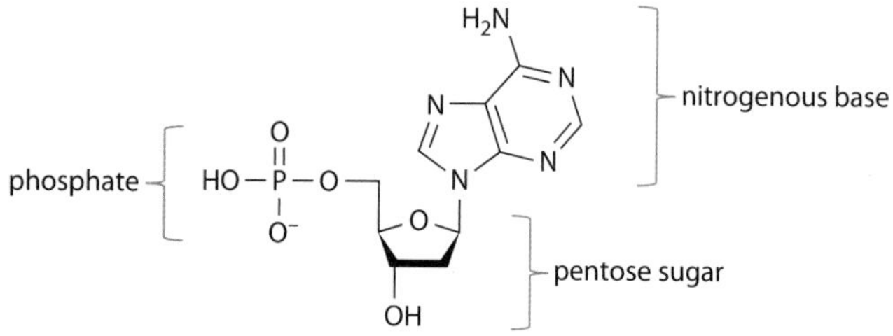

Figure B.19

Polynucleotides (including DNA and RNA) are formed through multiple **condensation reactions** between the phosphate group of one nucleotide and the pentose sugar of another (Figure **B.20**).

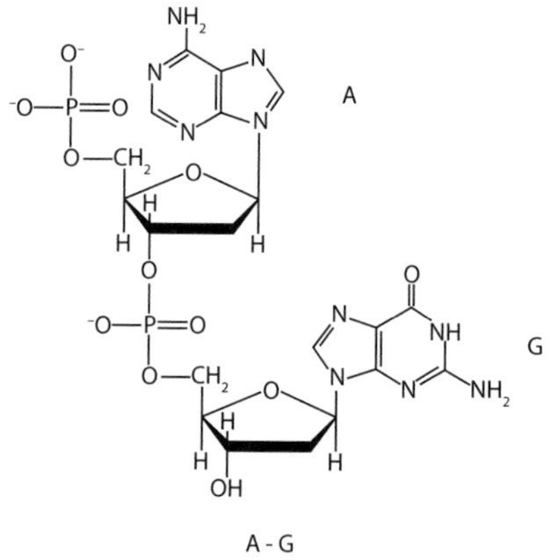

Figure B.20

DNA

DNA (deoxyribonucleic acid) is a **double helix** (Figure **B.21a**) of two anti-parallel polynucleotide chains held together by hydrogen bonding between the bases (A, C, T, G) on the different strands (Figure **B.21b**).

DNA is stabilised by:

- having hydrophobic bases on the inside away from water
- the hydrogen bonding between bases on opposite strands
- having hydrophilic phosphate and pentose sugar groups on the outside
- the extensive hydrogen bonding between water and the external phosphate and pentose sugar groups.

The phosphate groups give DNA molecules an overall negative charge. This enables them to associate with basic proteins called **histones** (containing many NH_3^+ groups) which are important for efficient packaging of DNA in the nucleus.

RNA (ribonucleic acid) is a condensation polymer of nucleotides in which thymine bases are replaced by **uracil** and the deoxyribose sugar is replaced by **ribose** (Figure **B.22**). Typically, it exists as a **single-stranded** molecule rather than a double helix.

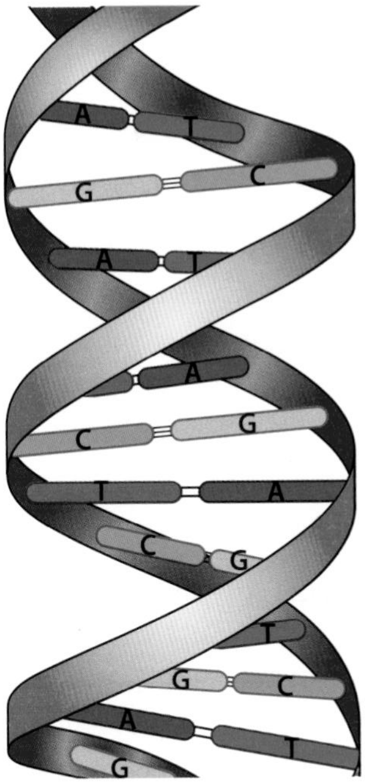

Figure B.21a

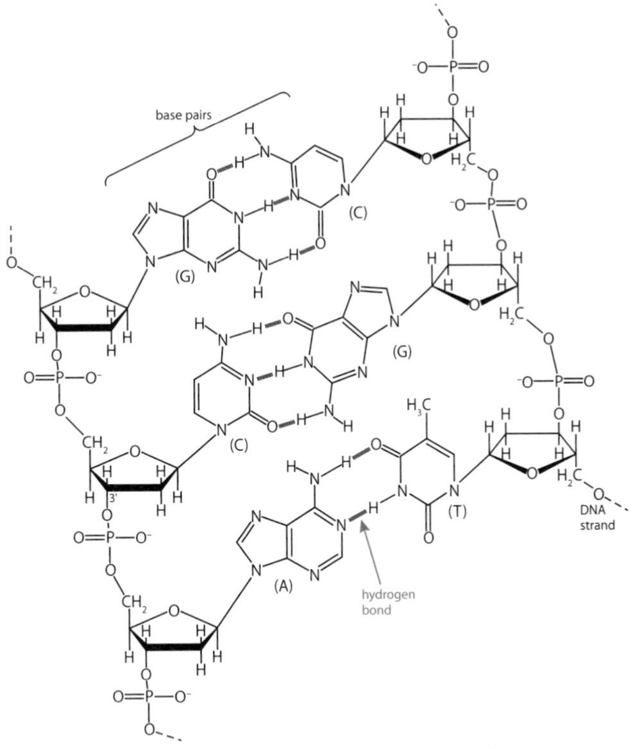

Figure B.21b

The genetic code

The sequence of bases along a DNA or RNA strand determines the sequence of amino acids in the synthesised proteins (primary structure). Three consecutive bases code for one amino acid (the **triplet code**), e.g. GCG → alanine, AAA → lysine, etc. This triplet **genetic code** is universal and does not differ between species.

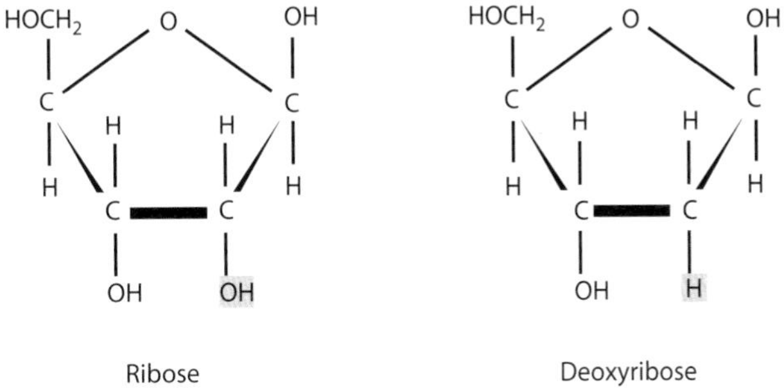

Figure B.22

Adenine always pairs (through hydrogen bonding) with thymine and cytosine always pairs with guanine. In RNA, all thymine bases are replaced with uracil (U). If the sequence of bases of one strand is known then it is possible to determine the sequence of bases of the other (**complementary**) strand.

Known sequence: 5′- A G T C T G G A C T G C A A T -3′

Complementary sequence (DNA): 3′- T C A G A C C T G A C G T T A -5′

Synthesised RNA strand: 3′- U C A G A C C U G A C G U U A -5′

Complementary base pairing is essential to the self-replication of DNA – the two strands of the double helix are separated and the complementary strand to each is synthesised creating two identical double-stranded molecules.

Original DNA: 5′- A G T C T G G A C T G C A A T -3′

3′- T C A G A C C T G A C G T T A -5′

Strands separated and complements synthesised:

5′- A G T C T G G A C T G C A A T -3′

3′- T C A G A C C T G A C G T T A -5′ (complement)

5′- A G T C T G G A C T G C A A T -3′ (complement)

3′- T C A G A C C T G A C G T T A -5′

TEST YOURSELF B.10

1 Give the complementary DNA strand to the following single strands of DNA:

a 5′- A G T T C A A A G T C G T C C -3′

b 5′- G G A T C G T A G T C T A C T -3′

2 Describe three ways in which the structure of DNA differs from that of RNA.

hint

You should learn 3/4 benefits and 3/4 concerns of genetic modification for the exam.

Genetic modification

A genetically modified organism (GMO) is one that has had its genetic material modified by genetic engineering techniques. It often involves the transfer of DNA between species. The use of GMOs in food production provides a number of benefits but there are several concerns about GMO usage.

Benefits of using GMOs:

- Increased yield and nutritional content of crops
- Improved crop resistance to disease, drought, pests and herbicides
- Enhanced taste, texture, quality and shelf life
- Improved animal health – GM animals could be developed to be more resistant to disease.

Concerns of using GMOs:

- Not enough is known about the long-term effects
- Escape of transgenic material into the 'wild'
- Damage to the environment and unbalancing of ecosystems
- Exploitation of farmers and monopolization by large companies producing GM seeds
- Ethical considerations.

hint

When a question ask you to 'suggest' reasons, e.g. for and against using GMOs, do **not** give explanations. Keep it short and simple!

hint

The bonds in a conjugated system do not have to be between carbon atoms but they must alternate with single bonds.

B.9 Biological pigments (HL only)

Biological pigments are coloured compounds produced by metabolic processes. They contain a **chromophore** – a portion of the molecule that is able to absorb visible light – making the compound coloured. In most biological pigments, the chromophore is a region of the molecule that is **extensively conjugated** (a long series of alternating single and double bonds) (Figure **B.23**). This conjugation creates a system of **delocalised electrons** and leads to intense **absorption bands in the visible region** of the electromagnetic spectrum. The longer the conjugated system the more likely a molecule is to be coloured.

e.g. lycopene (red pigment in tomatoes) has 11 conjugated double bonds:

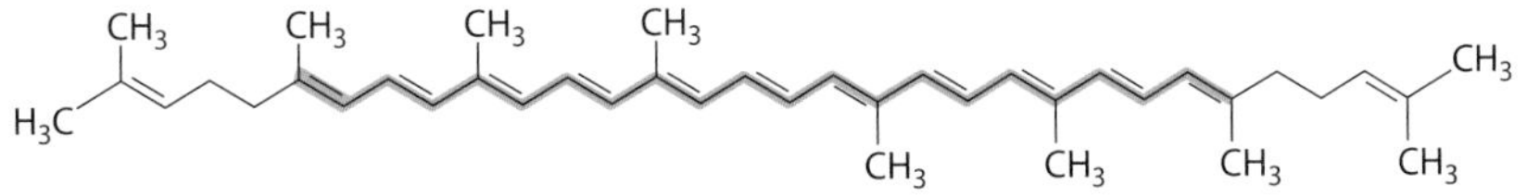

Figure B.23

Porphyrins

The porphyrins comprise a series of nitrogen-containing macrocyclic (large ring) conjugated ligands that are able to coordinate certain metal ions to form chelates (the ligand bonds to the metal ion through more than one donor atom). Examples you are required to know are: chlorophyll (Mg^{2+}), hemoglobin (Fe^{2+}), myoglobin (Fe^{2+}) and cytochromes.

Cytochromes contain heme groups in which the iron ion is able to interconvert between the +2 and +3 oxidation states during metabolic redox processes.

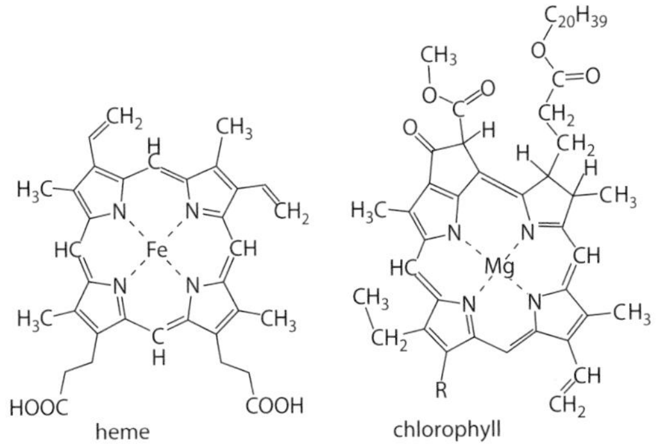

Figure B.24

Hemoglobin and myoglobin

Hemoglobin is a protein that transports oxygen in the blood. Oxygen binds **cooperatively** to hemoglobin resulting in a **sigmoidal (S-shaped) oxygen dissociation curve** (a plot of partial pressure of oxygen versus oxygen saturation of the protein) (Figure **B.25**).

Hemoglobin is a tetramer (four protein subunits) – binding of oxygen to one of the subunits causes a **conformational change** in the shape of the protein making it easier for oxygen to bind to the other subunits.

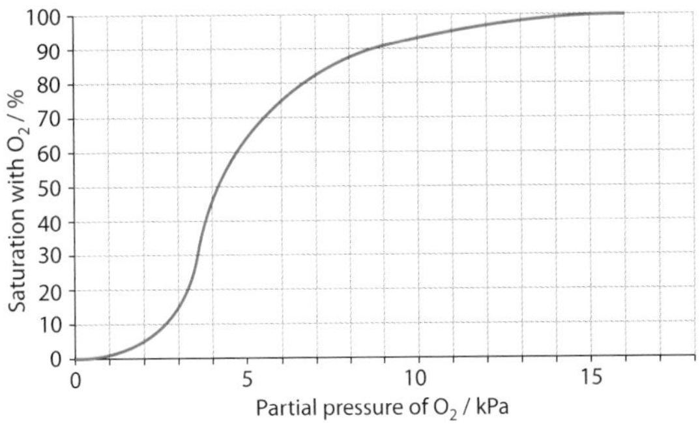

Figure B.25

The oxygen saturation (how much oxygen can bind) of hemoglobin is influenced by:

pH – hemoglobin can be protonated at low value of pH resulting in the dissociation of oxygen as the protein undergoes a conformational change.

$$HbO_2 + H^+ \rightleftharpoons HbH^+ + O_2$$

Carbon dioxide – forms carbonic acid lowering the pH and causing oxygen to dissociate. CO_2 is also able to react with some amine ($-NH_2$) groups in hemoglobin causing the protein to change shape and decreasing its affinity for oxygen.

Carbon monoxide (CO) – is a **competitive inhibitor** of oxygen binding – it binds to the same sites in hemoglobin as oxygen, therefore competes with oxygen for these binding sites and reduces the ability of hemoglobin to transport oxygen (CO is transported instead). **Hemoglobin has a higher affinity for CO than for O_2** – the lone pair of electrons of the carbon atom forms a coordinate covalent bond to the Fe^{2+} ion.

Temperature – oxygen saturation of hemoglobin decreases as temperature increases. The reaction:

$HbO_2 \rightleftharpoons Hb + O_2$ is an endothermic process ($\Delta H = +ve$)

The position of equilibrium shifts to the right as temperature is increased and less oxygen binds.

Fetal hemoglobin is a different form of hemoglobin present in the blood of unborn fetuses. It has a higher affinity for oxygen than maternal hemoglobin. This allows the transfer of oxygen from the mother's blood to the fetal blood in the placenta.

Anthocyanins

Anthocyanins are a class of **aromatic, water-soluble pigments** that share the same core structure (Figure **B.26**).

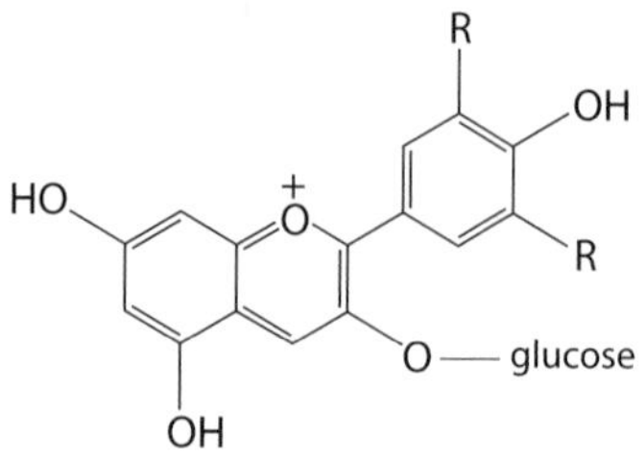

Figure B.26

Anthocyanin colour is affected by:

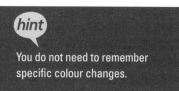

- **The presence of metal ions:** anthocyanin molecules can act as ligands to ions such as Fe^{3+} and Al^{3+} to form vividly-coloured complexes
- **pH**: a complex equilibrium exists in aqueous solution:

  $$\mathbf{A} \rightleftharpoons \mathbf{AH^+} \rightleftharpoons \mathbf{B} \rightleftharpoons \mathbf{C}$$

  quinoidal base ⇌ flavylium cation ⇌ carbinol ⇌ chalcone

  purple/blue ⇌ red ⇌ colourless ⇌ yellow

- Low pH favours the flavylium form – anthocyanins tend to appear red in acidic solutions.
- As the pH increases the purple quinoidal base and colourless carbinol predominate.
- High pH favours the chalcone form – anthocyanins tend to appear yellow in alkaline solutions.

Because the colour of anthocyanins depends on pH they can be used as acid–base indicators. The anthocyanin exists in different forms (with different colours) in acidic and alkaline solutions.

> **hint**
> You do not need to remember specific colour changes.

- **Temperature:** at low temperatures the red flavylium form is very stable but as temperature is increased the following equilibria shift to the right:

  flavylium cation ⇌ carbinol ⇌ chalcone

  red ⇌ colourless ⇌ yellow

  Low temp → High temp

Carotenoids

Carotenoids are a class of lipid-soluble pigments that contain long conjugated hydrocarbon chains, e.g. lycopene. They are involved in light harvesting during photosynthesis and transfer absorbed light energy to chlorophyll molecules.

The stability of carotenoid molecules is affected by:

- **Temperature:** above 50 °C, carotenoids isomerise from the all-*trans* form into a mixture of *cis* forms.
- **Light:** carotenoids are highly susceptible to photo-oxidation due to the presence of multiple C=C double bonds.
- **Metal ions and hydroperoxides:** both are able to cause oxidation of carotenoid molecules. Oxidation of a carotenoid causes the conjugation to be destroyed and the colour to be bleached.

Chlorophyll

The stability of chlorophyll is affected by:

- **pH:** chlorophyll is stable in alkaline conditions but in acidic conditions (pH < 3) it is unstable as the central Mg^{2+} ion is replaced by H^+ ions.
- **Temperature:** high temperatures cause chlorophyll to decompose.

In **photosynthesis**, chlorophyll absorbs light to promote electrons. These electrons then pass down an **electron-transport chain** (a series of redox reactions) which converts the energy of the excited electrons into chemical energy.

Chromatography

Paper and thin-layer chromatography (TLC) can be used to separate and identify pigments in a mixture. Paper chromatography has been described earlier in this chapter – TLC is a similar process that uses a stationary phase of silica particles on a glass/plastic plate. The mobile phase is an appropriate solvent.

TLC separates mixtures of pigments based on how strongly they are **ad**sorbed on the stationary phase/dissolve in the mobile phase. The greater the tendency of a pigment molecule to be adsorbed onto the stationary phase (higher affinity for silica particles), the more slowly it moves along the plate.

There is no need for a locating agent as the pigments are coloured. The **retardation factor (R_f)** for each pigment can be calculated and compared to literature values or samples of pure pigment to identify a particular pigment.

TLC is faster and gives better resolution than paper chromatography.

TEST YOURSELF B.11

1 Describe three ways in which paper chromatography and TLC differ.
2 State and explain what happens to the colour of **a** chlorophyll and **b** an anthocyanin in the presence of an acid.
3 Explain why carbon monoxide is highly toxic to humans.

B.10 Stereochemistry in biomolecules (HL only)

Due to the specific three-dimensional shape of most proteins and enzymes, many biological molecules exist in nature as a single enantiomer – so that they fit into an enzyme active site (also chiral).

- Amino acids (with the exception of glycine) are all chiral molecules and exist in the L-configuration in nature.
- Monosaccharide sugars that are chiral usually exist in the D-configuration in nature.

Stereoisomerism in monosaccharide sugars

When drawn in **Fischer (linear) projection** (Figure **B.27**), the position of the hydroxyl group on the **chiral** carbon **furthest** from the aldehyde/ketone group determines the D/L configuration:

- –OH group drawn on the right → D-configuration
- –OH on the left → L-configuration.

hint
Remember Left for L-configuration.

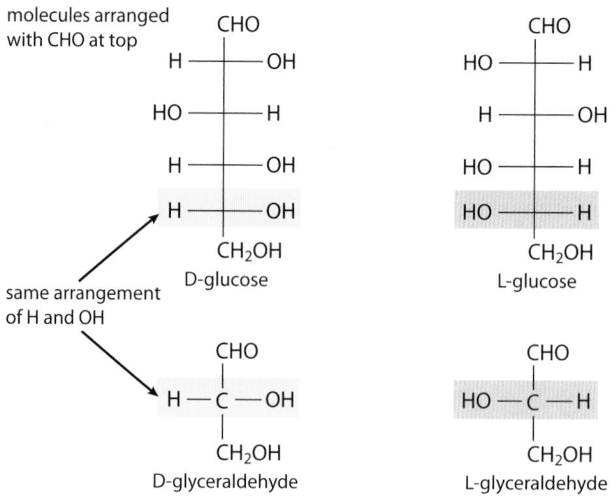

Figure B.27

TEST YOURSELF B.12

State the D/L configuration of the following monosaccharides:

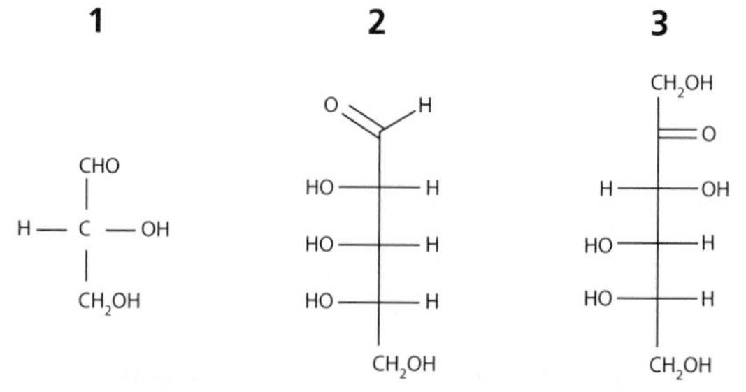

Figure B.28

---

hint

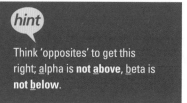

Think 'opposites' to get this right; alpha is **not above**, beta is **not below**.

When a monosaccharide cyclises to form a ring the sugar can be classified as α or β based on the position of a hydroxyl group – for glucose it is the –OH group on C1 and for fructose it is the –OH group on C2. If the –OH group lies below the plane the molecule is in the α-configuration; if the –OH group is above the plane then the molecule is a β-sugar.

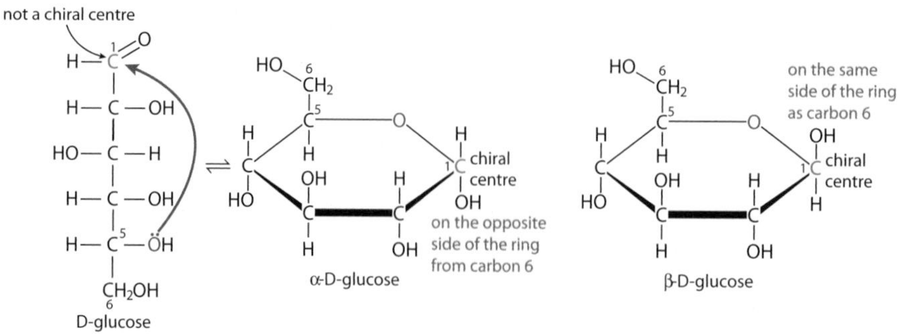

Figure B.29

Cellulose versus starch

Cellulose	Starch
Linear polymer chains only	Both linear and branched chains present
Polymer of β-glucose	Polymer of α-glucose
Only β-1,4-links between monomers	Both α-1,4 and α-1,6-links present
No coiling of polymer chains	Polymer chains form coils (helical)
Linear chains pack well forming extensive hydrogen bonding between –OH groups on neighbouring chains	Branching and coiling decreases the amount of hydrogen bonding between chains
Forms strong fibres	Does not form fibres and is not very strong
Insoluble in water	Soluble in hot water

Cellulose forms the strong **cell wall** of plant cells – the strength comes from the linearity of the chains and the extensive hydrogen bonding between them.

Cellulose provides **fibre** in our diets. It is insoluble in water and remains undigested – water binds to its surface which softens and bulks out the feces. This makes constipation and other intestinal disorders less likely.

Fatty acids

Carbon-carbon (C=C) double bonds in unsaturated fatty acids can exist in *cis* or *trans* forms. Most fatty acids exist as all-*cis* isomers in nature but food processing can convert them into the *trans* form.

Hydrogenation

Unsaturated lipids are reacted with **hydrogen gas** with the lipid at a high temperature (about **150 °C**) in the presence of a metal catalyst such as **nickel**. Hydrogen adds across the C=C double bonds to produce a saturated fatty acid containing C–C single bonds. **Partial hydrogenation** occurs when only some of the double bonds are saturated (converted to single bonds). During partial hydrogenation, some fatty acids convert to the *trans* configuration (Figure **B.30**)

Hydrogenation (and *cis*-to-*trans* conversion) increases the melting point of the lipid, turning it from a liquid into a semi-solid. It also increases the shelf-life of the product as saturated fatty acids are less likely to undergo oxidation.

Hydrogenation is important for the production of margarine from plant oils. However, saturated and *trans* fats increase cardiovascular disease – *trans* fats are not metabolised well in the body and accumulate as fatty deposits in arteries.

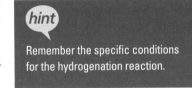

hint

Remember the specific conditions for the hydrogenation reaction.

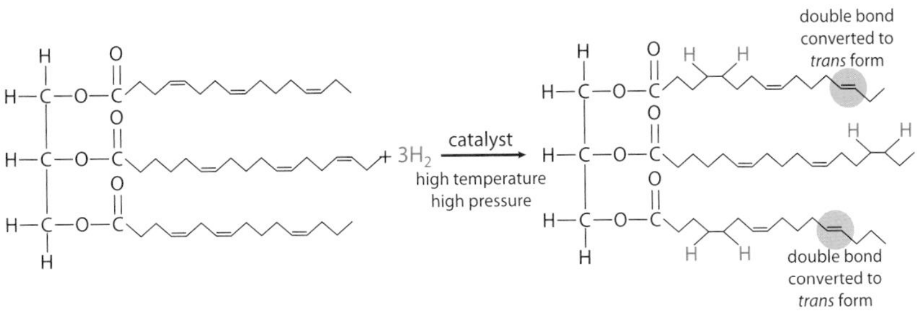

Figure B.30

Vision chemistry

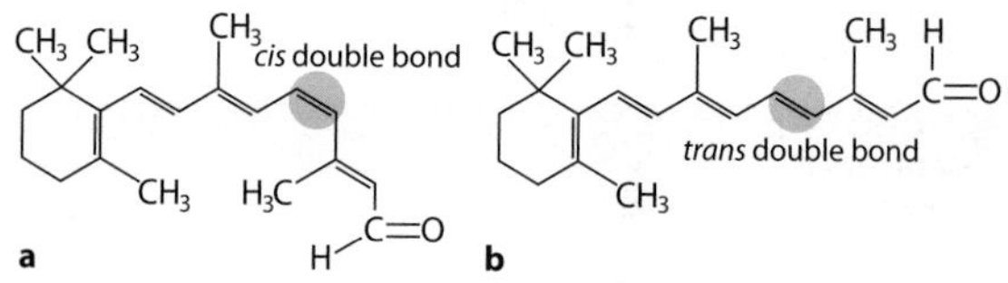

Figure B.31

Vision chemistry (the chemistry occurring in the eye) relies on the interconversion of the molecule **retinal** between *cis* and *trans* isomers.

Rod and cone nerve cells in the retina of the eye contain a protein called **rhodopsin**, comprising the protein **opsin** bound to a molecule of **retinal**. The absorption of light by the *cis* form of retinal (one *cis*-C=C) converts it to the all-*trans* configuration. This causes the protein to undergo a change in shape which leads to a cascade of signals resulting in nerve impulses to the brain.

Vitamin A

The molecule retinal is produced by the oxidation of vitamin A (retin**ol**). Vitamin A (see earlier in chapter for structure) contains a primary alcohol group that is oxidised via an enzyme to an aldehyde – retinal. Vitamin A deficiency can lead to night-blindness and poor vision because retinal is required for vision.

Checklist

At the end of this chapter you should be able to:

- ☐ Describe the basics of typical metabolic reactions such as anabolism, catabolism, condensation and hydrolysis.
- ☐ Explain the formation of structure in proteins and how this enables them to act as enzymes.
- ☐ Explain how changes in environment affect the function of enzymes.
- ☐ Describe the chemistry of amino acids including their physical properties.
- ☐ Describe the features of lipid molecules and their biological role.
- ☐ Calculate the iodine number of a lipid and relate this to unsaturation.
- ☐ Describe the chemistry of carbohydrates and explain how simple monosaccharides are joined to create more complex molecules.
- ☐ Describe what is meant by a vitamin and relate the structures of vitamins A, C and D to their physical properties.
- ☐ Explain what is meant by xenobiotics, host–guest chemistry and biomagnification, giving examples of each.

Higher Level only

- ☐ Determine enzyme kinetics from graphical data and explain how enzyme inhibition arises.
- ☐ Perform calculations to determine the pH of a buffer solution.
- ☐ Determine protein concentrations using the Beer–Lambert Law or calibration curve.
- ☐ Describe the structure of DNA and RNA and how DNA is replicated.
- ☐ Suggest reasons for and against genetic modification of organisms.
- ☐ Describe how the colour of biological pigments arises using chlorophyll, carotenoids and anthocyanins as examples.
- ☐ Explain how changes to the environment affect the colour of biological pigments.
- ☐ Describe how stereochemistry of amino acids and monosaccharides creates the diversity in the structures of their polymers and relate this to the different biological roles of those polymers.

ENERGY

C

In an increasingly technological society the availability, cost and environmental impact of energy sources are major political and economic issues. Scientists, politicians and economists can all make important contributions to the debate about energy for the future.

This chapter covers the following topics:

- ☐ Energy sources
- ☐ Fossil fuels
- ☐ Nuclear fusion and fission
- ☐ Solar energy
- ☐ Environmental impact – global warming
- ☐ Electrochemistry, rechargeable batteries and fuel cells (HL only)
- ☐ Nuclear fusion and nuclear fission (HL only)
- ☐ Photovoltaic cells and dye-sensitised solar cells (DSSC) (HL only)

C.1 Energy sources

Energy

Energy is the ability to do work. To be useful, a source must produce energy at a reasonable rate and with minimal pollution. However, energy always goes from a concentrated to a dispersed form when work is done. The quality and amount of energy available to do work decreases as heat is transferred to the surroundings.

Renewable and non-renewable energy

- Renewable energy comes from sources that can be **replenished**.
- Non-renewable energy sources are **finite**.

Source of energy	Advantages	Disadvantages
Coal	Deposits should last hundreds of years. Geographically widespread. Can be converted to gas and liquid fuels. Relatively cheap.	Produces large amounts of CO_2, SO_2 and particulates when burnt. More difficult to transport than oil and gas. Mining creates pollution.
Oil	Easily transported (ship or pipeline). Wide variety of compounds for petrochemical industry. Yields fuels suitable for internal combustion engine.	Produces large amounts of CO_2 and SO_2 when burnt. Supplies may last only decades. Uneven geographical distribution.
Natural gas	Clean fuel (little sulfur). Easy to transport. Higher specific energy than coal or oil. Less CO_2 per kJ produced compared to coal and oil. Cheap to produce.	Produces CO_2 when burnt. Supplies may last only 100 years. Geographical distribution uneven. Risk of explosion due to leaks. Storage requires pressurisation or liquefaction.
Nuclear fission	Extremely high specific energy. No greenhouse gases produced.	Non-renewable. Waste is radioactive and difficult to dispose of. Expensive. Risk of nuclear disaster.
Electrochemical cells	Portable. Rechargeable batteries can be used multiple times.	Non-renewable. Expensive. Limited lifetime. Require specialist disposal.

Source of energy	Advantages	Disadvantages
Solar energy (photovoltaic cells)	Free and renewable source. Non-polluting. Can be used in remote locations, e.g. deserts.	Diffuse form of energy. Expensive. Dependent on weather. Do not work at night.
Wind power	Free and renewable source. Non-polluting. Farms can be placed on or offshore.	Diffuse form of energy. Expensive. Dependent on weather.
Nuclear fusion	Potentially huge energy production. Non-polluting. Does not produce nuclear waste.	Currently not viable. Energy input is vastly greater than output.
Fuel cells (hydrogen)	Portable. Hydrogen can be produced from water. Non-polluting.	Hydrogen is flammable. Hydrogen also produced from fossil fuels.
Biomass	Renewable source. Can use up waste material.	Requires large areas of land to grow plants. May create food shortages/price inflation.

Other sources of renewable energy include geothermal, tidal and hydroelectric.

TEST YOURSELF C.1

Suggest advantages and disadvantages of using **1** biomass, **2** tidal, **3** hydroelectric and **4** geothermal energy sources.

renewable, uses up waste material
requires land, food shortage
free, renewable
dependent on weather
free and renewable, non polluting
dependent on weather

Energy density and specific energy

DEFINITIONS

$$\text{ENERGY DENSITY} = \frac{\text{energy released from fuel}}{\text{volume of fuel used}}$$

$$\text{SPECIFIC ENERGY} = \frac{\text{energy released from fuel}}{\text{mass of fuel used}}$$

You should be able to calculate the energy density or specific energy of a fuel.

Model answer C.1

Calculate the specific energy and energy density of hexane, C_6H_{14}, giving your answers in kJ g^{-1}, and kJ dm^{-3}, respectively. (The density of hexane is 0.654 g cm^{-3}.)

ΔH_c [hexane] = –4163 kJ mol^{-1}

4163 kJ are released when 1 mole of hexane is burnt

Mass of 1 mol of hexane = 86.20 g

$$\text{Specific energy} = \frac{\text{energy released from fuel}}{\text{mass of fuel used}} = \frac{4163}{86.20} = 48.3 \text{ kJ g}^{-1}$$

hint

You can ignore the sign on the enthalpy change when putting the value into the equation.

$$\text{Density} = \frac{\text{mass}}{\text{volume}}$$

$$\text{volume of 1 mol hexane} = \frac{\text{mass of 1 mol of hexane (g)}}{\text{density of hexane (g cm}^{-3})} = \frac{86.20}{0.654} = 131.8\ \text{cm}^3 = 0.1318\ \text{dm}^3$$

$$\text{Energy density} = \frac{\text{energy released from fuel}}{\text{volume of fuel used}} = \frac{4163}{0.1318} = 31\ 600\ \text{kJ dm}^{-3}$$

TEST YOURSELF C.2

By using appropriate data from the *IB Data Booklet*, determine both the specific energy (in kJ g^{-1}) and energy density (in kJ dm^{-3}) of:

1 ethanol (density = 0.79 g cm^{-3}) 29.7, 23400
2 carbon (density = 2.27 g cm^{-3}) 32.8, 74 500
3 hydrogen (density = 89.9 g m^{-3}) 146.6, 12.7

hint

Take extra care to check the units required for the final answer.

Fuels with high specific energy such as hydrogen are preferred when the weight of fuel has to be kept low, e.g. in rockets. Fuels with high energy density such as gasoline are preferred when storage volume needs to be minimised, e.g. car fuel tank.

Energy transfer efficiency

DEFINITIONS

$$\text{ENERGY TRANSFER EFFICIENCY} = \frac{\text{useful output energy}}{\text{total input energy}} \times 100$$

Model answer C.2

The fuel in the internal combustion engine of a car releases 20.0 kJ of energy, of which 0.4 kJ is converted to sound and 14.0 kJ is converted into heat. Determine the efficiency of the energy transfer in the engine.

Heat and sound are not useful in propelling the car, therefore:

Useful energy = 20.0−(14.0+0.4) = 5.6 kJ

$$\text{Energy transfer efficiency} = \frac{\text{useful output energy}}{\text{total input energy}} \times 100 = \frac{5.6}{20.0} \times 100 = 28\%$$

C.2 Fossil fuels

Fossil fuels are produced via the reduction of biological molecules (contained within dead organisms) made from carbon, hydrogen, nitrogen, sulfur and oxygen. The result of this reduction is a carbon-rich fuel, such as crude oil or coal.

Petroleum/crude oil

Crude oil (petroleum) is a complex mixture of different hydrocarbons. During crude oil refining, the components of this mixture undergo **fractional distillation** – a physical process – which separates them based on boiling point.

- The crude oil is heated until it boils.
- The vapour passes into a large tower (fractionating column).
- The temperature is higher at the bottom of the column than at the top.
- Different fractions condense and are drawn off at different heights in the tower.
- The hydrocarbons with the highest boiling points condense towards the bottom of the tower.
- The smaller hydrocarbons travel further up the column until they condense and are drawn off.

Fraction	Uses
refinery gases	Fuel for cooking and heating. Bottled gas. Used for fuel in the refinery.
gasoline	Fuel for cars.
naphtha	Feedstock for petrochemical industry. Converted by catalytic reforming into gasoline. Used as a solvent.
kerosene (paraffin)	Jet fuel. Household heaters and lamps. May also be cracked to produce more gasoline.
diesel oil (gas oil)	Diesel fuel for cars, lorries etc. May be cracked to produce more gasoline.
fuel oil	Fuel for ships and industry. Fuel for home central heating systems.
lubricating oil	Lubricant in engines and machinery. May be cracked.
wax	Candles, petroleum jelly, waxed paper and cardboard in the food industry.
bitumen/asphalt	Tarmac for roads and waterproofing roofs.

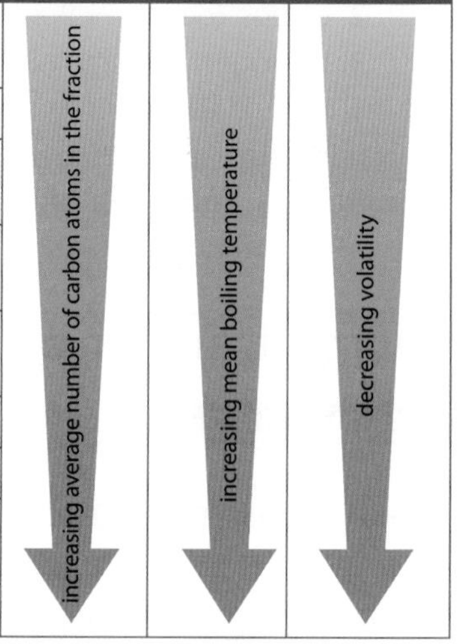

Fractions obtained from crude oil. Other names are sometimes used for fractions. Volatility refers to how readily a substance evaporates.

Octane numbers and auto-ignition

DEFINITIONS

AUTO-IGNITION occurs when the fuel in the cylinder of an engine burns instantly ('explodes') rather than burning smoothly. This leads to knocking, which can cause excessive engine wear and damage.

OCTANE NUMBER an experimentally-derived value that provides a measure of the tendency of a fuel to auto-ignite. The **higher** the octane number, the **lower** the tendency of a fuel to auto-ignite.

Octane numbers are dependent on the molecular structure of the fuel:

- **Carbon chain length:** the longer the chain, the lower the octane number.
- **Chain branching:** branching increases octane number (for the same number of carbon atoms).
- **Cyclisation:** cyclisation increases octane number (for the same number of carbon atoms).
- **Unsaturation:** alkenes tend to have a **higher** octane number than alkane equivalents.

Cracking and catalytic reforming

Octane numbers can be increased by **cracking** and **catalytic reforming**:

Cracking: Long-chain alkanes are passed over an aluminosilicate catalyst heated to 500 °C under pressure. Each alkane molecule decomposes into a shorter-chain alkane and one or more alkene molecules. Some alkane molecules may also cyclise or become branched during the process.

e.g. $C_{12}H_{26} \rightarrow C_8H_{18} + C_4H_8$ or $C_{20}H_{42} \rightarrow C_8H_{18} + 2C_4H_8 + 2C_2H_4$

Reforming: Hydrocarbons are passed over a platinum catalyst at 500 °C. The molecules become branched, cyclised or aromatised.

e.g. $CH_3(CH_2)_6CH_3 \rightarrow CH_3CH(CH_3)CH(CH_3)CH_2CH_2CH_3$ (two methyl branches formed)

$CH_3(CH_2)_4CH_3 \rightarrow C_6H_6 + 4H_2$ (formation of benzene)

TEST YOURSELF C.3

1 For each of the following pairs of molecules, deduce which has the higher octane number.
   a nonane (C_9H_{20}) hexane
   b heptane (C_7H_{16}) 2,2-dimethylpentane
   c 1,4-dimethylbenzene octane

2 Complete the equations for the cracking of an alkane.
   a $C_{10}H_{22} \rightarrow C_8H_{18} + ..$ C_2H_4
   b $C_{14}H_{30} \rightarrow C_8H_{18} + 2 .$ C_3H_6
   c $C_{13}H_{28} \rightarrow C_6H_{14} + C_3H_6 + 2C_2H_4$

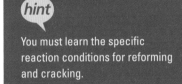

hint
You must learn the specific reaction conditions for reforming and cracking.

Coal gasification and liquefaction

In coal gasification, solid coal is converted into a gaseous fuel (**syngas**, a mixture of CO and H_2). The process involves heating coal to temperatures above **1000 °C** in the presence of **oxygen** and **steam**.

The main reaction occurring is:

$C(s) + H_2O(g) \rightarrow CO(g) + H_2(g)$

Syngas can be converted to methane by heating it with additional hydrogen:

$CO(g) + 3H_2(g) \rightarrow \mathbf{CH_4(g)} + H_2O(g)$

In the **coal liquefaction** process, coal is converted to liquid hydrocarbons. One method involves mixing coal with a **solvent and hydrogen**, and then heating to about **400 °C** at a **high pressure** and in the presence of a **catalyst** such as **red mud** (from bauxite purification).

Reactions such as $6C(s) + 7H_2(g) \rightarrow C_6H_{14}(l)$ may occur.

It is also possible to produce liquid hydrocarbon from syngas using the **Fischer–Tropsch** process. At a temperature of **300 °C**, **high pressure** and in the presence of a **ruthenium catalyst**, the CO reacts with H_2 to form liquid alkanes:

e.g. $15H_2(g) + 7CO(g) \rightarrow C_7H_{16}(l) + 7H_2O(g)$

hint
You will be expected to know the conditions for coal liquefaction/gasification and be able to deduce the equations.

Carbon footprints

DEFINITION

A CARBON FOOTPRINT is the total amount of greenhouse gas produced during the course of a human activity. It is typically expressed in equivalent tonnes of carbon dioxide (CO_2e).

Carbon footprints can be calculated for a variety of things: a product, an activity, a person, even a whole country. Direct production of CO_2 results from activities such as burning a fuel, e.g. driving a vehicle or heating a home. Indirect production is trickier to determine, such as how much CO_2 is produced via the consumption of an avocado pear.

Model answer C.3

Calculate the carbon footprint for a car journey of 250 km if the car consumes 22 dm³ of gasoline [assume pure octane, C_8H_{18} (density 0.703 g cm^{-3})].

mass of octane consumed = density × volume = 0.703 (g cm^{-3}) × 22 000 (cm³) = 15 466 g

moles of octane consumed = $\frac{15\,466}{114.26}$ = 135.4 mol

$C_8H_{18}(l) + 12.5O_2(g) \rightarrow 8CO_2(g) + 9H_2O(l)$

moles of carbon dioxide produced = moles of octane × 8 = 135.4 × 8 = 1082.9 mol

mass of carbon dioxide produced = 1082.9 × 44.01 = 47 700 g = 47.7 kg of CO_2

C.3 Nuclear fusion and fission

Nuclear fusion

Nuclear fusion is an ideal source of energy for the future; the fuel is cheap and abundant; and no radioactive waste or greenhouse gas is produced. Nuclear fusion occurs when two **light nuclei**, e.g. hydrogen nuclei, fuse (join) together to make a larger one releasing large quantities of energy; the **binding energy per nucleon** increases during the reaction.

DEFINITION

BINDING ENERGY is the amount of energy required to break a nucleus into individual protons and neutrons (nucleons), or the amount of energy released when the nucleons come together to form a nucleus – the greater the binding energy per nucleon, the more stable the nucleus.

Binding energy per nucleon increases from hydrogen to iron and then decreases as the nuclei get larger.

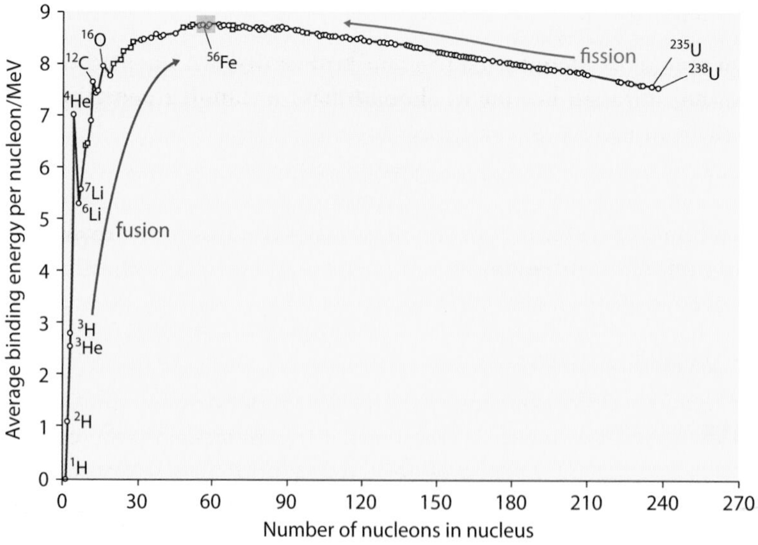

Figure C.1

Fusion equations

A typical fusion reaction could be:

$${}^{2}_{1}H + {}^{2}_{1}H \rightarrow {}^{3}_{2}He + {}^{1}_{0}n$$

The sum of the atomic masses and atomic numbers on left and right sides of the equation must balance.

> **hint**
> ${}^{2}_{1}H$ is called deuterium.

The fusion reaction generates a larger, more stable nucleus (helium-3) in which the binding energy per nucleon is greater than in deuterium. The extra binding energy is released as heat.

TEST YOURSELF C.4

Write an equation for the fusion of a nucleus of deuterium (hydrogen-2) and tritium (hydrogen-3) to form a helium nucleus and a neutron. ${}^{2}_{1}H + {}^{3}_{1}H \rightarrow {}^{4}_{2}H + {}^{1}_{0}n$

Absorption spectra

An absorption spectrum is produced when specific frequencies of light are absorbed from a continuous spectrum (Figure **C.2** lower panel). When an atom absorbs light, its electrons are promoted to higher energy levels. The **visible absorption spectrum of hydrogen** is created when electrons in main energy level 2 are promoted to higher energy levels – it appears as a series of black bands within a continuous spectrum that get closer together at higher frequencies.

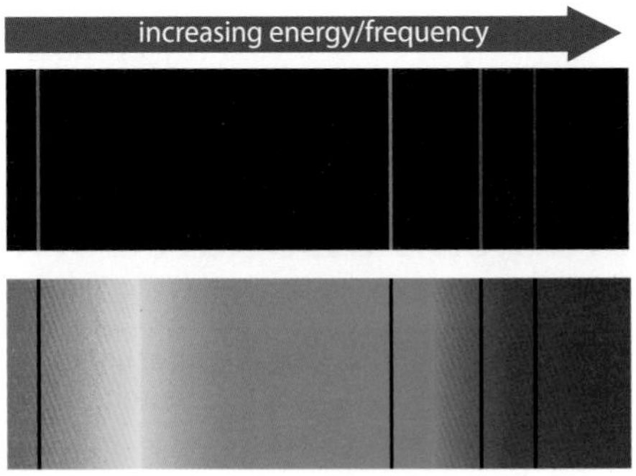

Figure C.2

As light from the core of a star travels through the cooler outer regions, specific frequencies of light are absorbed depending on which elements are present. Astronomers look for certain absorbed frequencies in the spectrum that correspond to specific elements in order to determine the star's elemental composition.

Nuclear fission

Nuclear fission involves the **splitting of heavy nuclei** into smaller nuclei of comparable mass. **Binding energy per nucleon increases in a fission reaction**, e.g. a heavy nucleus, such as uranium, has a lower binding energy per nucleon than the lighter products of its fission, e.g. barium and krypton (see Figure **C.1**).

> **hint**
> In hydrogen, promotion of an electron from n=1 requires the absorption of **ultraviolet** light.

The typical fuel in a fission reactor is uranium-235. When bombarded with neutrons, it becomes unstable and splits into two smaller nuclei. The increase in binding energy per nucleon causes the reaction to release a lot of energy in the form of heat.

$${}^{235}_{92}U + {}^{1}_{0}n \rightarrow {}^{236}_{92}U \rightarrow {}^{141}_{56}Ba + {}^{92}_{36}Kr + 3{}^{1}_{0}n$$

> **hint**
> The mass numbers and atomic numbers must balance on each side of the equation.

Uranium does not always split into the same two nuclei. You may be asked to write nuclear equations for the fission of uranium-235 into any two smaller nuclei.

Neutrons released in the reaction are able to cause the fission of further uranium-235 nuclei which releases more neutrons which causes more fission. This is called a **chain** reaction (Figure **C.3**).

If only a small quantity of uranium-235 is present, many neutrons escape from the sample without causing fission – a chain reaction does not occur. Above a certain mass of uranium-235 – the **critical mass** – there is enough material to ensure that at least one neutron from each fission process is captured by another uranium nucleus to cause another fission process and a chain reaction occurs – the reaction is self-sustaining.

TEST YOURSELF C.5

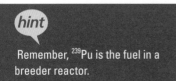

Complete and balance the following nuclear equations for the fission of uranium-235.

1 $^{235}_{92}U + ^{1}_{0}n \rightarrow ^{236}_{92}U \rightarrow ^{134}_{54}Xe + ^{Z}_{A}X + 3^{1}_{0}n$ 99 38 Sr

2 $^{235}_{92}U + ^{1}_{0}n \rightarrow ^{236}_{92}U \rightarrow ^{100}_{44}Ru + ^{Z}_{A}X + 3^{1}_{0}n$ 133 38 Cd

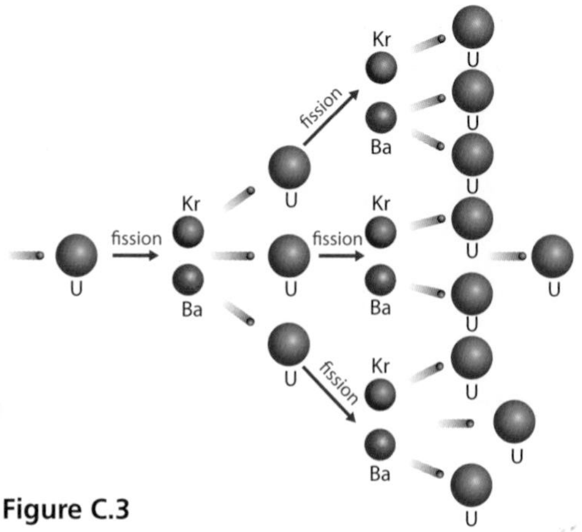

Figure C.3

hint

Remember, ^{239}Pu is the fuel in a breeder reactor.

Plutonium and breeder reactors

A breeder reactor is a nuclear reactor that produces more fissionable material than it consumes.

A naturally-occurring sample of uranium contains mostly ^{238}U, which does not undergo fission. In a breeder reactor this can be converted (by bombardment with fast neutrons) to ^{239}Pu, which does undergo fission.

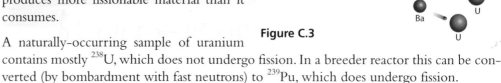

$$^{238}_{92}U + ^{1}_{0}n \rightarrow ^{239}_{92}U \rightarrow ^{239}_{94}Pu + 2^{0}_{-1}e$$

Problems associated with nuclear power

Health issues	Exposure to radioactivity can cause radiation poisoning, cancer and birth defects in children.
Meltdown	If the chain reaction becomes uncontrollable or cooling systems malfunction, the core can overheat and fuel rods may melt causing radioactive material to escape into the environment.
Nuclear weapons	Plutonium from breeder reactors and fuel reprocessing can be used to construct nuclear weapons that have the potential to cause mass destruction and severe loss of life.

Radioactive waste

A major problem of nuclear power is nuclear waste.

Low-level waste	High-level waste
Lower activity – usually contains isotopes with shorter half-lives.	Higher activity – usually contains isotopes with longer half-lives.
Items that have been contaminated with radioactive material or have been exposed to radioactivity, e.g. gloves, tools.	For example, spent fuel rods.
May be stored on site until it has decayed or buried underground (near surface disposal) in individual concrete canisters.	Stored in cooling ponds before being buried deep underground in thick concrete containers.

Half-life

DEFINITION

The **HALF-LIFE** ($t_{1/2}$) of a radioisotope is the time taken for half of the radioactive nuclei present in a sample to decay; the number of radioactive nuclei present halves.

Consider a sample of radioactive ^{32}P (half-life 14.3 days). If we start with x g of ^{32}P then after 14.3 days we would have $x/2$ g remaining. After another 14.3 days we would have $x/4$ g etc.

$$\begin{array}{ccccc} & 14.3\text{ d} & 28.6\text{ d} & 42.9\text{ d} & 57.2\text{ d} \\ x/\text{g} \rightarrow & x/2\text{ g} \rightarrow & x/4\text{ g} \rightarrow & x/8\text{ g} \rightarrow & x/16\text{ g}\end{array}$$

hint

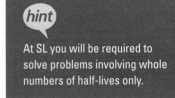

At SL you will be required to solve problems involving whole numbers of half-lives only.

Worked example C.1

A sample of radioactive ^{22}Na has an initial mass of 3.4 g. Calculate the mass of ^{22}Na remaining after it has decayed through three half-lives.

The radioactive mass halves three times and therefore $(½)^3$ (i.e. one-eighth) of the starting mass remains.

Mass of ^{22}Na remaining = original mass $\times (½)^3 = 3.4 \times 0.125 = 0.425$ g

Alternatively: $3.4\text{ g} \xrightarrow{\text{half-life}} 1.7\text{ g} \xrightarrow{\text{half-life}} 0.85\text{ g} \xrightarrow{\text{half-life}} 0.425\text{ g}$

Model answer C.4

A 2.40 mg sample of neodymium-147 decays to 0.15 mg in 44 days. Calculate the half-life of neodymium-147.

$2.40\text{ mg} \xrightarrow{\text{half-life}} 1.20\text{ mg} \xrightarrow{\text{half-life}} 0.60\text{ mg} \xrightarrow{\text{half-life}} 0.30\text{ mg} \xrightarrow{\text{half-life}} 0.15\text{ mg}$

This will take four half-lives, therefore 4 half-lives = 44 days and the half-life is 11 days.

TEST YOURSELF C.6

1 Uranium-235 decays with a half-life of 7.03×10^8 years. How long would it take for the radioactivity in 16.0 kg of uranium-235 to decay to 250 g? 4.22 x 10^9 years

2 120 g of radioactive ^{141}Ba decays to 15 g in 38.1 days. Determine the half-life of ^{141}Ba. 12.7 days

C.4 Solar energy in biological systems

Photosynthesis

DEFINITION

CONJUGATED SYSTEM a series of alternating single and double bonds in a molecule.

The conjugated system is highlighted in blue.

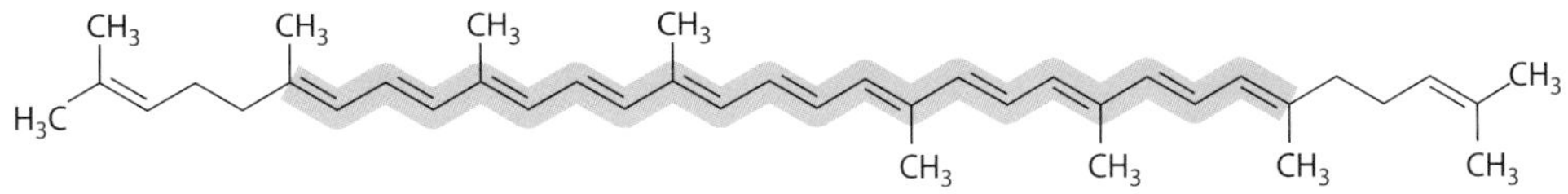

Figure C.4

In order to absorb light in the visible region of the electromagnetic spectrum the molecule must contain a sufficiently long conjugated system. Molecules such as chlorophylls and carotenoids, contain an **extensive conjugated system** and can **absorb visible light**.

hint

See Option B (Biochemistry) for the molecular structures of these two types of molecule.

Energy from the Sun's light is used in **photosynthesis** to produce carbohydrates such as glucose:

$$6CO_2(g) + 6H_2O(l) \rightarrow C_6H_{12}O_6(aq) + 6O_2(g)$$

Photosynthesis therefore converts light energy into chemical energy.

Biofuels

The sugar/carbohydrate produced in photosynthesis can be fermented using yeast to make ethanol.

$$C_6H_{12}O_6(aq) \rightarrow 2C_2H_5OH(aq) + 2CO_2(g)$$

Ethanol made via the fermentation of glucose from sugar cane can be used as a **biofuel**.

Transesterification

In transesterification, an ester is mixed with an alcohol and either a strong base (e.g. NaOH) or strong acid (e.g. H_2SO_4) catalyst – the part of the ester from the alcohol is exchanged with the new alcohol:

e.g.

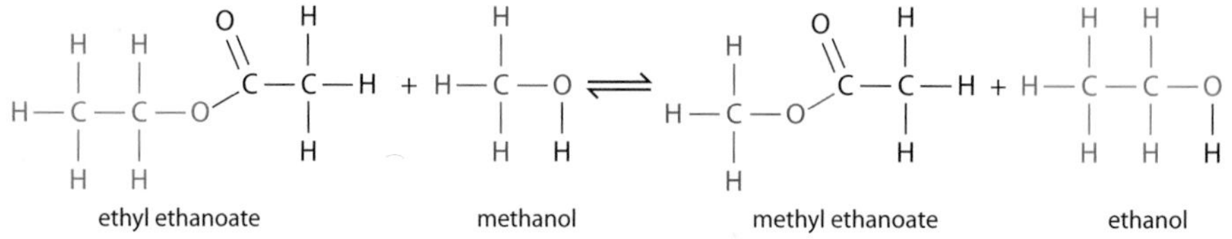

Figure C.5

Transesterification of a triglyceride molecule (e.g. in vegetable oil) with methanol or ethanol yields a mixture of methyl/ethyl esters, glycerol and some free fatty acids.

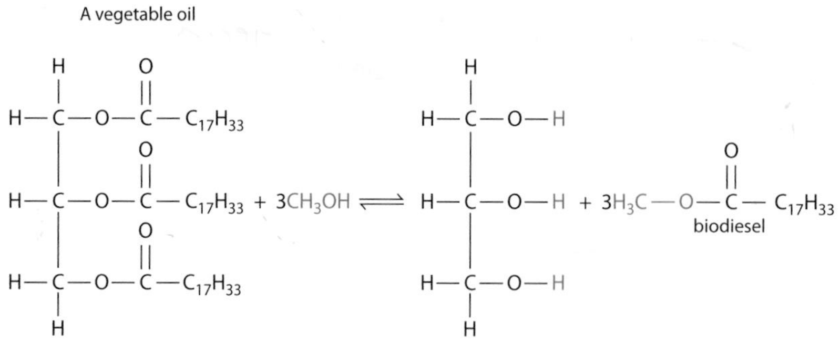

Figure C.6

Biodiesel

Vegetable oils contain a similar amount of energy per gram as conventional diesel fuel. However, vegetable oils are not used in internal combustion engines due to their very **high viscosity** (they are resistant to flowing). **Transesterification** turns the triglyceride molecules (vegetable oil) into methyl and ethyl esters which have a much lower viscosity than the original oil – these are used as **biodiesel**.

Biodiesel molecules have lower viscosity than vegetable oils because they have lower relative molecular masses and therefore weaker London forces between molecules. The weaker forces between the molecules allow them to flow more easily.

TEST YOURSELF C.7

1 Complete the equations for the following transesterification reactions:

a $C_4H_9COOC_2H_5 + C_3H_7OH \rightleftharpoons$ $C_4H_9COOC_3H_7 + C_2H_5OH$

b $CH_3CH(CH_3)COOCH_3 + CH_3CH(OH)CH_3 \rightleftharpoons$ $CH_3CH(CH_3)COOCH(CH_3)_2 + CH_3OH$

2 Draw structural formulae for the esters produced in the following transesterification reactions:

a $CH_3CH_2COOC_2H_5 + CH_3OH \rightleftharpoons$

b $CH_3COOCH_3 + CH_3CH_2CH(OH)CH_3 \rightleftharpoons$

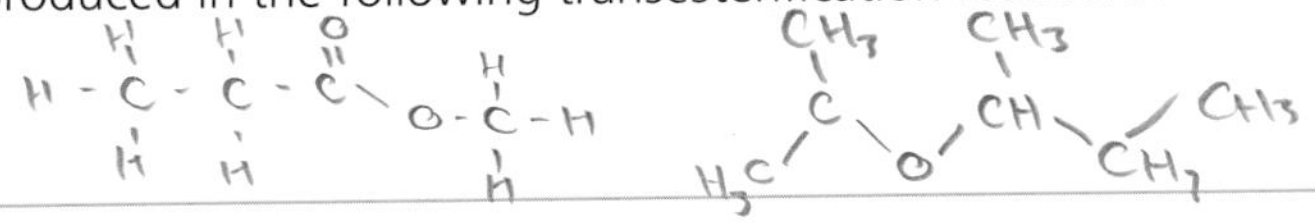

Advantages and disadvantages of biofuels

Advantages	Disadvantages
Produced from renewable feedstocks.	More expensive then petroleum-derived fuels.
Contain no sulfur so do not produce sulfur dioxide when burnt.	Land used to grow biofuel crops not used to produce food crops – creates food shortages and a rise in food prices.

Biofuels are sometimes described as **carbon neutral** – the carbon atoms in the biofuel came originally from CO_2 in the atmosphere (photosynthesis) and only these are put back into the atmosphere when the fuel is burnt – there is no net change in the number of CO_2 molecules in the atmosphere.

C.5 Environmental impact – global warming

The greenhouse effect

- Short-wavelength radiation from the Sun passes through the atmosphere without being absorbed.
- The planet's surface is warmed and emits longer wavelength infrared (IR) radiation.
- Some of this IR radiation is absorbed by greenhouse gases such as CO_2.
- Some absorbed radiation is re-emitted into the atmosphere.
- Atmospheric temperature is warmer than without the effect.

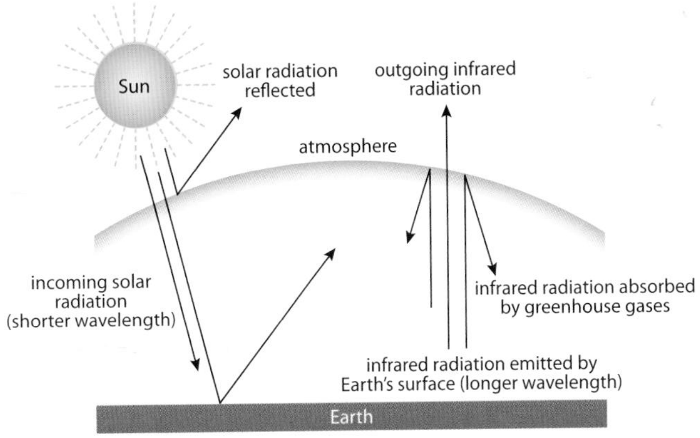

Figure C.7

Absorption of infrared radiation

All molecules vibrate but different molecules vibrate in different ways. Molecules with more than two atoms have different vibrational modes. A molecule will absorb infrared radiation if it possesses a **vibrational mode** (bend or stretch) which leads to a **change in the dipole moment** (polarity) of the molecule. Absorption of IR radiation promotes the molecule from a lower to a higher vibrational energy level.

Greenhouse gases (e.g. water vapour, carbon dioxide and methane) absorb IR radiation because they possess bending and/or stretching vibrations that lead to a change in dipole moment:

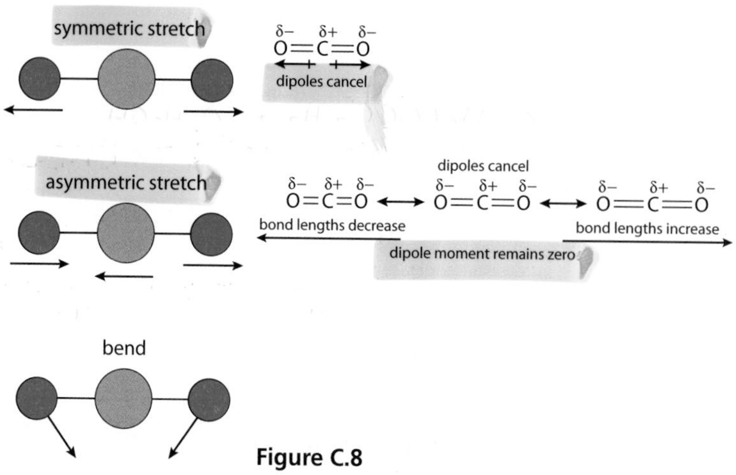

Figure C.8

Molecules such as O_2 and N_2 are non-polar and as they vibrate they remain non-polar (no change in dipole moment) – they do not absorb IR radiation.

Annotated exemplar answer C.1

Explain why molecules such as CO_2 and H_2O absorb infrared radiation. **[3]**

When the molecule vibrates it becomes polar and so can absorb radiation.

You will get a mark for mentioning vibration, but you also need to mention stretching and bending modes to get the second mark.

'becomes polar' will not gain a mark as H_2O is already polar – 'change in dipole moment' is the key term.

1/3

TEST YOURSELF C.8

Identify which of the following molecules are able to absorb infrared radiation.

H_2 HF Cl_2 CH_3Cl SO_2 H_2O Br_2 CO

Evidence for the effect of greenhouse gases on global warming

- Atmospheric CO_2 levels were fairly constant between the end of the last ice age and the start of the industrial revolution.
- In the last 250 years, CO_2 levels have risen 200 times faster.
- Average global temperatures rose by 0.7 °C in the last 100 years.

Is this coincidence or a causal effect?

- Computer models tested against past data provide compelling evidence for a causal relationship.
- Solar and volcanic activity have been proposed as alternatives.

Sources and effects of greenhouse gases

Greenhouse gas	Source	Relative effect	Explanation
Water vapour	Evaporation from oceans and lakes	Largest natural effect	Poor absorber of IR radiation but present in huge quantities compared to other greenhouse gases. Levels not influenced by humans to any real extent.
Carbon dioxide	Burning of fossil fuels and biomass	Largest anthropogenic effect	Weak absorber of IR radiation; large quantities present.
Methane	Anaerobic decomposition of vegetation; animals in agriculture	Smaller effect than CO_2	Good absorber of IR radiation; lower abundance than CO_2.

Controlling carbon dioxide emissions

There are many ways to control carbon dioxide emissions:

- Reduce burning of fossil fuels and move to cleaner or renewable energy sources.
- Reduce deforestation – more trees to absorb excess CO_2.
- Carbon capture and storage – CO_2 is piped underground for storage. CO_2 can also be reacted with minerals to make insoluble carbonates (stored underground).

hint

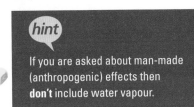

If you are asked about man-made (anthropogenic) effects then **don't** include water vapour.

Global dimming

Global dimming occurs when solar radiation is reflected back into space by particulates such as soot, ash or dust in the atmosphere as well as by clouds. Global dimming is responsible for a decrease in global temperatures.

Ocean acidification

The following heterogeneous (different physical states) equilibrium exists between the atmosphere and the oceans:

$$CO_2(g) \rightleftharpoons CO_2(aq)$$

Rising atmospheric levels of CO_2 shift this equilibrium to the right.

Dissolved carbon dioxide in the oceans reacts with water to form carbonic acid which then dissociates:

$$CO_2(aq) + H_2O(l) \rightleftharpoons H_2CO_3(aq) \qquad 1$$

$$H_2CO_3(aq) \rightleftharpoons HCO_3^-(aq) + H^+(aq) \qquad 2$$

$$HCO_3^-(aq) \rightleftharpoons CO_3^{2-}(aq) + H^+(aq) \qquad 3$$

An increase in $[CO_2(aq)]$ causes equilibrium 1 to shift to right – this leads to an increase in the concentration of H_2CO_3 – this causes equilibrium 2 to shift to the right (to use up the H_2CO_3), which increases the concentration of H^+ in the oceans causing a decrease in pH. Acidification may lead to the extinction of many marine species.

C.6 Electrochemistry, rechargeable batteries and fuel cells (HL only)

Batteries

DEFINITION

A **BATTERY** is a portable electrochemical source made up of one or more voltaic (galvanic) cells connected in series.

hint

Make sure you refer back to Chapter 9 for detailed information on voltaic cells.

Voltaic cells

In voltaic cells, charge is carried by free moving ions, which take a **finite time to diffuse** through the cell to the electrodes – this causes **internal resistance**, which limits the maximum current.

The maximum **voltage** that a cell can generate depends on the nature of the materials from which it is constructed (see examples later).

The total amount of work (energy) that can be obtained depends on the quantity of these materials.

Types of cell

Primary cells: cells in which the redox reaction is non-reversible. They cannot be recharged.

Rechargeable cells: cells in which the redox reaction can be reversed using an input of electricity allowing the cell to be used multiple times.

Fuel cells: cells that have a continuous supply of reactants from an external source. Chemical energy contained in a fuel, e.g. hydrogen, is converted directly to electricity using an oxidising agent, e.g. oxygen.

Fuel cells

Typical fuels used in fuel cells are hydrogen and methanol, with oxygen gas as the oxidising agent. The electrolyte between the two electrodes may be alkaline (e.g. KOH) or acidic (e.g. H_3PO_4), involving the movement of hydroxide ions or protons respectively. Current is generated by the flow of the electrons from the anode (–) through an external circuit to the cathode (+).

hint

You need to know the half-equations occurring at both electrodes in all three fuel cells mentioned in this chapter.

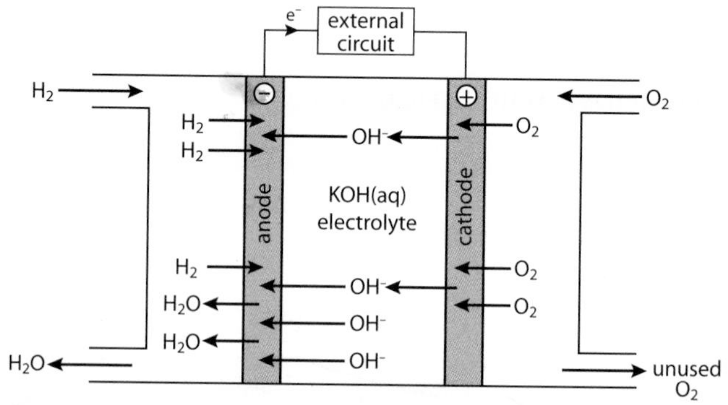

Figure C.9

Type of fuel cell	Hydrogen	Methanol
Oxidation half equation (anode)	Alkaline: $H_2(g)+2OH^-(aq) \rightarrow 2H_2O(l)+2e^-$ Acidic: $H_2(g) \rightarrow 2H^+(aq)+2e^-$	$CH_3OH(aq)+H_2O(l) \rightarrow CO_2(g)+6H^+(aq)+6e^-$
Reduction half equation (cathode)	Alkaline: $\frac{1}{2}O_2(g)+H_2O(l)+2e^- \rightarrow 2OH^-(aq)$ Acidic: $\frac{1}{2}O_2(g)+2H^+(aq)+2e^- \rightarrow H_2O(l)$	$\frac{3}{2}O_2(g)+6H^+(aq)+6e^- \rightarrow 3H_2O(l)$
Overall equation	$H_2(g)+\frac{1}{2}O_2(g) \rightarrow H_2O(l)$	$CH_3OH(aq)+\frac{3}{2}O_2(g) \rightarrow CO_2(g)+2H_2O(l)$

Proton-exchange membrane (PEM) fuel cells

Hydrogen gas passes through a porous carbon anode containing platinum where it is oxidised to H^+ ions. The electrons produced flow through the external circuit and the protons pass through a proton-exchange membrane (PEM) to the cathode (also contains platinum) combining with oxygen and electrons to form water.

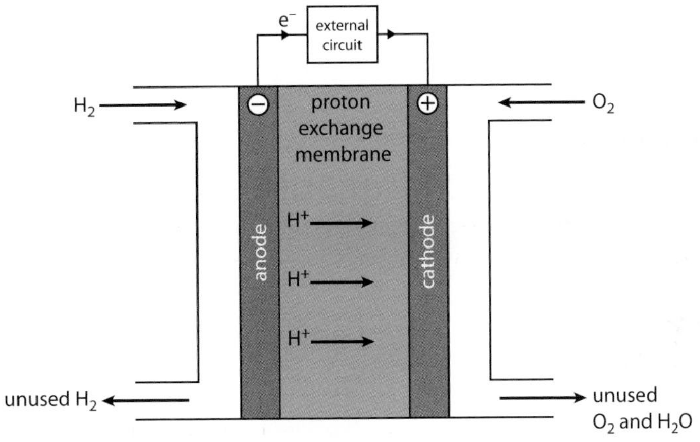

Figure C.10

Microbial fuel cells (MFCs)

The MFC is essentially a PEM fuel cell where bacteria, such as *Geobacter* species, are attached to the anode. Carbohydrates and other substrates present in waste water are oxidised by the bacteria at the anode to carbon dioxide, protons and electrons (which flow through the external circuit to the cathode).

hint

The PEM fuel cell is a type of hydrogen fuel cell.

For example, ethanoate ions can be oxidised by bacteria:

anode: $CH_3COO^-(aq) + 2H_2O(l) \rightarrow 2CO_2(g) + 7H^+(aq) + 8e^-$

cathode: $2O_2(g) + 8H^+(aq) + 8e^- \rightarrow 4H_2O(l)$

Overall: $CH_3COOH(aq) + 2O_2(g) \rightarrow 2CO_2(g) + 2H_2O(l)$

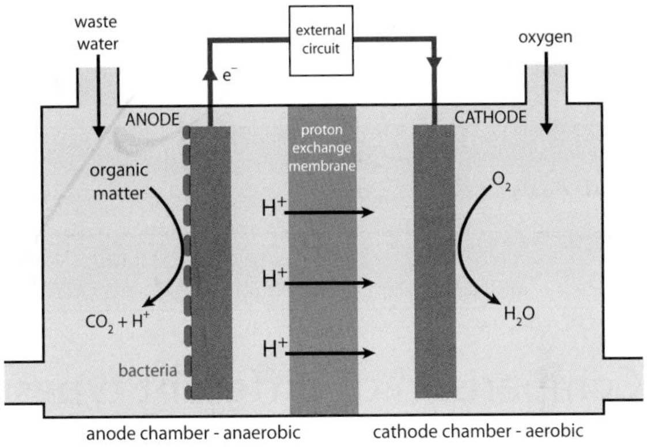

Figure C.11

Thermodynamic efficiency of a fuel cell

DEFINITION

THERMODYNAMIC EFFICIENCY $= \frac{\Delta G}{\Delta H}$

The value for ΔG can be calculated using the equation: $\Delta G^\ominus = -nFE^\ominus$ (see Chapter 9).

Model answer C.5

A hydrogen fuel cell has a maximum voltage of 1.23 V. The standard enthalpy change of combustion of hydrogen is –286 kJ mol⁻¹. Use these data to calculate the thermodynamic efficiency of a hydrogen fuel cell.

$\Delta G^\ominus = -nFE^\ominus = -2 \times 96\,500 \times 1.23 = -237\,000 \text{ J mol}^{-1} = -237 \text{ kJ mol}^{-1}$

Thermodynamic efficiency $= \frac{\Delta G}{\Delta H} = \frac{-237}{-286} = 0.829$

Converted to a percentage, the thermodynamic efficiency of a hydrogen fuel cell is **82.9%**.

hint

Thermodynamic efficiency can be greater than 100%.

TEST YOURSELF C.9

Calculate the thermodynamic efficiency for a methanol-oxygen fuel cell using data from the *IB Data Booklet* and given that the maximum voltage of the cell is 1.21 V.

96.5%

Rechargeable batteries

Type of battery	
Oxidation half equation (anode)	**Nickel-cadmium:** $Cd(s)+2OH^-(aq) \rightarrow Cd(OH)_2(s)+2e^-$ **Lead-acid:** $Pb(s)+SO_4^{2-}(aq) \rightarrow PbSO_4(s)+2e^-$ **Lithium-ion:** $Li_xC_6 \rightarrow xLi^+ + xe^- + 6C$
Reduction half equation (cathode)	**Nickel-cadmium:** $2NiO(OH)(s)+2H_2O(l)+2e^- \rightarrow 2Ni(OH)_2(s)+2OH^-(aq)$ **Lead-acid:** $PbO_2(s)+4H^+(aq)+SO_4^{2-}(aq)+2e^- \rightarrow PbSO_4(s)+2H_2O(l)$ **Lithium-ion:** $CoO_2+xLi^+ + xe^- \rightarrow Li_xCoO_2$
Overall equation	**Nickel-cadmium:** $Cd(s)+2NiO(OH)(s)+2H_2O(l) \rightarrow Cd(OH)_2(s)+2Ni(OH)_2(s)$ **Lead-acid:** $Pb(s)+PbO_2(s)+4H^+(aq)+2SO_4^{2-}(aq) \rightarrow 2PbSO_4(s)+2H_2O(l)$ **Lithium-ion:** $Li_xC_6+CoO_2 \rightarrow Li_xCoO_2+6C$

Fuel cells versus rechargeable batteries

Similarities	Differences
Both convert chemical energy to electrical energy.	Fuel cells have an external source of fuel; rechargeable batteries contain the energy source.
Both comprise an anode, cathode and an electrolyte.	Fuel cells will not run out if the supply of fuel and oxidant is maintained. Rechargeable batteries have to be recharged by connecting to a supply of electricity.
Both use separated reduction and oxidation reactions to drive the flow of electrons through an external circuit.	Fuel cells can generate more electricity.

Comparison of different types of cell

Cell type	Size/mass	Voltage
Fuel Cells	Size depends on use; single cells are light but stacks can be heavy and bulky; high power-to-mass ratio.	0.6–0.8 V per cell – used in a stack with several joined in series
Lead-acid batteries	Large and heavy – not portable; low power-to-mass ratio	2 V per cell – usually used as a battery with 6 cells in series
Nickel-cadmium batteries	Small and light – used in portable devices. Intermediate power-to-mass ratio	1.2 V per cell
Lithium-ion batteries	Small and light – used in a variety or portable devices. Highest power-to-mass ratio of rechargeable batteries.	about 3.7 V per cell

hint

You will need to learn all of the half-equations in this section and examples for the uses of each type of battery.

The effect of concentration on electrode potentials

The Nernst equation

The Nernst equation is used to determine the electrode potential of a half-cell when the conditions are non-standard.

$$E = E^{\ominus} - \left(\frac{RT}{nF}\right) \ln Q$$

E is the electrode potential

$E^{\ominus}$ is the standard electrode potential of the cell

R is the gas constant ($8.31\ J\ K^{-1}\ mol^{-1}$)

T is the temperature in kelvin

n is the number of electrons transferred

F is the Faraday constant ($96\ 500\ C\ mol^{-1}$)

Q is the reaction quotient (see Chapter 7).

hint

The Nernst equation is in the *IB Data Booklet.*

hint

The concentration of a pure solid is 1.

Worked example C.2

Calculate the cell potential, at 298 K, when a $Cu^{2+}|Cu$ half-cell with $[Cu^{2+}(aq)] = 0.10\ mol\ dm^{-3}$ is connected to a $Mg^{2+}|Mg$ half-cell with $[Mg^{2+}(aq)] = 0.25\ mol\ dm^{-3}$.

$Cu^{2+}(aq) + 2e^{-} \rightarrow Cu(s) \quad E^{\ominus} = +0.34\ V$

2 electrons are transferred, therefore $n = 2$

$$= \frac{[Cu(s)]}{[Cu^{2+}(aq)]} = \frac{1}{0.10} = \mathbf{10}$$

Substitute into the equation with the other known values:

$$E = E^{\ominus} - \left(\frac{RT}{nF}\right)\ln Q = 0.34 - \left(\frac{8.31 \times 298}{2 \times 96\ 500}\right)\ln 10 = \mathbf{+0.31\ V}$$

$Mg^{2+}(aq) + 2e^{-} \rightarrow Mg(s) \quad E^{\ominus} = -2.37\ V$

$$= \frac{[Mg(s)]}{[Mg^{2+}(aq)]} = \frac{1}{0.25} = \mathbf{4}$$

$$E = E^{\ominus} - \left(\frac{RT}{nF}\right)\ln Q = -2.37 - \left(\frac{8.31 \times 298}{2 \times 96\ 500}\right)\ln 4 = \mathbf{-2.39\ V}$$

The more negative electrode potential is reversed (see Chapter 9)

Cell potential, $E_{cell} = +0.31 + 2.39 = \mathbf{+2.70\ V}$

TEST YOURSELF C.10

1 Use data from the *IB Data Booklet* to determine the electrode potential for the following half-cells at 298 K:

  a $Zn^{2+}|Zn$ half-cell; $[Zn^{2+}] = 0.1\ mol\ dm^{-3}$ - 0.79V

  b $MnO_4^{-}|Mn^{2+}$ half-cell; $[MnO_4^{-}] = 0.3\ mol\ dm^{-3}$ and $[Mn^{2+}] = 0.9\ mol\ dm^{-3}$ +1.50V

2 Determine the concentration of aqueous ions in the following non-standard cells at 298 K:

  a $Fe^{2+}(aq)$ in an $Fe^{2+}|Fe$ half-cell coupled to a standard $Cu^{2+}|Cu$ half-cell; $E_{cell} = +0.77\ V$ 4.75

  b $Pb^{2+}(aq)$ in a $Pb^{2+}|Pb$ half-cell coupled to a standard $Ni^{2+}|Ni$ half-cell; $E_{cell} = +0.11\ V$ 0.210

Concentration cells

It is possible to create a voltaic cell using two half-cells containing the same materials as long as the concentration of the electrolyte is different in each – this is a **concentration cell**.

Consider a cell made from the following $Cu^{2+}|Cu$ half-cells:

$Cu^{2+}|Cu$ half-cell with $[Cu^{2+}(aq)] = 1.00\ mol\ dm^{-3}$ $\qquad E^{\ominus} = +0.34\,V$

$Cu^{2+}|Cu$ half-cell with $[Cu^{2+}(aq)] = 0.10\ mol\ dm^{-3}$ $\qquad E^{\ominus} = +0.31\,V$ (calculated using the Nernst equation)

The half-cell with the less positive (more negative) value undergoes oxidation and the sign of $E^{\ominus}$ is changed. The overall cell potential in this case is:

$E_{cell} = +0.34 - 0.31 = +0.03V$

TEST YOURSELF C.11

Use the Nernst equation and the *IB Data Booklet* to determine the electrode potentials of the following concentration cells at 298 K:

1 an $Fe^{2+}|Fe$ standard half-cell coupled to an $Fe^{2+}|Fe$ half-cell with $[Fe^{2+}(aq)] = 0.01\ mol\ dm^{-3}$ 0·06

2 a $Zn^{2+}|Zn$ half-cell with $[Zn^{2+}(aq)] = 0.2\ mol\ dm^{-3}$ coupled to a $Zn^{2+}|Zn$ half-cell with $[Zn^{2+}(aq)] = 0.05\ mol\ dm^{-3}$ 0·02

C.7 Nuclear fusion and nuclear fission (HL only)

Nuclear binding energy and mass defect

DEFINITIONS

The **MASS DEFECT (ΔM)** is the difference between the mass of a nucleus and the sum of the masses of the individual nucleons (neutrons and protons).

NUCLEAR BINDING ENERGY (ΔE) is the energy required to break apart a nucleus into individual protons and neutrons.

The **UNIFIED ATOMIC MASS UNIT (u)** is equivalent to $\frac{1}{12}$ of the mass of a carbon-12 atom.

hint

amu may be used in the exam instead of u.

The masses of subatomic particles in atomic mass units are:

Particle	Mass/u
proton	1.007276
neutron	1.008665

The **binding energy per nucleon** is worked out by dividing the nuclear binding energy by the number of nucleons – it provides a measure of the stability of a nucleus – the higher the binding energy per nucleon, the more stable the nucleus.

Worked example C.3

A boron-10 nucleus has a mass of 10.012937 u. Calculate the mass defect (Δm), nuclear binding energy (ΔE) and binding energy per nucleon for a boron-10 nucleus.

Boron-10 contains 5 protons and 5 neutrons

Mass of nucleons $= (5 \times 1.007276) + (5 \times 1.008665) = 10.079705$ u

Mass defect = mass of nucleons − mass of nucleus $= 10.079705 - 10.012937 =$ **0.066768 u**

Use **Einstein's mass–energy equivalence relationship**: $E = mc^2$

where c is the speed of light and m is the mass in kg (1 u = 1.66×10^{-27} kg).

Mass defect = $0.066768\times1.66\times10^{-27}$ = $\mathbf{1.11\times10^{-28}}$ **kg**

The mass has been converted to energy, which is released mainly as heat.

Nuclear binding energy = $\Delta m\times c^2 = 1.11\times10^{-28}\times(3.00\times10^8)^2 = \mathbf{9.98\times10^{-12}}$ **J**

There are 10 nucleons therefore the binding energy per nucleon = $9.98\times10^{-12}/10$ = $\mathbf{9.98\times10^{-13}}$ **J**

TEST YOURSELF C.12

Determine the mass defect (in u), nuclear binding energy and binding energy per nucleon for the following nuclei:

1 ^{12}C 0.098938, 1.48×10^{-11}, 1.23×10^{-12}
2 ^{79}Br 0.736784, 1.10×10^{-10}, 1.39×10^{-12}
3 ^{235}U 1.915048, 2.86×10^{-10}, 1.22×10^{-12}

Particle	Mass/u
proton	1.007276
neutron	1.008665
^{12}C nucleus	11.996708
^{79}Br nucleus	78.899136
^{235}U nucleus	234.993439

Calculating the energy released during fusion and fission reactions

You need to determine the difference in the mass of the nuclei of the reactants and products (Δm). In fusion and fission reactions, the mass of the nuclei of the products will be lower than that of the reactants. The lost mass is converted to energy according to $E = mc^2$.

Annotated exemplar answer C.2

Calculate the energy released, in kJ mol^{-1}, during the fusion of two deuterium (^{2}H) nuclei according to the equation:

$^2_1H + ^2_1H \rightarrow ^3_2He + ^1_0n$ **[3]**

Particle	Mass /u
^{2}H nucleus	2.013553
^{3}He nucleus	3.014932
neutron	1.008665

1/3

Δm = mass of products − mass of reactants

= (3.014932+1.008665) − (2×2.013553) = −0.003509 u

Mass must be converted to kg – multiply by 1.66×10^{-27} – should be 5.82494×10^{-30} kg

Answer should be 5.24×10^{-13} J

$E = mc^2 = 0.003509\times(3.00\times10^8)^2 = 3.16\times10^{14}$ J

Think about the size of your answers – this is a lot more energy than the Sun produces each year!

$E = 3.16\times10^{14}\times6.02\times10^{23} = 1.90\times10^{38}$ J mol^{-1} = 1.90×10^{41} kJ mol^{-1}

The answer must be divided by 1000 to convert to kJ – not multiplied.

Final answer should be 3.16×10^8 kJ mol^{-1}

The method for working out the energy released in a fission reaction is identical to that for the fusion reaction shown previously.

TEST YOURSELF C.13

1 Calculate the energy released, in kJ mol^{-1}, in the following reactions.

a $^{8}_{4}Be + ^{4}_{2}He \rightarrow ^{12}_{6}C$ 2·37×10^8

b $^{12}_{6}C + ^{4}_{2}He \rightarrow ^{12}_{6}O$ 2·30×10^8

Particle	Mass/u
^{4}He nucleus	4.001506
^{8}Be nucleus	8.003111
^{12}C nucleus	11.996708
^{16}O nucleus	15.990556

2 Calculate the energy released, in kJ mol^{-1}, for the following reaction.

$^{239}_{94}Pu + ^{1}_{0}n \rightarrow ^{100}_{42}Mo + ^{134}_{52}Te + 6^{1}_{0}n$

5·70×10^9

Particle	Mass/u
neutron	1.008665
^{239}Pu nucleus	239.000595
^{100}Mo nucleus	99.884436
^{134}Te nucleus	133.882842

Radioactive decay and half-life

Radioactive decay is a first order kinetic process – it has constant half-life, $t_{½}$.

hint

Equations relevant to this subtopic are given in the *IB Data Booklet*.

Half-life is related to the decay constant, λ, by the equation: $\lambda = \frac{\ln 2}{t_{½}}$

The number of undecayed nuclei remaining at any time (N) can be worked out using the equation: $N = N_0 e^{-\lambda t}$

where N_0 is the initial number of undecayed nuclei.

Worked example C.4

Plutonium-239 has a half-life of 24 100 years. What percentage of the original radioactive material remains after 100 000 years?

Determine the value of λ: $t_{½} = \frac{\ln 2}{\lambda}$ Therefore: $\lambda = \frac{\ln 2}{t_{½}}$

$$\lambda = \frac{0.693}{24\,100} = 2.88 \times 10^{-5}\,y^{-1}$$

The proportion of material remaining is equal to $\frac{N}{N_0}$

$N = N_0 e^{-\lambda t}$ therefore: $\frac{N}{N_0} = e^{-\lambda t} = e^{-2.88 \times 10^{-5} \times 100000} = 0.056$

Therefore, the percentage of material remaining after 100 000 years is $0.056 \times 100 =$ **5.6%**.

TEST YOURSELF C.14

1 Calculate the proportion of radioactive nuclei remaining in each of the following samples:
 a ^{32}P (half-life = 14.3 d) after 30 days 0.234
 b ^{235}U (half-life = 7.04×10^8 y) after 4.54×10^9 years 0.0114
 c ^{289}Fl (half-life = 2.6 s) after 1 minute 1.15×10^{-7}
2 Determine after how long the following amounts of radioactive material will remain:
 a 10% of radioactive ^{36}Cl (half-life = 3.01×10^5 y) 1×10^6
 b 0.234 g of radioactive tritium, ^3H (half-life = 12.3 y; original mass 3.00 g) 45.3
3 Determine the half-life of the following radioisotopes:
 a 30% of radioactive material remains after 7 days 4.03
 b 8.0 g of radioactive material from an original mass of 50.0 g remains after 32 minutes 12.1

hint

Think about whether your answer is sensible – the time is just over four half-lives. In four half-lives the nuclei decay from 100% → 50% → 25% → 12.5% → 6.25%, so this answer looks about right.

hint

You could be asked to use the equations to work out the half-life of a substance or how long it would take for a certain amount of material to decay. If you are unsure about how to use logs to simplify equations then you could learn a modified version of the key equation:

$$\ln\left(\frac{N}{N_0}\right) = -\lambda t$$

Graham's law of effusion

DEFINITIONS

EFFUSION the process by which a gas escapes through a very small hole in a container.

GRAHAM'S LAW the rate of effusion of a gas is **inversely proportional** to the **square root** of the relative molecular mass of the gas.

At the same temperature, the average kinetic energies of two gases are the same. Given, kinetic energy = $\frac{1}{2}mv^2$, the particles of the lighter gas will, on average, be moving faster and therefore 'collide' more often with the hole and escape more quickly.

Graham's law can be applied to two gases in a container:

$$\frac{\text{rate of effusion of gas 1}}{\text{rate of effusion of gas 2}} = \sqrt{\frac{\text{molar mass of gas 2}}{\text{molar mass of gas 1}}}$$

hint

This equation is given in an abbreviated form in the *IB Data Booklet*

$$\frac{\text{rate}_1}{\text{rate}_2} = \sqrt{\frac{M_2}{M_1}}$$

☆ Model answer C.6

a A vessel with a small hole contains an equal number of moles of sulfur hexafluoride (SF_6) and ammonia (NH_3); what is the relative rate of effusion of ammonia compared to sulfur hexafluoride?

b A vessel with a small hole contains two noble gases, one of which is xenon. The xenon effuses from the vessel at 1.05 cm^3 h^{-1}. The other, unknown gas effuses from the vessel at 6.02 cm^3 h^{-1}. Deduce the identity of the second noble gas present in the vessel.

a $$\frac{\text{rate}_{NH_3}}{\text{rate}_{SF_6}} = \sqrt{\frac{M_{SF_6}}{M_{NH_3}}} = \sqrt{\frac{146.07}{17.04}} = 2.93$$

The rate of effusion of ammonia is 2.93 times faster than that of sulfur hexafluoride.

b $$\frac{\text{rate}_{\text{unknown}}}{\text{rate}_{Xe}} = \sqrt{\frac{M_{Xe}}{M_{\text{unknown}}}} \quad \frac{6.02}{1.05} = \sqrt{\frac{131.29}{M_{\text{unknown}}}} \quad M_{\text{unknown}} = \frac{131.29}{\left(\frac{6.02}{1.05}\right)^2} = 3.99$$

Relative molecular mass of the unknown gas is approximately 4, therefore the gas is **helium**.

TEST YOURSELF C.15

1 Calculate the relative rates of effusion of the following pairs of gases:

a Argon relative to neon. 0.712

b Chloromethane (CH_3Cl) relative to tetrachloromethane (CCl_4). 1.75

2 Calculate the rate of effusion of the named gas:

a Hydrogen; if methane contained in the same vessel effuses at a rate of 5.4×10^{-5} mol h^{-1}. 1.52×10^{-4}

b Silane, SiH_4; if hydrogen bromide in the same vessel effuses at a rate of 2.3 cm^3 h^{-1}. 3.65

3 Calculate the relative molecular mass of an unknown gas if:

a The relative rate of effusion of phosphine (PH_3) is 1.2 times greater. 48.96

b Ethane effuses from a vessel at a rate of 0.96 cm^3 h^{-1} and the unknown gas effuses from the same vessel at 1.04 cm^3 h^{-1}. 25.63

Uranium enrichment

Uranium has two main isotopes, ^{238}U (99.27% abundance) and ^{235}U (0.72% abundance). Only ^{235}U is suitable for fission reactions in a nuclear reactor. Uranium fuel rods require at least 3% ^{235}U to achieve a critical mass; nuclear weapons require >85% ^{235}U.

There are two enrichment techniques used to increase the percentage of ^{235}U in a sample of uranium. Both rely on the conversion of uranium ores containing UO_2 to the volatile UF_6. The UF_6 gas enriched in ^{235}U is then converted back into UO_2 to be turned into fuel rods for the reactors.

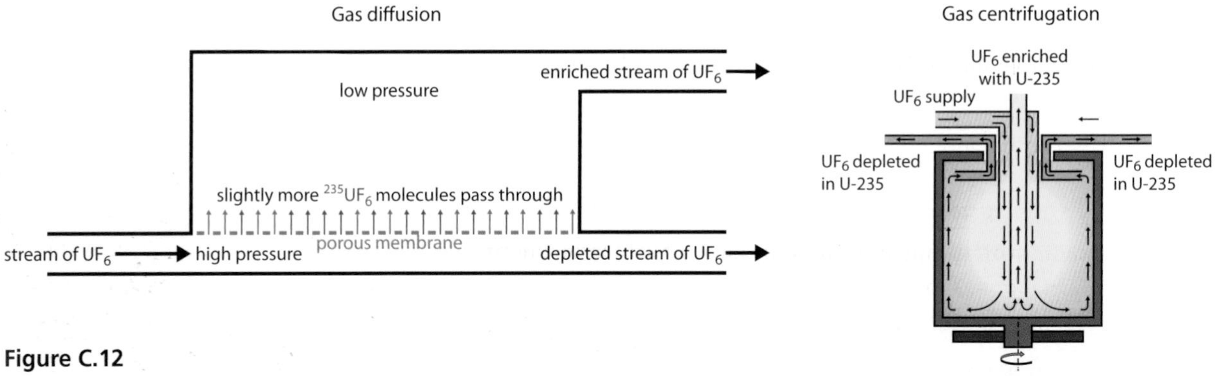

Figure C.12

Gas diffusion	Gas centrifugation
• UF_6 gas is forced at high pressure through a container with walls made of a porous membrane • $^{235}UF_6$ effuses through the pores slightly faster (1.004×) than $^{238}UF_6$ due to having a lower relative molecular mass • Effused gas is slightly richer in $^{235}UF_6$ than before • Process is repeated hundreds of times with effused gas to achieve enrichment of a few percent	• UF_6 is pumped into a large centrifuge spinning at high speed • The slightly heavier $^{238}UF_6$ molecules move towards the outside more than $^{235}UF_6$ • Gas slightly richer in $^{235}UF_6$ is found in the centre of the centrifuge and pumped out into a second centrifuge • The process is repeated many times more to produce gas enriched in ^{235}U by a few percent

Properties of UF_6 and UO_2

UF_6	UO_2
• White crystalline solid • Octahedral shaped molecule • Non-polar • Sublimes at 56 °C • Strong covalent bonds between uranium and fluorine atoms • Weak London forces between molecules • Little energy required to separate molecules from each other	• Dark brown crystalline solid • Giant ionic lattice • Very high melting point (>2800 °C) • Strong electrostatic forces of attraction between U^{4+} and O^{2-} ions • A lot of energy is required to break lattice apart

Dangers of nuclear energy

Nuclear fuel and nuclear waste are radioactive – they contain isotopes that can undergo radioactive decay by the emission of alpha particles, beta particles or gamma rays (called ionising radiation as they cause the formation of ions when they interact with matter).

Ionising radiation can kill living cells by damaging DNA and enzymes directly or by the production of highly reactive oxygen-containing free radicals (e.g. hydroxyl radical, HO•; superoxide radical, $•O_2^-$).

C.8 Photovoltaic cells and dye-sensitised solar cells (HL only)

Effect of conjugation on light absorption

In order for a molecule to be coloured it must absorb light in the visible region of the spectrum.

Molecules containing a conjugated system of alternating single and double bonds are able to absorb light in the ultraviolet-visible region when an electron moves from a low energy level to a higher energy level.

The longer the conjugated system the longer the wavelength of light absorbed.

Molecules with short conjugated systems have a fairly large energy gap between lower and upper energy levels and absorb UV light (appear colourless), but molecules with longer conjugated systems have a smaller energy gap and can absorb longer wavelengths of light in the visible region – they, therefore, appear coloured.

TEST YOURSELF C.16

For each pair of electrons, deduce which absorbs light of the longer wavelength.

1

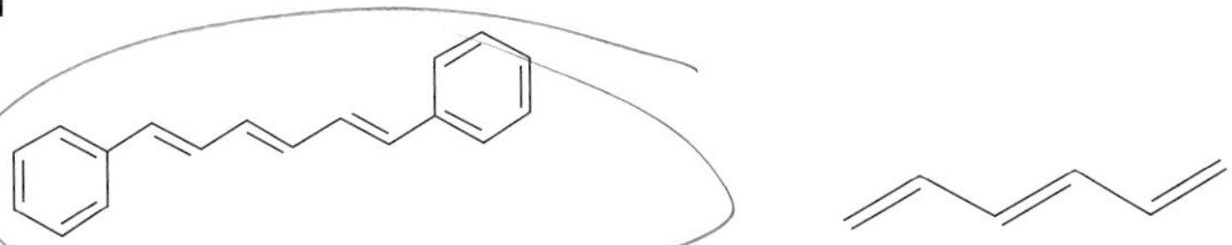

2

Solar cells

Semiconductors

A semiconductor is a solid with intermediate electrical conductivity between that of a full conductor (e.g. a metal) and an insulator. Semiconductors are able to conduct electricity under particular conditions such as high temperatures or in the presence of sunlight.

Metals – resistance increases as temperature increases.	Metals conduct electricity well because the delocalised electrons are free to move throughout the structure. Resistance in metals arises because these electrons collide with the positive ions in the lattice. As the temperature increases, the metal ions vibrate more and so there is essentially a larger cross-section for the electrons to collide with and the electrical conductivity decreases (resistance increases).
Semiconductors – resistance decreases as temperature increases.	At higher temperatures more electrons are promoted to the conduction band and are free to move (Figure **C.13**).

Metals tend to have lower ionisation energies than semiconductors in the same period. This means it requires less energy to remove electrons from metals to generate free-moving electrons. Semiconductors require more energy input to ionise and free the electrons.

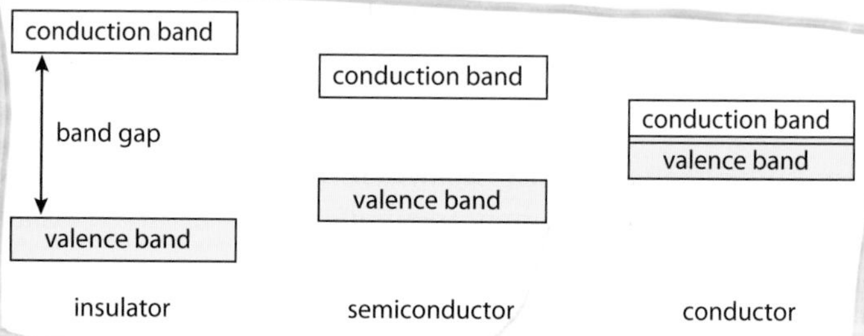

Figure C.13

Silicon doping

The semiconductor silicon can be modified by 'doping' in order to increase its electrical conductivity.

1 **Adding a Group 13 element**, e.g. boron: only three valence electrons are supplied instead of four, creating positive 'holes' in the valence band. Electrons are able to move into these holes and so electrical conductivity increases. As the charge-carriers are the **positive** holes, these materials are called **p-type semiconductors**.

2 **Adding a Group 15 element**, e.g. phosphorus: five valence electrons are supplied. Four are used in bonding and the fifth is promoted into the conduction band increasing the electrical conductivity. As the charge-carriers are **negative** electrons, these materials are called **n-type semiconductors**.

Photovoltaic cells

When an n-type and a p-type semiconductor are joined together a potential difference is built up at the junction between the two (Figure **C.14**).

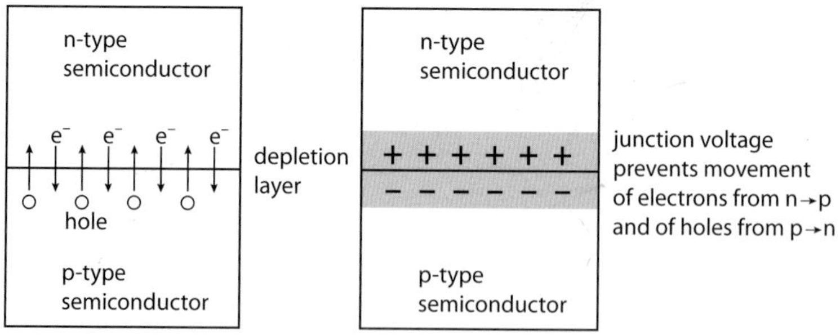

Figure C.14

When light hits the n-type semiconductor, electrons are promoted to the conduction band. These electrons are not able to move directly from the n-type to the p-type, because the voltage at the p–n junction prevents this. They must travel from the n-type to the p-type through an external circuit, and this flow of electrons can be used to power an electronic device.

Dye-sensitised solar cells (DSSCs)

A DSSC (Figure **C.15**) is designed to imitate the way biological molecules, such as chlorophyll, absorb and harness sunlight. In a typical Grätzel DSSC, a dye is used to absorb sunlight, promoting electrons to higher energy levels. These excited electrons are then 'injected' into the conduction band of a titanium(IV) oxide, TiO_2, semiconductor.

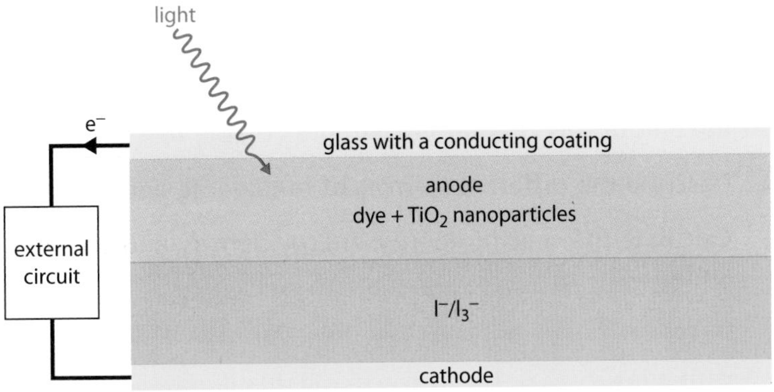

Figure C.15

1 The dye-sensitiser (D) absorbs energy from sunlight ($h\nu$) to promote an electron to a higher energy level:

$$D + h\nu \rightarrow D^{\star}$$

2 The promoted electron is passed to the conduction band of the TiO_2 semiconductor oxidising the dye-sensitiser:

$$D^{\star} \rightarrow D^{+} + e^{-} \qquad e^{-} \text{ passed to } TiO_2$$

3 The delocalised electrons flow from the TiO_2 through an external circuit to the cathode where they reduce I_3^- ions in the electrolyte:

$$I_3^- + 2e^- \rightarrow 3I^-$$

4 The dye-sensitiser is reduced to its original state by accepting electrons from the iodide ions, I^-, in the electrolyte:

$$2D^+ + 3I^- \rightarrow 2D + I_3^-$$

In a DSSC, dye-coated nanoparticles of TiO_2 are used. This increases the surface area available for light harvesting enormously and increases the cell's efficiency.

hint

You need to know the equations that explain how a DSSC works.

Photovoltaic cells versus DSSCs

DSSCs have a number of advantages over photovoltaic cells:

- The materials to make a DSSC are more abundant and cheaper so the technology should be less expensive.
- DSSC efficiency is unaffected by temperature – the efficiency of photovoltaic cells decreases as temperature increases. Placement of a DSSC in direct sunlight (where it gets hot) has no significant effect on the cell's efficiency – unlike a photovoltaic cell.
- DSSCs work better in low light conditions such as an overcast day.

In a photovoltaic cell the absorption of a photon of light causes the movement of electrons directly; in a DSSC the processes of absorption of photons and movement of electrons are separated.

Checklist

At the end of this chapter you should be able to:

- ☐ Describe the different sources of renewable and non-renewable energy.
- ☐ Calculate the specific energy, energy density and energy efficiency of various fuels and processes.
- ☐ Describe the advantages and disadvantage of using fossil fuels.
- ☐ Explain the processes of fractional distillation, cracking, reforming, coal gasification and liquefaction.
- ☐ Describe what is meant by an octane number and how these numbers can be altered.
- ☐ Explain why nuclear fission and fusion result in the release of energy and be able to construct equations for nuclear reactions.
- ☐ Explain how emission spectra can be used to analyse the composition of stars.
- ☐ Perform calculations involving whole numbers of half-lives.
- ☐ Explain how energy can be extracted from plant material via fermentation and transesterification.
- ☐ Discuss the advantages and disadvantages of using biofuels.
- ☐ Explain how certain molecules are able to absorb infrared radiation and relate this to their role in the greenhouse effect.
- ☐ Describe the relative effects of carbon dioxide, water and methane on the greenhouse effect.
- ☐ Describe the phenomena of global dimming and ocean acidification.

Higher Level only

- ☐ Explain the functioning of primary cells, fuel cells and rechargeable batteries.
- ☐ Construct redox equations for the chemical reactions occurring within these cells.
- ☐ Use the Nernst equation to solve problems involving non-standard conditions within an electrochemical cell.
- ☐ Use mass defect and Mass-Equivalence relationship to determine the energy released during nuclear fission and fusion reactions.
- ☐ Describe the techniques used to enrich uranium.
- ☐ Perform calculations on gas effusion using Graham's law.
- ☐ Perform radioactive decay calculations using non-integer values of half-life.
- ☐ Describe how silicon and dye-sensitised solar cells function.

MEDICINAL CHEMISTRY

D

Medicinal chemistry is a discipline which studies the design, synthesis and development of pharmaceutical compounds (drugs and medicines). It utilises knowledge of chemistry (principally organic) and pharmacology (the study of drugs and medicines).

This chapter covers the following topics:

- ☐ Pharmaceutical products and drug action
- ☐ Aspirin and penicillin
- ☐ Opiates
- ☐ pH regulation of the stomach
- ☐ Antiviral medications
- ☐ Environmental implications of some medications
- ☐ Taxol – a chiral auxillary case study (HL only)
- ☐ Nuclear medicine (HL only)
- ☐ Drug detection and analysis (HL only)

D.1 Pharmaceutical products and drug action

Drug dosing

DEFINITIONS

TD_{50}, LD_{50}, ED_{50} the dosage of a drug that is toxic (TD), lethal (LD) or effective (ED) in 50% of a test population.

THERAPEUTIC INDEX (TI) is defined as:

$$TI = \frac{LD_{50}}{ED_{50}} \text{ in animal studies}$$

$$TI = \frac{TD_{50}}{ED_{50}} \text{ in human studies}$$

THERAPEUTIC WINDOW is the range of dosages (or plasma concentrations) of a drug for which it is effective without causing toxic side-effects, i.e. between the minimum effective dose and the minimum toxic dose.

t

…ust learn the definitions in …ection.

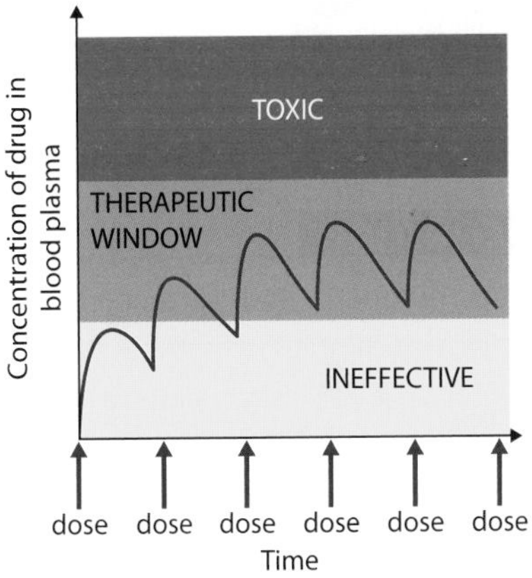

Figure D.1

Annotated exemplar answer D.1

Explain why therapeutic indices are defined in different terms for humans and animals. **[1]**

Therapeutic index for humans uses TD_{50} whereas for animals it uses LD_{50}.

The question asks 'why' rather than 'in what way'. The answer does not address the question – refer to the fact that it is unethical to kill humans so an LD_{50} value cannot be established.

0/1

TEST YOURSELF D.1

Using your understanding of the therapeutic index, suggest one reason why it may be considered unethical to use animals in drug testing.

Drug administration

Method	Description
Injection – intravenous	Hypodermic needle used to inject drug directly into blood. Rapid onset of effects.
Injection – intramuscular	Hypodermic needle used to inject drug into a large muscle.
Injection – subcutaneous	Hypodermic needle used to inject bolus of drug into the layer of fat underneath the skin.
Oral – ingestion	Standard ingestion of a drug, usually in the form of a pill, capsule or liquid.
Oral – inhalation (pulmonary)	Drug is inhaled directly into the lungs using a pocket inhaler or nebuliser.
Rectal	Suppositories are inserted through the anus into the rectum. Drug is absorbed through the rectum wall.
Topical	Direct application of a drug, usually a cream, ointment, lotion or gel, directly on to the location of the problem.

hint

Intravenous injection will generally results in a more rapid therapeutic effect than oral ingestion because the drug is injected directly into the blood stream.

TEST YOURSELF D.2

Suggest which method(s) of administration:

1 provides the quickest onset of drug-related effects,

2 does **not** involve a drug being introduced directly **into** the body,

3 might be used at home (rather than in hospital) by a patient unable to swallow.

Bioavailability

DEFINITION

BIOAVAILABILITY is the fraction of the administered dose which reaches the drug target in the body.

Factors affecting bioavailability:

- **Method of administration**

  Intravenous injection has the highest bioavailability because it is injected directly into the blood stream.

  Oral ingestion – lower bioavailability: depends on the solubility of the drug in water/fat and whether it is broken down in the liver.

- **Functional groups present in the molecule**
- **Polarity of the molecule.**

The bioavailability of an orally-administered drug depends strongly on its solubility in water – water solubility increases absorption through the intestine. If a molecule is more polar and contains (several) functional groups that can hydrogen bond to water (e.g. –OH and $–NH_2$) it will be more soluble in water and have higher bioavailability.

Drugs that are fat-soluble pass through cell membranes (lipids) more easily – this increases bioavailability if the drug is administered topically or rectally.

Tolerance

DEFINITION

TOLERANCE occurs when the body becomes less responsive to the effects of a drug, and so larger and larger doses are needed to produce the same effect.

Developing tolerance means that the patient may be at higher risk of toxic side-effects – the risk of overdosing is increased.

> **hint**
> **Tolerance** is **not** the same as **addiction**. Addiction occurs when the user has developed a physical dependence on the drug which may produce craving or negative effects upon withdrawal. This is not true for tolerance.

Side-effects

DEFINITION

SIDE-EFFECTS are the negative effects of administration of a drug.

Side-effects vary hugely – from mild to life-threatening. The potential side-effects of a drug increase as the dose increases and must be considered when working out the dosage. It may be acceptable to give potentially toxic doses of a drug to a patient with, for example, aggressive cancer but not to someone with a mild headache.

> **hint**
> No specific side-effects are mentioned in the IB syllabus so you will not be expected to know any.

Drug development

The main stages in the development of a drug compound that involve chemists are:

1. Determination of the **need** for the drug: is there a market, significantly-sized population affected by the condition, are there already excellent treatments available?
2. Elucidating the **structure**: analytical and spectroscopic methods can be used to work out the structure of any active compounds.
3. Drug **synthesis**: may be a long, multi-stage process that is time-consuming and has poor overall **yield**. Organic chemists look for ways to improve the yields and/or reduce the number of steps required for a synthesis.
4. **Extraction** of the final product: the drug must be extracted from the reaction mixture, solvents removed and the drug purified.

> **hint**
> Questions on drug development are likely to be concerned with the process from the viewpoint of a synthetic chemist rather than a holistic view. You should avoid talking about clinical trials and animal testing unless directed.

> **hint**
> The bold terms are those stated in the IB syllabus – if possible you should use them in your answers to exam questions.

Annotated exemplar answer D.2

Discuss the main steps in the development of a new **medicine**. **[4]**

First, the drug has to be synthesised by chemists. It is then tested by biologists to see if it works. It might be used in animal testing to see if it is toxic. If it is not toxic then it is tested on humans in a series of clinical trials before being approved by the appropriate agency.

Reference to 'synthesis of drug' scores one mark

Reference to further biological testing is not relevant to the chemistry specification and no mark is gained by including this – avoid wasting time and effort including irrelevant information.

Use specific terms from the syllabus to gain full marks:

- *a need is identified*
- *the structure of an active compound is determined*
- *a synthesis with good yield is designed*
- *the active drug is extracted and purified*

1/4

> **hint**
>
> Annotated Exemplar Answer D.2 illustrates the importance of having your own copy of the syllabus which you have read and annotated. The examiners will be looking at the syllabus when designing a mark scheme – they will often use the syllabus as the basis for the mark scheme.

Drug–receptor interactions

Signals can be carried to cells by small molecules (substrates) binding to specific sites in protein **receptors**. Many drugs are designed to target **receptor** proteins. A receptor implicated in a disease can be targeted by drugs that mimic the natural substrate by having key functional groups in similar positions.

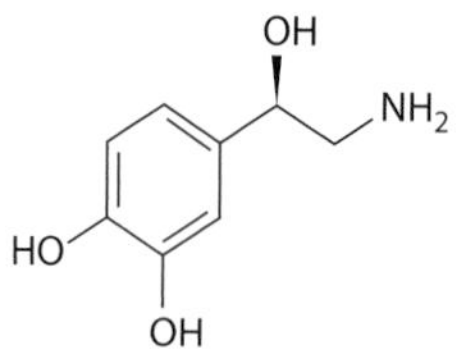

noradrenaline
natural

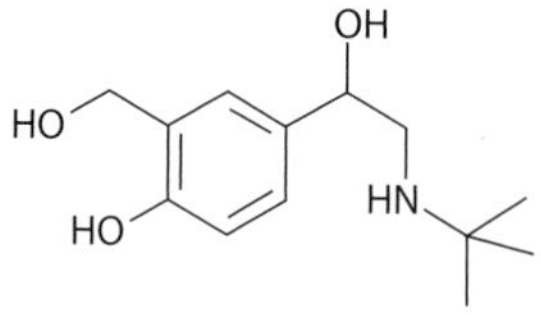

salbutamol
used to treat asthma

Figure D.2

Drugs can have different effects on their target protein:

- **Agonistic:** drug acts in a similar way to the normal substrate, activating the receptor and the associated response.
- **Antagonistic:** drug blocks the substrate-binding site and prevents the natural substrate from activating the receptor.
- **Allosteric:** drug binds to a site other than the substrate-binding site. This action may increase or decrease the receptor's response to the natural substrate.

D.2 Aspirin and penicillin

Aspirin

DEFINITION

A **MILD ANALGESIC** is a drug which works by intercepting the pain stimulus at the source. It does this by preventing the synthesis of substances that cause pain.

> **hint**
>
> Aspirin and penicillin are mentioned specifically in the IB syllabus. Their structures are given in the *IB Data Booklet*.

> **hint**
>
> Ethanoic acid **cannot** be used as an acylating agent in this reaction.

Aspirin is a mild **analgesic** which prevents the synthesis of **prostaglandins** by the enzyme **cyclooxygenase (COX)**. COX is also responsible for synthesising the natural coagulant **(causes clotting of blood)** thromboxane and this explains why aspirin is used as a blood-thinning agent (**anticoagulant**) in the prevention (as a **prophylactic**) of strokes and heart attacks.

Aspirin is synthesised from salicylic acid (2-hydroxybenzoic acid). An acylating agent such as **ethanoyl chloride** or **ethanoic anhydride** reacts with the 2-hydroxy group of salicylic acid to form an **ester** (Figure **D.3**). The carboxylic acid group of salicylic acid is unchanged.

This synthesis involves an **addition-elimination** reaction.

Aspirin is purified by **recrystallisation:**

- The aspirin is dissolved in the minimum amount of hot solvent (e.g. ethyl ethanoate).
- The solution is filtered while still hot to remove any insoluble impurities.

- As the solution cools, crystals of aspirin form.
- The solid aspirin is separated from the solvent by vacuum filtration and washed with cold solvent.

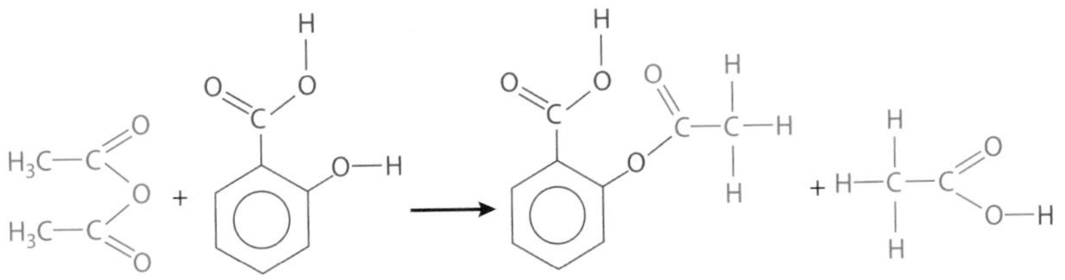

Figure D.3

The aspirin is then dried.

The purity of the final product can be checked by **melting point determination** – a pure substance will melt at a specific temperature but the presence of impurities lowers the melting point and causes the solid to melt over a wider range of temperatures.

Infrared spectroscopy can be used to determine the functional groups present in the final product and to distinguish between the starting material (salicylic acid) and the final product.

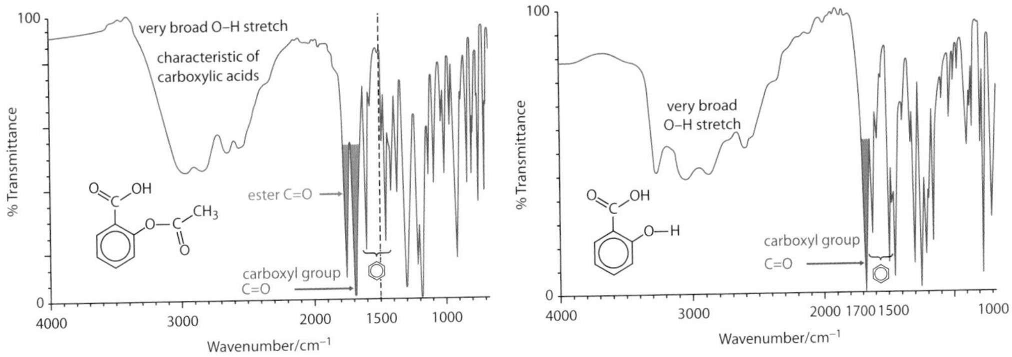

Figure D.4

There are three possible sources of O–H stretches (Figure **D.4**): carboxyl hydroxyl (both compounds), phenolic hydroxyl (salicylic acid only), water (wet sample). There are two sources for the carbonyl absorptions: carboxyl carbonyl (both compounds), ester carbonyl (aspirin only).

hint

If asked to create a balanced equation for the synthesis of aspirin through the reaction of salicylic acid with ethanoyl chloride or ethanoic anhydride, do not forget the small molecules produced: HCl in the case of ethanoyl chloride; ethanoic acid with ethanoic anhydride.

hint

Remember that both impurities and desired product dissolved in the hot solvent during recrystallisation but only the desired product crystallises upon cooling as it is present at **far higher concentrations** than the impurities.

hint

There are two C=O groups in aspirin but only one in salicylic acid, therefore there is an extra absorption in the 1700–1750 cm^{-1} region for aspirin due to the **additional ester carbonyl group (C=O)**.

TEST YOURSELF D.3

1 State the names of functional groups present in both aspirin and salicylic acid.

2 State the name of a functional group that is present in aspirin but not in salicylic acid.

Aspirin exhibits low solubility in water due to the hydrophobic phenyl group. However, it will react with strong bases, such as sodium hydroxide, to form a salt. This **ionic** form of aspirin increases the molecule's water-solubility enormously. Such 'soluble' aspirin can be drunk as a solution and has a higher bioavailability than the molecular form.

Synergy with alcohol

When taken in combination with alcohol, the side-effects of aspirin are increased – stomach bleeding and the formation of peptic ulcers are more likely to occur.

Nature of Science. Aspirin is an excellent early example of the chemical modification of a natural product to enhance pharmacological properties. Salicylic acid was isolated from willow bark after it was noted that indigenous populations chewed the bark to relieve mild pain. Modifications were made to minimise the unpleasant side-effects of taking salicylic acid.

Penicillin

Nature of Science. The discovery of penicillin by Sir Alexander Fleming is an excellent example of a serendipitous (by chance) discovery.

Penicillins are a group of antibiotic compounds produced naturally by fungi (though marketed penicillins are entirely synthetic). The compound works by preventing bacteria from cross-linking their peptidoglycan cell wall, thus weakening it. When the cell absorbs water the osmotic pressure inside the cell increases. The weakened cell wall can no longer prevent the cell from rupturing and being destroyed.

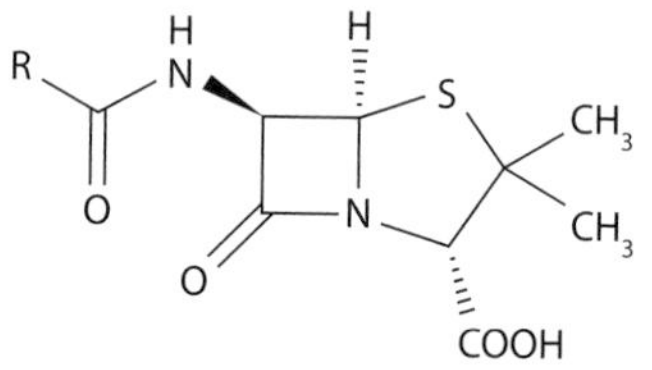

Figure D.5

hint

The ring does not contain separate amine and carbonyl functional groups – a carbonyl group next to a nitrogen atom makes this a **carboxamide** functional group.

Penicillins (Figure **D.5**) contain two key structural components: (1) a beta-lactam ring (Figure **D.6**) and (2) a variable side group, R.

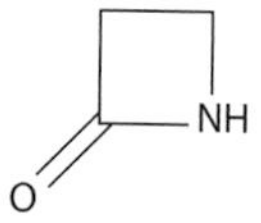

Figure D.6

The beta-lactam ring is the essential feature of penicillins – in a bacterial cell, the ring opens and the carbonyl carbon covalently bonds to the active site of the enzyme, **transpeptidase**, which is involved in cell wall formation.

Some bacteria are penicillin-resistant as they produce an enzyme called **penicillinase**, which hydrolyses the ring before it reaches its target. Modification of the R group can prevent a penicillin molecule being recognised by penicillinase, thus overcoming the penicillin-resistance in some bacteria.

Modification of the R group may also have other effects such as increasing bioavailability and stability and decreasing side-effects.

Resistance and patient compliance

Bacteria mutate rapidly and strains resistant to penicillin (and penicillinase-resistant compounds) quickly developed. Methicillin-resistant *Staphylococcus aureus* (MRSA) is a major problem in modern hospitals. The overprescribing of penicillins and other antibiotics for minor complaints has worsened the problem. Bacteria are able to develop resistance to new drugs quickly as they are constantly exposed to low levels of the compounds.

hint

You will need to be able to describe, not define, the term compliance.

DEFINITION

COMPLIANCE describes a patient's willingness and ability to take a medication in the correct doses, at the correct times for the whole length of the course prescribed.

It is of vital importance that a course of antibiotics is completed by a patient and according to the doctor's instructions. Symptoms of infection often disappear before the infection has fully gone; patients who stop taking antibiotics at this stage risk the infection returning. The infections are often more difficult to treat as the bacteria have reproduced from the hardy survivors of the first treatment.

TEST YOURSELF D.4

1 Suggest two ways in which human activity may lead to antibiotic-resistance in bacteria.
2 State the name of the structural feature of penicillins which gives them biological activity. Describe how this feature acts to confer an antibiotic effect.

D.3 Opiates

DEFINITIONS

A **STRONG ANALGESIC** is a drug that works by temporarily bonding to opioid receptors in the brain and thereby prevents the transmission of pain impulses without depressing the central nervous system.

OPIATES are naturally-occurring narcotic compounds derived from the opium poppy, e.g. morphine.

OPIOIDS is a more general term used to describe all compounds with structures similar to morphine, e.g. codeine and diamorphine (heroin).

Morphine, codeine and diamorphine are examples of commonly-used strong opioid analgesics for the treatment of acute pain, e.g. caused by bone breaks, amputations, etc. Due to their numbing and euphoric effects, they are also commonly abused recreationally.

hint

These three molecules are all specific examples in the IB syllabus.

Structure

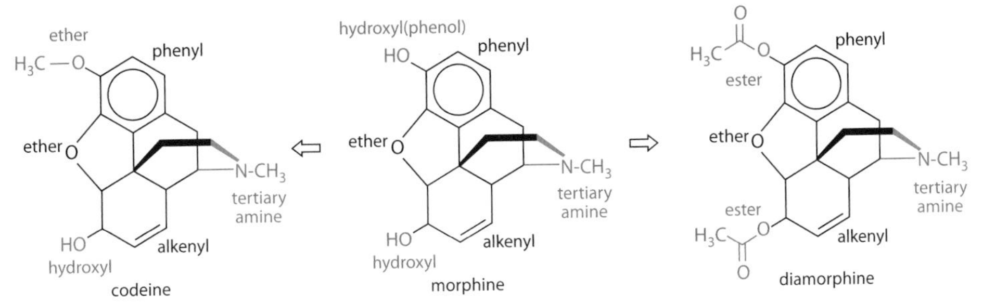

Figure D.7

hint

There is another important functional group common to all three molecules; this is the tertiary amine present in the backbone structure (in blue).

The structures of all three have the same basic skeleton but with some simple changes to one or two side groups. Morphine is used as the main organic starting material to make codeine and diamorphine.

Compound	Side group functionality (red in figure)	Structure	Synthesis from morphine
Morphine	**hydroxyl** (phenol)	R—OH	
Codeine	(methyl) **ether**	R—OCH_3	react with CH_3I in KOH
Diamorphine	**ester**	R—$OCOCH_3$	react with ethanoic anhydride

hint

Questions on these opioid molecules are likely to concentrate on changes to the hydroxyl groups.

Solubility

Opioids must cross the blood–brain barrier (BBB) in order to work. In general, the more hydrophobic (non-polar) a compound, the more rapidly it penetrates the BBB. Although all three molecules mentioned previously are able to cross the BBB, the hydroxyl groups in morphine form hydrogen bonds to water molecules, reducing BBB permeability compared to codeine and diamorphine. The ester groups in diamorphine do not form hydrogen bonds and increase BBB permeability significantly. This explains why diamorphine has **very rapid** onset of action.

Advantages and disadvantages of opioid use

Advantages: rapid onset of analgesia, powerful analgesic effect, cough suppression, anti-diarrhoeal.

Disadvantages: euphoria, vomiting, addiction, constipation, sedation.

Addiction occurs when drug users become dependent on a compound. This may be:

- psychological – they crave the euphoric high a drug such as diamorphine creates,
- physical – withdrawal from the drug produces intolerable symptoms.

Opiate compounds are derived from the opium poppy. Opium and related compounds have been used as painkillers and narcotics for thousands of years.

hint

When describing amine groups, it is always safest to identify whether the amine is primary, secondary or tertiary and include this in the description.

TEST YOURSELF D.5

1 Calculate the difference in relative molecular mass between codeine and diamorphine.
2 State the names of three functional groups present in morphine, codeine **and** diamorphine.
3 Other than pain relief, suggest one other medical use of opioids.
4 Describe how the structure of diamorphine differs from that of morphine. Explain how this difference increases the rate of onset of the effects of diamorphine compared to morphine.

D.4 pH regulation of the stomach

hint

You must be able to construct stoichiometric equations for the reactions of each of these compounds with hydrochloric acid.

Dyspepsia, or indigestion, occurs when excess acid is created in the stomach and/or there is reflux of this acid into the oesophagus. The simplest treatment is to neutralise the excess acid using a non-specific base. Common bases used are: calcium carbonate, sodium hydrogencarbonate, magnesium hydroxide and aluminium hydroxide.

e.g. $CaCO_3 + 2HCl \rightarrow CaCl_2 + CO_2 + H_2O$

$Al(OH)_3 + 3HCl \rightarrow CaCl_2 + 3H_2O$

hint

Notice that aluminium hydroxide neutralises three moles of acid per mole of base compared to two moles of acid neutralised per mole of the carbonate.

TEST YOURSELF D.6

1 Construct stoichiometric equations for the reactions of sodium hydrogencarbonate and, separately, calcium hydroxide with hydrochloric acid.
2 Calculate how many moles of hydrochloric acid can be neutralised by 0.1 mol of magnesium hydroxide.
3 State and explain which of magnesium hydroxide or aluminium hydroxide is the more effective antacid.

Model answer D.1

What volume of stomach acid can be neutralised by 0.500 g of aluminium hydroxide? Assume that stomach acid is 0.050 mol dm^{-3} HCl(aq).

Moles of aluminium hydroxide $= \dfrac{\text{mass}}{\text{mass of 1 mole}} = \dfrac{0.500}{26.92+(3\times(16.00+1.01))}$

$= 0.00641\ldots\text{mol}$

Moles of HCl neutralised $= 3\times$moles of aluminium hydroxide $= 3\times0.00641\ldots$

$= 0.0192\ldots\text{mol}$

Volume of hydrochloric acid neutralised $= \dfrac{\text{moles}}{\text{concentration}} = \dfrac{0.0192\ldots}{0.050}$

$= 0.39\ \text{dm}^3 = 390\ \text{cm}^3$

hint

What is the least number of significant figures used in the question? Use this amount for your answer. However, don't forget to use **unrounded** values throughout your calculations.

Annotated exemplar answer D.3

Suggest a reason why sodium hydroxide is not used as a treatment for indigestion but sodium hydrogencarbonate is. **[1]**

It is only able to neutralise one mole of H+ per mole of base and is therefore not very effective compared to other remedies.

$NaHCO_3$ is also only able to neutralise 1 mol H+ so this is not relevant and gains no marks.

Think what makes NaOH different from $NaHCO_3$ – NaOH is highly corrosive and therefore is dangerous to human health whereas $NaHCO_3$ is non-corrosive and does no damage.

0/1

Buffer solutions

Option B (Biochemistry) covered buffer calculations using the Henderson–Hasselbalch equation. Option D requires the same understanding and application, so you should refer to Option B for worked examples.

hint

You may have to work out the concentration of HCl yourself from a pH value. Remember to use the equation $[H^+] = 10^{-pH}$.

Specific inhibition of stomach acid production

Excessive production of stomach acid can be inhibited using drugs such as **ranitidine** (Zantac^{TM}), which **antagonise** (prevent the proper functioning of) **the histamine-H$_2$ protein receptor**, which is responsible for promoting stomach acid production.

Other drugs such as **omeprazole** (Prilosec^{TM}) and **esomeprazole** (Nexium^{TM}) selectively **inhibit the proton pumps** that physically move H^+ ions into the stomach. Omeprazole itself is not able is to inhibit proton pumps but in the highly acidic medium of the cells that produce stomach acid it is converted to a different compound (the **active metabolite**) that can inhibit proton pumps.

DEFINITION

ACTIVE METABOLITES are the forms of a drug responsible for its pharmacological action after it has undergone chemical processing by the body.

TEST YOURSELF D.7

1 Calculate the neutralising power, in mol g^{-1}, of magnesium hydroxide.

2 Determine the volume of stomach acid, pH 1.0, that could be neutralised by 2.00 g of calcium carbonate.

3 Equimolar amounts of ethanoic acid ($pK_a = 4.75$) and sodium ethanoate are mixed to create a buffer solution. Determine the pH of the resulting solution.

D.5 Antiviral medications

Ways in which viruses differ from bacteria

Compared to bacteria, viruses:

- possess no classical cell structure:
  - no cell organelles
  - no cell membrane
  - no cell wall
- are much smaller than bacterial cells
- need a living cell to multiply
- exhibit no metabolic functions or ability to grow or feed
- contain only simple genetic information (DNA or RNA)
- multiply and mutate far more rapidly
- are always parasitic.

Challenges associated with creating antiviral medication

Viruses are far more difficult to target than bacteria because:

- The lack of a cellular structure means there are fewer targets for drug to act upon.
- It is difficult to target virus without also targeting – and damaging – the host cell.
- Rapid mutation of viruses means drugs quickly become obsolete.

Mechanisms of antiviral drug action

It is difficult to target viruses, and antiviral drugs rely on preventing the host cell from replicating the virus particles or preventing them from being released into the body.

- Alteration of the host cell's genetic information so that the virus cannot use it to multiply:

  e.g. aciclovir (used to treat cold sores – herpes virus) is mistaken for a DNA building block during replication of viral DNA and inhibits the enzyme involved.
- Inhibition of enzyme activity preventing the multiplication of virus particles:
  - AZT (for treating HIV) inhibits the **reverse-transcriptase** enzyme responsible for converting viral RNA into DNA.
  - Oseltamivir and zanamivir (used in the treatment of influenzas) inhibit an enzyme called **neuraminidase**, which virus particles need in order to be released from the host cell. Inhibition of the enzyme prevents the release of virus particles from the cell.

Oseltamivir (Tamiflu™) and zanamivir (Relenza™)

The structures of oseltamivir and zanamivir are provided in the *IB Data Booklet*. You will need to be able to recognise the key functional groups present in both compounds as you may be asked to compare them.

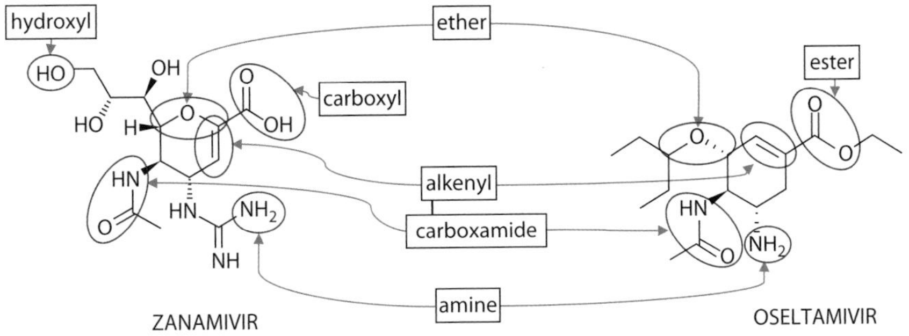

Figure D.8

AIDS

There are several issues related to tackling the AIDS problem:

- HIV mutates extremely rapidly meaning it develops drug-resistance quickly.
- As with other viruses, HIV cannot be targeted without targeting and harming the host cell as well.
- HIV destroys the helper T-cells that are part of the immune system.
- Social and economic barriers exist:
  - AIDS and HIV infection rates are high among certain populations due to lifestyle choices, e.g. sharing needles during drug taking, unprotected sex, etc.
  - Countries with high rates of infection are some of the poorest and cannot afford medication.
  - Education related to treatment and prevention of infection is minimal in certain areas.

hint

Does the question ask you to state the **names** or **formulae** of the functional groups? You will not be credited for giving names where formulae are required and *vice versa*.

TEST YOURSELF D.8

1 Describe two ways in which viruses differ from bacteria.
2 Explain why viruses are generally harder to target than bacteria.
3 Suggest two reasons why tackling the AIDS crisis is challenging.
4 State the formulae of two functional groups present in both oseltamivir and zanamivir. State the formulae of one functional group present only in oseltamivir and one present only in zanamivir.

D.6 Environmental impact of some medications

Nuclear waste

The use of radioactive isotopes in medicine such as ^{60}Co and ^{131}I creates several difficulties. They must be stored and used appropriately, and when their usefulness has expired, they must be disposed of safely without posing any danger to the public or the environment.

Low-level waste (LLW)	High-level waste (HLW)
Waste which gives off small amounts of ionising radiation for short periods of time. Contains isotopes with short half-lives and few nuclei decay each second (low activity).	Waste that gives off large amounts of ionising radiation for a long period of time. Contains isotopes with long half-lives and many nuclei decay each second (high activity).
Items that have been contaminated with radioactive material or have been exposed to radioactivity. e.g. gloves, tools.	E.g. spent fuel rods and other material from nuclear reactors.
May be stored on site until it has decayed or buried underground (near-surface disposal) in individual concrete canisters.	Stored in cooling ponds before being buried deep underground in thick concrete containers.

Improper disposal of radioactive waste can expose humans and animals to radioactivity unnecessarily. Prolonged exposure to low levels of radioactivity can increase rates of cancer, mutation and birth defects. Short or prolonged exposure to high levels of radiation can lead to radiation poisoning, which can be fatal.

Antibiotics

Small quantities of antibiotics enter the water supply from the waste of treated livestock or through improper disposal of medicines. Mutations can allow some bacteria to become resistant to antibiotics. Exposure of bacteria to low concentrations of antibiotics in the environment kills those without resistance but allows those with antibiotic resistance to thrive and multiply further.

Solvents

Many solvents used in industry in the production of drugs are toxic or harmful. To reduce environmental impact:

- the minimum quantity of solvent should be used
- left-over solvent should be reused
- a less toxic alternative should be employed
- disposal should be controlled.

Waste solvents can be incinerated or injected underground – both have the potential to contaminate surrounding water or land if carried out improperly.

Green chemistry (sustainable chemistry)

The key principles underpinning green chemistry are:

- Using renewable resources
- Minimising waste
- Reducing hazards (processes, reagents, products and waste)
- Reducing energy usage
- Improving atom economy (use of catalysts).

hint

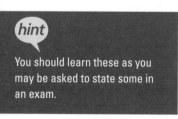

You should learn these as you may be asked to state some in an exam.

Green synthesis of oseltamivir

Total synthesis of oseltamivir (Tamiflu) from petrochemical feedstock (chemicals from crude oil) is energy intensive, produces large quantities of waste and relies on non-renewable resources. A greener synthesis of oseltamivir has been developed starting from shikimic acid (extracted from renewable Chinese star anise or made by genetically modified bacteria) – this synthesis involves fewer steps and is overall more energy efficient.

However, supplies of star anise are limited and shikimic acid is hard to extract from it. The production of shikimic acid by **fermentation of glucose** by **genetically modified bacteria** uses low temperatures and an aqueous solvent, making it a longer-term and greener solution.

TEST YOURSELF D.9

1. Describe the principal differences between high-level and low-level radioactive waste.
2. Explain why it is important to dispose of antibiotics appropriately.
3. Use your understanding of the principles of green chemistry to describe how this approach has been adopted in the synthesis of oseltamivir.

D.7 Taxol – a chiral auxiliary case study (HL only)

Taxol (paclitaxel) is an optically active anti-cancer drug used in chemotherapy. Taxol is used to treat breast cancer and lung cancer and works by preventing cell division.

The original source of Taxol was from the bark of the **Pacific yew tree**, removal of which killed the trees, causing the species to become endangered. A more sustainable approach to Taxol synthesis uses compounds extracted from yew tree needles. These are a renewable resource and produce minimal waste. However, the synthesis of Taxol is a multi-stage process that has low overall yield.

DEFINITION

A **CHIRAL AUXILIARY** is an optically active substance that is temporarily incorporated into an organic synthesis so that it can be carried out asymmetrically with the selective formation of a single enantiomer.

Taxol contains 11 chiral carbons and has a very large number of stereoisomers, only one of which is the desired product. The stereochemistry has to be controlled carefully during the synthesis. Chiral auxiliaries are used to control the enantiomer produced.

- This chiral auxiliary is a pure enantiomer and combines with a non-chiral group to form a chiral intermediate.
- The physical presence of the chiral auxiliary allows the reagent in the next stage of the synthesis to approach from one side of the molecule only, so forcing the formation of just one enantiomer.
- Once the reaction is complete, the chiral auxiliary is removed to leave the desired enantiomer.
- The chiral auxiliary can then be recycled for use in other reactions.

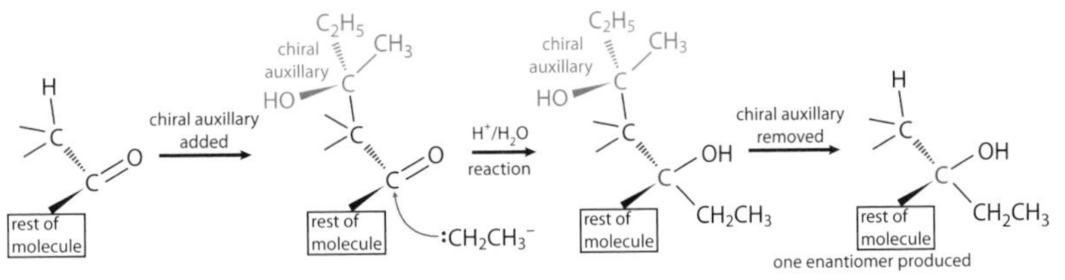

Figure D.9

The enantiomers of a chiral drug may interact differently with protein targets in the body. This means they may exhibit different biological activity, potentially causing physiological harm or exhibiting no activity whatsoever.

Polarimetry

Enantiomers can be distinguished using a **polarimeter** because they **rotate the plane of plane-polarised light by the same angle but in opposite directions** to each other (clockwise and counter-clockwise).

Plane-polarised light is passed through a known amount of the substance in a polarimeter and the angle and direction of rotation of the light recorded.

The direction of rotation allows identification of which enantiomer is present. The angle of rotation can be used to work out the specific rotation, which can aid in identification of an unknown substance or provide an indication of the purity of the sample.

If the sample does not rotate the light then it either contains no optically active compounds or contains equimolar amounts of a pair of enantiomers (a **racemic mixture**) in which both rotations cancel each other.

TEST YOURSELF D.10

1 Outline the challenges associated with the production and synthesis of Taxol.
2 Outline how a chiral auxiliary could be used to create a single enantiomer product.
3 Describe how a polarimeter could be used to distinguish between two enantiomers.

hint

In IB exams, 'outline' means write a brief summary whereas 'describe' requires a more detailed answer.

D.8 Nuclear medicine (HL only)

Elements that emit specific types of radiation can be used in medical imaging or treatment of cancer. **Ionising radiation** (alpha and beta particles and gamma rays) causes damage to DNA in cells – this can cause the death of the cell and tumours can be selectively destroyed.

Use of emissions

- **Alpha ($^{4}_{2}$He)** – cancer radiotherapy
- **Beta ($^{0}_{-1}$e)** – cancer radiotherapy
- **Gamma** – cancer radiotherapy and imaging
- **Proton** – proton-beam therapy
- **Neutron** – neutron-beam therapy
- **Positron** – imaging and diagnosis

Radiotherapy can be given **internally**; radioisotopes are ingested or injected and accumulate at the site of the action, e.g. ^{131}I in the treatment of thyroid tumours. **Externally** applied radiotherapy uses a focused beam of ionising radiation directed at the tumour, e.g. ^{60}Co.

Magnetic Resonance Imaging (MRI) is an application of Nuclear Magnetic Resonance spectroscopy (NMR). The hydrogen atoms (protons) in water molecules in cells in different organs are in slightly different environments, so the various organs in the body can be distinguished to produce a three-dimensional picture of the body. MRI does **not** involve the emission of radioactive particles/ionising radiation and is perfectly safe.

Side-effects of radiotherapy

Common side-effects include: hair loss, nausea, fatigue and sterility. Ionising radiation causes damage to DNA in cells – the biggest effect is on growing/regenerating cells.

Commonly used radioisotopes

Technetium-99m: used in medical imaging, is the most commonly used radioisotope in medicine. It emits penetrating **gamma rays** (they can pass out through the body), has a short **six-hour** half-life that minimises patient exposure to radiation and has **water-soluble compounds**, e.g. **technetate(VII)**, TcO_4^-, for easy transport around the body.

Lutetium-177: used internally in radiotherapy – it emits **highly ionising β-particles** that can kill tumour cells. **Low tissue penetration** (the β-particles do not travel very far in the body) minimises the death of healthy cells. ^{177}Lu also emits gamma rays, making it useful in imaging.

Yttrium-90: is also a β-particle emitter used internally for radiotherapy.

Lutetium-177 and yttrium-90 both have relatively short half-lives.

Targeted alpha therapy (TAT)

Alpha particles lose their energy rapidly in the body and cause a great deal of damage to cells over a very small distance – they do not penetrate very far and have a high probability of killing cells along the path they travel.

Monoclonal antibodies or **peptides** labelled with an alpha emitter can be used to transport the alpha emitter around the body to target specific tumour cells – the monoclonal antibodies/peptides bind to specific receptors only present on certain cells. Lead-212 is a radioisotope that can be used in TAT.

> **hint**
> Alpha decay reduces the proton number by two. Only alpha emission changes the mass number, e.g. $^{253}_{99}Es \rightarrow ^{249}_{97}Bk + ^{4}_{2}\alpha$

> **hint**
> Alpha particles can be represented as $^{4}_{2}He$ or $^{4}_{2}\alpha$; beta particles as or $^{0}_{-1}e$ or $^{0}_{-1}\beta$.

Boron neutron capture therapy (BNCT)

Compounds containing non-radioactive boron-10 are taken up by tumour cells. A beam of neutrons is focused on the tumour. Boron-10 atoms absorb the neutrons to form boron-11 which immediately releases an alpha particle that is able to ionise nearby tumour cells and kill them.

Balancing nuclear equations

Worked example D.1

Lead-212 emits a beta particle when it decays. Construct a nuclear equation for this process.

$$^{212}_{82}Pb \rightarrow ^{212}_{83}Bi + ^{0}_{-1}e$$

The mass numbers must balance (212 = 212 + 0) as must the atomic numbers (82 = 83 − 1). Beta decay increases the atomic number by one, creating a new element. The chemical symbol of the new element can be found using a periodic table.

TEST YOURSELF D.11

Create balanced nuclear equations for:

1 The capture of a neutron by boron-10 leading to emission of an alpha particle.
2 The emission of an alpha particle by bismuth-212.
3 The production of zirconium through the radioactive decay of yttrium-90.

> **hint**
> N_t and N_0 are proportional to the mass of the substance so masses can be used instead of N_t and N_0 in this equation.

hint

This equation is given in the *IB Data Booklet*.

hint

k and t must have the same units.

Nuclear half-life equation

Radioactive decay has constant half-life, $t_{½}$.

The ratio between the initial amount of undecayed nuclei (N_0) and the number present at time $t(N_t)$ is given by:

$N_t = N_0(0.5)^{t/k}$

k is the half-life ($t_{1/2}$)

Model answer D.2

Calculate what mass, in milligrams, of radioactive ^{177}Lu (half-life = 6.71 days) remains after 4 weeks if an initial dose of 1.20 g is administered.

$N_0 = 1.20$ g

$t = 4 \times 7 = 28$ d (convert weeks to days)

$k = 6.71$ d

Using the equation $N_t = N_0(0.5)^{\frac{t}{k}}$

$N_t = 1.20(0.5)^{\frac{28}{6.71}} = 0.0665$ g, i.e. 66.5 mg

hint

A different approach to solving problems on radioactive decay is considered in Option C.

Worked example D.2

Calculate what percentage of yttrium-90, with a half-life of 64 hours, decays over a period of 24 hours?

$\frac{N_t}{N_0}$ represents the fraction of material remaining and $1 - \frac{N_t}{N_0}$ represents the fraction of material that has decayed.

Rearranging the nuclear equation we get: $\frac{N_t}{N_0} = (0.5)^{\frac{t}{k}}$

$t = 24$ h and $k = 64$ h

Place values into equation: $\frac{N_t}{N_0} = (0.5)^{\frac{24}{64}} = 0.771$ (this is the fraction of material **remaining**).

The fraction of material that has decayed is $1 - 0.771 = 0.229$. Multiply by 100 to convert to percent = **23%**

hint

Read the question carefully. Are you asked for the amount/percentage of material remaining or that has decayed?

TEST YOURSELF D.12

1 Calculate what mass, in milligrams, of the original 0.21 g of iodine-131 (half-life = 8.0 days) administered to a patient has decayed after 14 days.

2 Calculate what percentage of technetium-99m (half-life = 6.0 hours) remains in the blood of a patient after a 15-hour stay in hospital.

D.9 Drug detection and analysis (HL only)

Drug extraction and purification

After a drug has been synthesised it must be **extracted** from the reaction mixture (containing solvents, unreacted material and side products) and **purified**.

> *hint*
> This section builds on knowledge gained in Chapter 11.

Methods of extraction

- Adding **ice-cold water to the reaction mixture:** if the product is a solid and not soluble in water it might precipitate out and can be removed from the reaction mixture by filtration. Aspirin can be extracted this way.
- **Solvent extraction:** The organic product partitions between aqueous and organic solvents depending on its solubility in each. A separating funnel is used to drain off each phase. The drug (an organic compound) is most likely to be in the organic layer.
  A drying agent (e.g. **anhydrous** magnesium sulfate) is added to remove traces of water in the organic phase and then filtered off. The solvent is then evaporated off to leave the required compound.

> *hint*
> Normally, the organic layer is the top layer as it has a lower density than water.

Solubility

If a compound contains ionic groups (e.g. COO^-Na^+ or $NH_4^+Cl^-$) or it contains several groups that are able to hydrogen bond to water (e.g. –OH or $-NH_2$) it may be soluble in water. If the compound contains mostly non-polar groups (e.g. phenyl groups, long hydrocarbon chains) it will generally be insoluble in water and more soluble in organic solvents.

> *hint*
> The relationship between the structure of organic compounds and their solubility in water and non-polar solvents is addressed in Chapter 4 – you should refer back for additional information.

TEST YOURSELF D.13

For each of the compounds published in Table 37 of the *IB Data Booklet*, suggest which of water or hexane would be the better solvent.

> *hint*
> Remember the saying **like dissolves like.**

Purification

Recrystallisation, chromatography and (fractional) distillation can be used to purify the products of a synthesis. This section covers fractional distillation as the other methods are discussed elsewhere. Fractional distillation is used to separate a mixture of liquids that have similar boiling points.

The column is hotter at the bottom and cooler at the top.

Consider a mixture of two liquids, A and B, with different boiling points – A has the lower boiling point. The mixture is heated until it boils. The vapour is richer in the more volatile component (A). The vapour travels up the column to the cooler parts where it condenses. This liquid (richer in A than the original) trickles down the column to the warmer parts and boils again – the vapour is again richer in the more volatile component. Each time this process of boiling/condensing is repeated the vapour becomes richer in the more volatile component (A) and travels further up the column until it condenses. Eventually pure A will be obtained at the top of the column where it is condensed to form liquid A. The glass beads in the column provide a large surface area for condensation of the vapour.

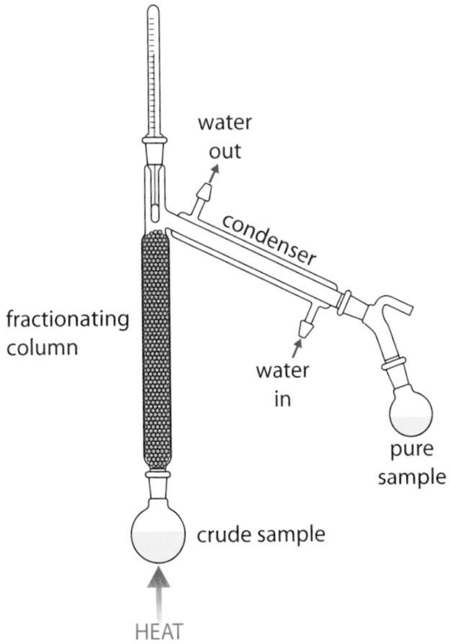

Figure D.10

hint

The amine, carboxamide, ester and phenyl functional groups are more likely to come up in this option.

Raoult's law

A liquid with lower boiling point (more volatile) has a higher vapour pressure (the pressure exerted by a vapour in equilibrium with a liquid). Raoult's law states that the partial vapour pressure of a component in the vapour above a liquid mixture depends on the amount of that component in the liquid mixture and the vapour pressure of the pure liquid. This essentially means that the vapour will be richer in the more volatile component than the liquid mixture.

hint

Look at the structures of the molecules in the relevant section of the *IB Data Booklet* as these are highly likely to be used as examples.

Spectroscopic analysis

The techniques of ^{1}H-NMR, mass spectrometry and infrared spectroscopy are used to analyse samples in the context of medicinal chemistry. In the section on aspirin, previously, the IR spectra of aspirin and salicylic acid were discussed.

In the exam, you may be required to interpret combinations of spectra to deduce information about the structure of an organic compound.

Worked example D.3

Paracetamol ($C_8H_9NO_2$) is a commonly used mild analgesic. The mass spectrum and infrared spectrum of paracetamol are given here.

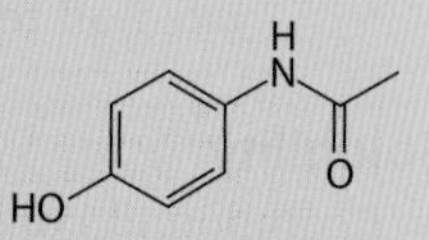

a State the formula of the fragment that produces the peak at $m/z = 43$ in the mass spectrum.

b Identify the **bonds** responsible for the peaks X and Y in the infrared spectrum.

c State and explain the splitting of the ^{1}H-NMR signal produced by the hydrogen atoms in the methyl group.

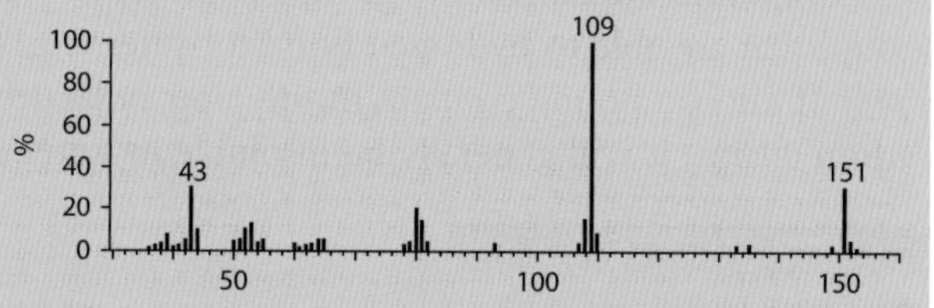

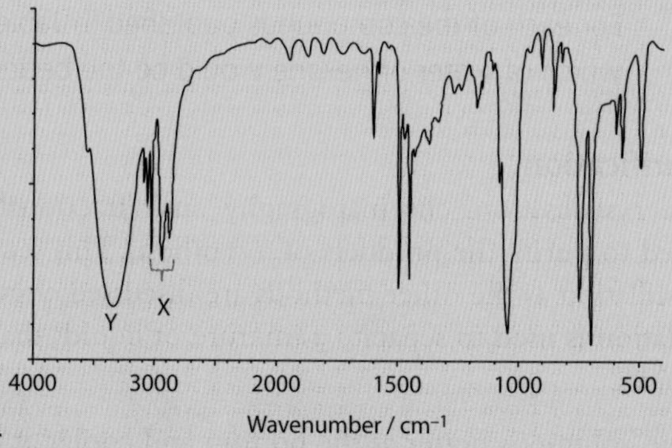

a Work inwards from either end of the molecule counting up the atomic masses of the atoms until you find a fragment that gives you the desired total. On the right of the structure the CH_3 and C=O groups give a total mass of 43.

$$CH_3CO^+$$

b Use the table of IR wavenumbers in the *IB Data Booklet*.

Peak X is at 2900–3000 cm^{-1} which corresponds to the **C–H bond**.

Peak Y is broad and at 3100–3600 cm^{-1} which corresponds to the **O–H bond**.

c The carbon atom adjacent to the methyl group is attached to **no hydrogen atoms** and therefore the methyl hydrogens produce a **singlet**.

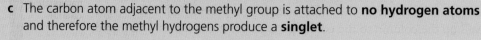

hint

Do not forget to **include a positive charge** on the formula of any fragment **responsible** for a peak in a mass spectrum. Remember to **omit the positive charge** if you are giving the identity of the fragment **lost**.

TEST YOURSELF D.14

1 Draw the structure of the fragment that produces the large peak at $m/z = 109$ in the mass spectrum of paracetamol.

2 Explain why it may not be possible to determine the identity of the bond responsible for the broad peak at 3350 cm^{-1} in the IR spectrum of paracetamol.

3 State how many different hydrogen environments are present in a molecule of paracetamol.

Steroid detection

The detection of steroids in a urine or blood sample is carried out by the combined technique of gas chromatography–mass spectrometry (GC-MS).

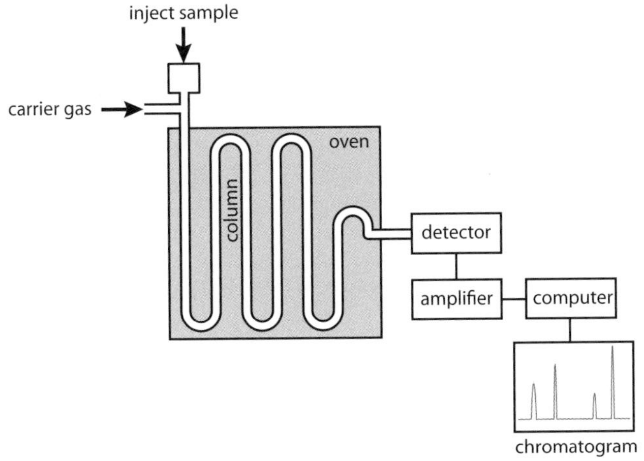

Figure D.11

> **hint**
> If you are asked to identify the bond responsible for an IR absorption then draw the bond, e.g. O–H not OH.

> **hint**
> Look for symmetry when determining the number of hydrogen environments in a molecule. Remember that a molecule may not be drawn symmetrically but may still show symmetry.

> **hint**
> Don't expect the peaks in an IR spectrum to match up perfectly with the values in the *IB Data Booklet*. However, there will be enough information for you to assign them correctly.

The basics of gas chromatography

Gas chromatography can be used to separate the components of a mixture.

1. The sample is injected and vaporised.
2. The vaporised sample is carried through the column by an inert gas (the mobile phase), such as helium or nitrogen.
3. The column is heated by a temperature-controlled oven.
4. The column contains particles of an inert solid, e.g. silica, coated in a non-volatile liquid hydrocarbon (the **stationary phase**).
5. The components of the sample partition between the two phases depending on their solubility in the stationary phase.
6. Components of the mixture that are more volatile or less soluble in the stationary phase travel through the column more quickly – lower **retention time**.

Each separated component enters a mass spectrometer allowing its relative molecular mass to be determined to a high precision. Steroids have the same basic carbon skeleton but different side groups. This means they each have a unique molecular mass which enables them to be identified via a database search of known compounds.

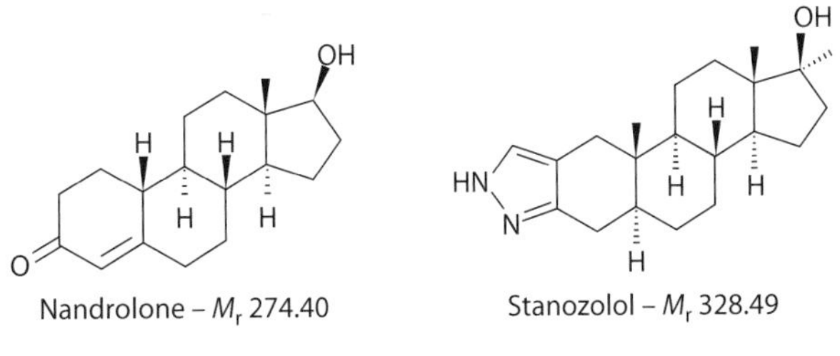

Figure D.12

Alcohol detection

The breathalyser tests used by police are based on redox chemistry. Two forms of breathalyser test exist.

1. Reduction of dichromate(VI) ions

The ethanol in a person's breath is a good measure of the amount of ethanol in their blood plasma. The recipient of the test breathes into the device which contains crystals or a solution of acidified sodium or potassium dichromate(VI), $Na_2/K_2Cr_2O_7$. The ethanol in the sample reduces the dichromate(VI) ions to chromium(III) ions causing them to change colour **from orange to green**. In the process, the ethanol is oxidised to ethanoic acid.

You may be asked to construct balanced ionic half-equations for the oxidation and reduction processes and an equation for the overall reaction. It is unlikely that state symbols will be required but be sure to check.

Reduction: $Cr_2O_7^{2-}(aq) + 14H^+(aq) + 6e^- \rightarrow 2Cr^{3+}(aq) + 7H_2O(l)$

Oxidation: $CH_3CH_2OH(g) + H_2O(l) \rightarrow CH_3COOH(aq) + 4H^+(aq) + 4e^-$

TEST YOURSELF D.15

1 Combine the two half-equations into an overall redox equation for the process occurring in the breathalyser.
2 State the change in oxidation state of the chromium atoms present.
3 State the change in the average oxidation state of the carbon atoms present.

2. Fuel cell breathalyser

The dichromate(VI) breathalyser has been superseded by the fuel cell breathalyser that uses atmospheric oxygen to oxidise the ethanol in the breath sample. The oxidation and reduction reactions are physically separated in a small fuel cell and this allows current to be generated. The current produced gives a more quantitative measure of the ethanol content of the breath – the higher the current, the more ethanol present.

The overall equation for the reaction that occurs is: $CH_3CH_2OH + O_2 \rightarrow CH_3COOH + H_2O$

TEST YOURSELF D.16

Construct a balanced ionic half-equation for the reduction of atmospheric oxygen to water.

Checklist

At the end of this chapter you should be able to:

- ☐ Describe details of pharmaceutical properties and drug action.
- ☐ Describe the chemistry of aspirin and penicillin and give details of some of their biological properties.
- ☐ Describe the uses of opiates and understand the chemical differences between different opiates.
- ☐ Understand how bases can be used to regulate stomach pH.
- ☐ Perform calculations involving buffer solutions.
- ☐ Explain how antiviral medications work and why viruses are difficult to target.
- ☐ Explain the chemical differences between oseltamivir and zanamivir.
- ☐ Describe the environmental issues caused by nuclear, antibiotic and solvent waste.
- ☐ Describe the basics of green/sustainable chemistry.

Higher Level only

- ☐ Describe the basics of Taxol synthesis and how to use a polarimeter to distinguish between enantiomers.
- ☐ Describe the use of radioisotopes in medicine and perform calculations involving half-lives.
- ☐ Interpret spectra to determine the structure of a simple compound.
- ☐ Describe the extraction and purification of drug compounds.
- ☐ Explain how steroids and alcohol can be detected.

GLOSSARY

acid deposition a more general term than acid rain; refers to any process in which acidic substances (particles, gases and precipitation) leave the atmosphere to be deposited on the surface of the Earth; can be divided into wet deposition (acid rain, fog and snow) and dry deposition (acidic gases and particles).

addition polymerisation alkenes undergo addition polymerisation, in which a large number of monomers are joined together into a polymer chain; no other groups/molecules are lost in the process.

addition reaction in organic chemistry, a reaction in which a molecule is added to a compound containing a multiple bond without the loss of any other atoms/groups.

amphoteric a substance that can act as an acid and a base.

analgesics drugs that reduce pain.

atoms the smallest part of an element that can still be recognised as the element.

Aufbau principle the process of putting electrons into atoms to generate the electron configuration.

biofuel a fuel produced from organic matter obtained from plants, waste material, etc.

chain reaction one initial event causes a large number of subsequent reactions – the reactive species is regenerated in each cycle of reactions.

chemical properties how a substance behaves in chemical reactions.

common ion effect a substance AB will be less soluble in an aqueous solution containing A^+ or B^- ions than it is in water.

continuous spectrum a spectrum consisting of all frequencies/wavelengths of light.

convergence limit the point in a line emission spectrum where the lines merge to form a continuum; may be used to determine the ionisation energy.

dipole moment the product of one of the charges making up a dipole and the distance between the charges; non-polar molecules have a zero dipole moment.

distillation a separation technique that can be used to separate the components in a mixture of liquids or extract a liquid from a solution. The mixture is heated and the vapour condensed and collected in a separate container.

drugs substances that, when applied to or introduced into a living organism, bring about a change in biological function through their chemical action.

emission spectrum electromagnetic radiation given out when an electron falls from a higher energy level to a lower one; only certain frequencies of electromagnetic radiation are emitted; each atom has a different emission spectrum.

Fischer Projection a way of representing the structure of a 3D organic molecule by projection onto a plane – bonds are shown as horizontal or vertical lines.

fossil fuels fuels formed from things that were once alive and have been buried underground for millions of years, e.g. coal, oil and gas.

genetic code how the four-base code in DNA determines the sequence of 20 amino acids in proteins; each three-base codon codes for only one amino acid and this code is the same in all organisms – it is universal.

hydrogen bonding an intermolecular force resulting from the interaction of a lone pair on a very electronegative atom (N/O/F) in one molecule with an H atom attached to N/O/F in another molecule [these forces may also occur intramolecularly].

ideal gas a theoretical model that approximates the behaviour of real gases; it can be defined in terms of macroscopic properties (a gas that obeys the equation $PV = nRT$) or in terms of microscopic properties (the main assumptions that define an ideal gas on a microscopic scale are that the molecules are point masses – their volume is negligible compared with the volume of the container – and that there are no intermolecular forces except during a collision).

ion a charged particle that is formed when an atom loses or gains electron(s); a positive ion is formed when an atom loses (an) electron(s) and a negative ion is formed when an atom gains (an) electron(s).

lattice structure regular 3D arrangement of particles.

limiting reactant the reactant that is used up first in a chemical reaction; when the number of moles of each species is divided by their coefficient in the stoichiometric equation, the limiting reactant is the one with the lowest number; all other reactants are in excess.

line spectrum the emission spectrum of an atom consists of a series of lines that get closer together at higher frequency; only certain frequencies/wavelengths of light are present.

lipoprotein an assembly of phospholipids and proteins that transports lipids around the body.

London (dispersion) force intermolecular forces resulting from temporary (instantaneous) dipole – induced dipole interactions.

Glossary

matter something that has mass and occupies space.

medicines things that treat, prevent or alleviate the symptoms of disease.

metabolic reactions chemical reactions that go on in cells – they involve the breakdown of molecules with the release of energy and the synthesis of molecules that are required by cells.

monomer a molecule from which a polymer chain may be built up, e.g. ethene is the monomer for polyethene.

noble gases the elements in group 18 of the periodic table; also sometimes called the 'inert gases'.

ore a sample of rock containing a metal/metal compound from which the metal can be economically extracted.

oxidation loss of electrons or increase in oxidation number.

oxidation state a purely formal concept that regards all compounds as ionic and assigns charges to the components accordingly; it provides a guide to the distribution of electrons in covalent compounds.

oxidising agent (oxidant) a species that oxidises other species and, in the process, is itself reduced; an oxidising agent takes electrons away from something.

physical properties properties such as melting point, solubility and electrical conductivity, relating to the physical state of a substance and the physical changes it can undergo.

polar molecule molecule in which one end is slightly positive relative to the other; whether a molecule is polar or not depends on the differences in electronegativity of the atoms and the shape of the molecule.

rancid a rancid fat is one that has gone off. Rancidity refers to the unpleasant odours and flavours that develop when fats go 'bad'.

reflux a way of heating organic liquids in an open system without significant loss of the liquids. The mixture is heated in a flask to which a vertical condenser is attached. The liquids evaporate and then condense and return to the reaction vessel.

spectrochemical series a series of ligands arranged in order of the extent to which they cause splitting of d orbitals in a transition metal complex ion.

standard hydrogen electrode the standard half-cell relative to which standard electrode potentials are measured.

steric effects effects associated with the physical size/bulkiness of atoms/groups.

steroids a class of lipid molecules – they are hydrophobic (mostly non-polar) molecules having a common structure, known as the **steroid backbone**. This is made up of three six-membered rings and a five-membered ring fused together.

titration a technique that involves adding measured amounts of a solution (from a burette) to another solution to determine the amounts that react exactly with each other.

volatile describing a substance that evaporates readily.

ANSWERS TO TEST YOURSELF QUESTIONS

1 Stoichiometric relationships

1.1

		sum of coefficients
1	$SF_4 + 2H_2O \rightarrow SO_2 + 4HF$	8
2	$Fe_2O_3 + 3CO \rightarrow 2Fe + 3CO_2$	9
3	$4NH_3 + 3O_2 \rightarrow 2N_2 + 6H_2O$	15

1.2 **1** 0.125 mol

2 1.36 g

1.3 **1** 9.99×10^{-23} g

2 7.31×10^{-23} g

1.4 **1** 3.61×10^{23} O atoms

2 3.61×10^{21} H atoms

1.5 **1** Empirical formula CH_2 N_2H_3 C_3H_5 NHO

Molecular formula C_2H_2 C_6H_6

2 PH_2

3 $KMnO_4$

1.6 **1** 1.39 dm^3

2 120 cm^3 340 cm^3

1.7 **1** 78.4%

2 0.420 tonne

1.8 **1** H_2SO_4

2 SO_2

1.9 **1** 20 cm^3

2 10 cm^3

1.10 400 cm^3

1.11 **1** 2.06×10^{-3} m^3 or 2.06 dm^3

2 43.2 g mol^{-1}

1.12 **1** 0.179 mol dm^{-3}

2 0.400 mol dm^{-3}

1.13 8.00 ppm

2 Atomic structure

2.1

	protons	neutrons	electrons
^{65}Cu	29	36	29
$^{15}N^{3-}$	7	8	10
$^{137}Ba^{2+}$	56	81	54

2.2 **1** 63.48

2 20.00% ^{10}B

80.00% ^{11}B

2.3 **1**

B	$1s^22s^22p^1$
P	$1s^22s^22p^63s^23p^3$
Ti	$1s^22s^22p^63s^23p^64s^23d^2$
Cr	$1s^22s^22p^63s^23p^64s^13d^5$
Cu	$1s^22s^22p^63s^23p^64s^13d^{10}$
Se	$1s^22s^22p^63s^23p^64s^23d^{10}4p^4$

2

Al	$[Ne]3s^23p^1$
As	$[Ar]4s^23d^{10}4p^3$

2.4

	1s	2s	2p	3s	3p	4s	3d
Si	↑↓	↑↓	↑↓ ↑↓ ↑↓	↑↓	↑ ↑ □		
Fe	↑↓	↑↓	↑↓ ↑↓ ↑↓	↑↓	↑↓ ↑↓ ↑↓	↑↓	↑↓ ↑ ↑ ↑ ↑
Cr	↑↓	↑↓	↑↓ ↑↓ ↑↓	↑↓	↑↓ ↑↓ ↑↓	↑	↑ ↑ ↑ ↑ ↑

2.5

Na^+	$1s^22s^22p^6$
Cl^-	$1s^22s^22p^63s^23p^6$
Fe^{3+}	$1s^22s^22p^63s^23p^63d^5$

2.6 $n = 2 \rightarrow n = 1$

2.7 **1** 4.57×10^{14} Hz

2 3.03×10^{-19} J

2.8 **1** First ionisation energy $K(g) \rightarrow K^+(g) + e^-$

Second ionisation energy $K^+(g) \rightarrow K^{2+}(g) + e^-$

2 B

3 Periodicity

3.1

	d block element	noble gas	s block element	halogen	f block element	lanthanoid
bromine				✓		
magnesium			✓			
europium					✓	✓
titanium	✓					
plutonium					✓	
iridium	✓					
radon		✓				
radium			✓			

3.2 **1** ionic radius

2 electronegativity, electron affinity becomes more exothermic

3.3 **1**

	Na_2O	SO_3	MgO	Al_2O_3	Cl_2O	SiO_2	P_4O_6	SO_2	NO_2	CaO
acidic		✓			✓	✓	✓	✓	✓	
basic	✓		✓							✓
amphoteric				✓						

2

Na_2O	$Na_2O(s) + H_2O(l) \rightarrow 2NaOH(aq)$
P_4O_{10}	$P_4O_{10}(s) + 6H_2O(l) \rightarrow 4H_3PO_4(aq)$
SO_3	$SO_3(g) + H_2O(l) \rightarrow H_2SO_4(aq)$

3.4 **1** $2Li(s) + 2H_2O(l) \rightarrow 2LiOH(aq) + H_2(g)$

2 $2K(s) + Br_2(l) \rightarrow 2KBr(s)$

3.5 **1** chlorine solution is added to potassium bromide solution – pale green/colourless to orange

bromine solution is added to potassium chloride solution – no change

2 $Cl_2(aq) + 2Br^-(aq) \rightarrow 2Cl^-(aq) + Br_2(aq)$

3.6 3–

3.7 $FeCl_2$ $MnCl_2$ CuO

3.8 **1** $[TiCl_5H_2O]^{2-}$

2 $[Cr(NH_3)_6]^{3+}$

4 Chemical bonding and structure

4.1

ionic	CaS	MgO	
covalent	CO_2	PCl_3	OF_2

4.2

lithium fluoride	LiF
magnesium chloride	$MgCl_2$
potassium carbonate	K_2CO_3
calcium hydroxide	$Ca(OH)_2$
ammonium sulfate	$(NH_4)_2SO_4$
sodium hydrogencarbonate	$NaHCO_3$
iron(II) phosphate	$Fe_3(PO_4)_2$

4.3

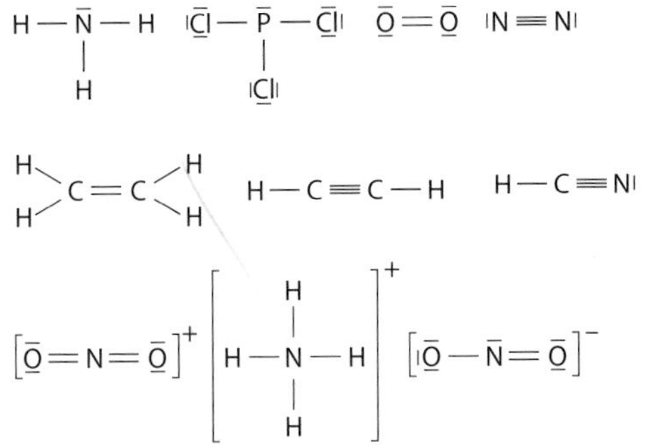

O_3

C_6H_6

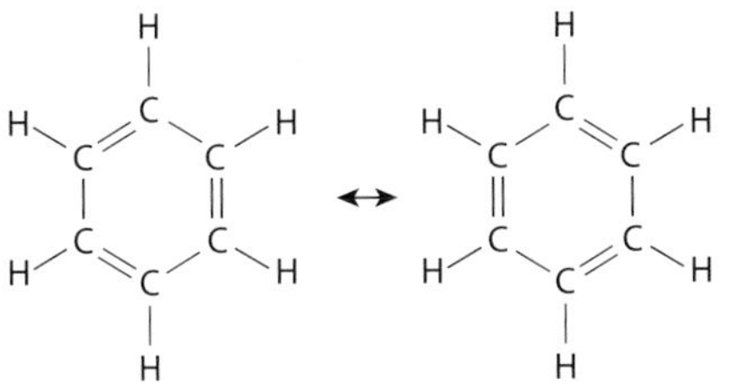

NO_3^-

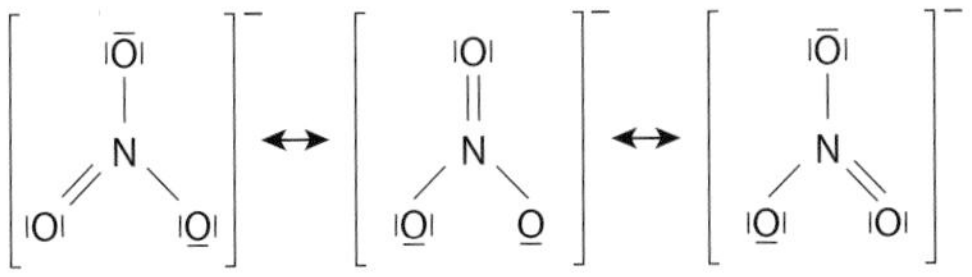

4.4

	shape	bond angle
CH_4	tetrahedral	109.5°
NH_3	trigonal pyramidal	107°
H_2O	bent	104.5°
CO_2	linear	180°
NO_2^+	linear	180°
NH_4^+	tetrahedral	109.5°
H_3O^+	trigonal pyramidal	100–108°
BF_3	trigonal planar	120°
CO_3^{2-}	trigonal planar	120°

4.5 **1** H–F

2 CO, NH_3, PCl_3

4.6 **1** NH_3 CH_3COOH

2 HF CH_3OH NH_3

4.7 The atoms shown have a formal charge, all other atoms have 0 formal charge.

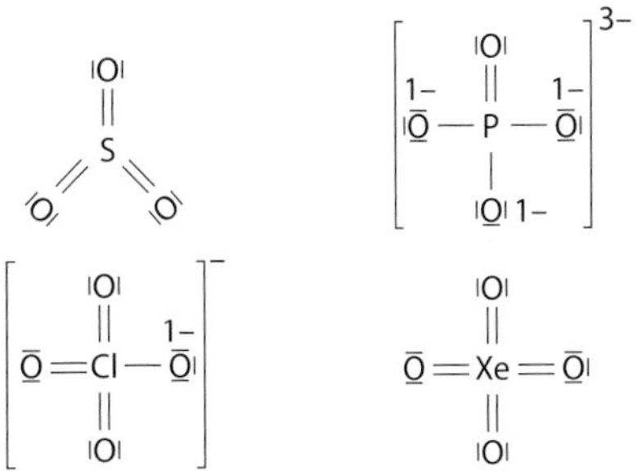

	σ bonds	π bonds
C_2H_4	5	1
HCN	2	2
C_2H_2	3	2

4.9

	shape	bond angle
XeF_4	square planar	90°
PCl_5	trigonal bipyramidal	120° and 90°
$XeOF_4$	square pyramidal	80–90°
BrF_5	square pyramidal	80–90°
SO_4^{2-}	tetrahedral	109.5°
ClF_3	T-shaped	80–90°
SF_6	octahedral	90°
XeO_2F_2	see-saw	110–120° and 80–90°

4.10

CH_4	sp^3
BF_3	sp^2
HCN	sp
H_2O	sp^3
SO_2	sp^2
CO_3^{2-}	sp^2
NH_3	sp^3
CO_2	sp
C_2H_4	sp^2

4.11 CO_3^{2-} C_6H_6 CH_3COO^- SO_4^{2-} NO_2^- NO_3^-

5 Energetics

5.1 $-4200\ kJ\ mol^{-1}$

5.2 $-46.5\ kJ\ mol^{-1}$

5.3 **1** $44\ kJ\ mol^{-1}$

2 $-1268\ kJ\ mol^{-1}$

5.4

$H_2O(l)$	$H_2(g) + \frac{1}{2}O_2(g) \rightarrow H_2O(l)$
$SO_2(g)$	$S(s) + O_2(g) \rightarrow SO_2(g)$
$H_2SO_4(l)$	$H_2(g) + S(s) + 2O_2(g) \rightarrow H_2SO_4(l)$

5.5 $598\ kJ\ mol^{-1}$

5.6 $-184\ kJ\ mol^{-1}$

5.7 $802\ kJ\ mol^{-1}$

5.8 $2067\ kJ\ mol^{-1}$

5.9 KBr LiF $CaCl_2$ CaO MgO

5.10

$NH_3(g) + HCl(g) \rightarrow NH_4Cl(s)$	decrease
$C_2H_5OH(g) + 3O_2(g) \rightarrow 2CO_2(g) + 3H_2O(l)$	decrease

5.11 $-433\ J\ K^{-1}\ mol^{-1}$

5.12 $-2110\ kJ\ mol^{-1}$ spontaneous

5.13 high temperature

6 Chemical kinetics

6.1

$(CH_3)_2CO(aq) + I_2(aq) \xrightarrow{H^+} CH_3COCH_2I(aq) + HI(aq)$	use a colorimeter to monitor the fading of the colour as iodine is used up with time
$2H_2O_2(aq) \xrightarrow{MnO_2} 2H_2O(l) + O_2(g)$	measure the volume of gas produced or the decrease in mass with time

6.2 **1** Rate = $k[A][B]^2$

$k = 2.0 \times 10^{-3}\ mol^{-2}\ dm^6\ s^{-1}$

2 Rate = $k[A]^2$

$k = 2.5\ mol^{-1}\ dm^3\ h^{-1}$

6.3 **1** Rate = $k[Q][P]$

2 C

[Hint – 3 particles colliding in step 2 of D]

6.4 **1** $41.6\ kJ\ mol^{-1}$

2 $43\ kJ\ mol^{-1}$

7 Equilibrium

7.1 **1**

$$K_c = \frac{[CO_2(g)]\ [H_2(g)]^4}{[CH_4(g)]\ [H_2O(g)]^2}$$

2

$$K_c = \frac{[SO_3(g)]^2}{[SO_2(g)]^2\ [O_2(g)]}$$

7.2 C

7.3

	1 effect on position of equilibrium	2 effect on value of K_c
increasing the pressure	shifts to left	no change
decreasing the temperature	shifts to left	decreases
adding hydrogen	shifts to left	no change
adding a catalyst	no change	no change

7.4 To the right because Q is less than K_c

7.5 **1** 2.37×10^{-4}

2 0.075 mol

7.6 **1** $-19\ kJ\ mol^{-1}$

2 1.4×10^{-6}

8 Acids and bases

8.1 **1**

	acids	bases
$NH_4^+(aq) + H_2O(l) \rightleftharpoons NH_3(aq) + H_3O^+(aq)$	NH_4^+ and H_3O^+	H_2O and NH_3
$HSO_4^-(aq) + H_2O(l) \rightleftharpoons SO_4^{2-}(aq) + H_3O^+(aq)$	HSO_4^- and H_3O^+	H_2O and SO_4^{2-}

2 H_2O

8.2

$CaCO_3 + H_2SO_4 \rightarrow CaSO_4 + CO_2 + H_2O$
$CaO + 2HNO_3 \rightarrow Ca(NO_3)_2 + H_2O$
$Mg + 2CH_3COOH \rightarrow (CH_3COO)_2Mg + H_2$
$2NH_3 + H_2SO_4 \rightarrow (NH_4)_2SO_4$

8.3

strong acid	weak acid	strong base	weak base
H_2SO_4 HNO_3 HCl	CH_3COOH H_2CO_3 ($CO_2(aq)$)	$CsOH$ $Ba(OH)_2$ $LiOH$	NH_3 $CH_3CH_2NH_2$

8.4 **1** increases by 1 unit

2 0.10 mol dm^{-3} H_2SO_4 < 0.10 mol dm^{-3} HCl < 0.010 mol dm^{-3} HCl < 0.010 mol dm^{-3} CH_3CH_2COOH < 0.001 mol dm^{-3} CH_3CH_2COOH < 0.0010 mol dm^{-3} NH_3 < 1.0 mol dm^{-3} NH_3 < 1.0 mol dm^{-3} NaOH

3 Decreases by a factor of 1000 from 1×10^{-2} mol dm^{-3} to 1×10^{-5} mol dm^{-3}

4 pH = 5

5 $[H^+(aq)] = 1.0 \times 10^{-9}$ mol dm^{-3}

8.5

0.010 mol dm^{-3} NaOH(aq)	12
0.200 mol dm^{-3} $Ba(OH)_2$(aq)	13.6

8.6 **1** HC > HD > HA > HB

2 2.94

8.7 9.25

8.8 **1**

	pH	pOH	$[H^+(aq)]$	$[OH^-(aq)]$
0.1 M HCl	1	13	0.1	1×10^{-13}
0.1 M NaOH	13	1	1×10^{-13}	0.1
0.0200 M $Ba(OH)_2$	12.6	1.4	2.50×10^{-13}	0.0400

2 11.3

8.9

pH	$[H^+(aq)]$	$[OH^-(aq)]$
6.77	1.71×10^{-7} mol dm^{-3}	1.71×10^{-7} mol dm^{-3}

8.10

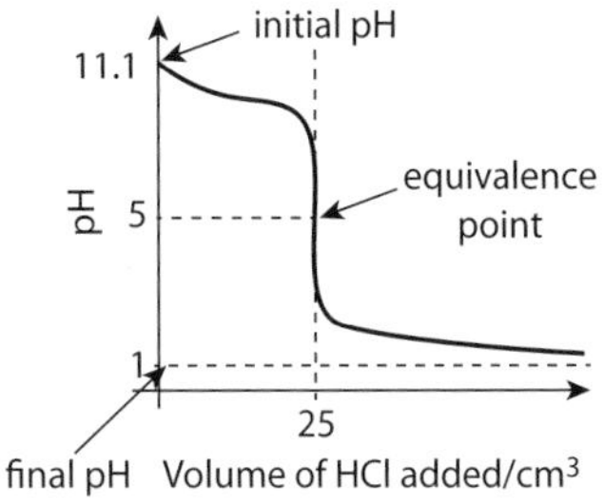

8.11

NaCl	neutral
CH_3COONa	basic
NH_4Cl	acidic
KNO_3	neutral

8.12 C

8.13 **1** Lewis acid – there are only 6 e$^-$ in the outer shell of B and therefore it has space to accept a pair of electrons.

2

OH^-	Lewis base
Cl^-	Lewis base
$(CH_3)_3C^+$	Lewis acid

9 Redox processes

9.1 **1**

NO_2	NO_3^-	HCl	$HClO_3$	ClO_4^-	CrO_4^{2-}	$Cr_2O_7^{2-}$	Na_2O_2	LiH
N +4 O −2	N +5 O −2	H +1 Cl −1	H +1 Cl +5 O −2	Cl +7 O −2	Cr +6 O −2	Cr +6 O −2	Na +1 O −1	Li +1 H −1

2

NO_2	nitrogen(IV) oxide
N_2O	nitrogen(I) oxide
$Cr_2O_7^{2-}$	dichromate(VI)
MnO_4^-	manganate(VII)
ClO_4^-	chlorate(VII)
ClO^-	chlorate(I)

9.2

redox reactions	oxidising agent	reducing agent
$Fe + Cu^{2+} \rightarrow Fe^{2+} + Cu$	Cu^{2+}	Fe
$C_2H_2 + 2H_2 \rightarrow C_2H_6$	C_2H_2	H_2

9.3 **1** $Fe^{2+} + 2e^- \rightarrow Fe$

$2Br^- \rightarrow Br_2 + 2e^-$

2 $Cr_2O_7^{2-} + 14H^+ + 6e^- \rightarrow 2Cr^{3+} + 7H_2O$

$H_3PO_3 + H_2O \rightarrow H_3PO_4 + 2H^+ + 2e^-$

3 $2MnO_4^-(aq) + 16H^+(aq) + 10I^-(aq) \rightarrow 2Mn^{2+}(aq) + 8H_2O(l) + 5I_2(aq)$

9.4 **1** 6.60 ppm

2 4.10 ppm

9.5 **1** $2Fe^{2+}(aq) + Cl_2(g) \rightarrow 2Fe^{3+}(aq) + 2Cl^-(aq)$
$E^{\ominus}_{cell} = 0.59\,V$

2 $Cl_2|Pt$ is the positive electrode

Electrons flow from Fe^{2+}/Fe^{3+} half-cell to $Cl_2|Cl^-$ half-cell

Salt bridge:

negative ions flow from $Cl_2|Cl^-$ half-cell to Fe^{2+}/Fe^{3+} half-cell

positive ions flow from Fe^{2+}/Fe^{3+} half-cell to $Cl_2|Cl^-$ half-cell

3 Fe^{2+} is oxidised, therefore the Pt electrode in this half-cell is the anode.

9.6 **1** $E^{\ominus}_{cell}$ is positive (+0.56 V), therefore the reaction is spontaneous.

Alternatively: $Cr_2O_7^{2-}/Cr^{3+}$ has a more positive electrode potential than Fe^{3+}/Fe^{2+}, therefore $Cr_2O_7^{2-}$ is a stronger oxidising agent than Fe^{3+} and will oxidise Fe^{2+} to Fe^{3+}.

2 strongest reducing agent is U^{3+}

strongest oxidising agent is Eu^{3+}

9.7 −357 kJ mol^{-1}

9.8

	product at anode	product at cathode
concentrated $MgCl_2(aq)$	chlorine	hydrogen
$Cu(NO_3)_2(aq)$ with platinum electrodes	oxygen	copper
$H_2SO_4(aq)$	oxygen	hydrogen

9.9 Volume of oxygen = 50.00 cm^3

Mass of copper = 0.2800 g

10 Organic chemistry

10.1

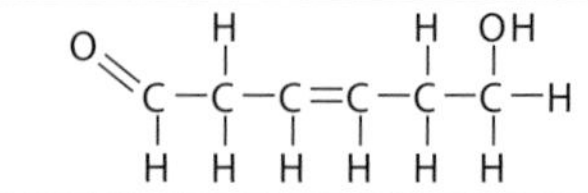	**1** carbonyl (aldehyde), alkenyl, hydroxyl
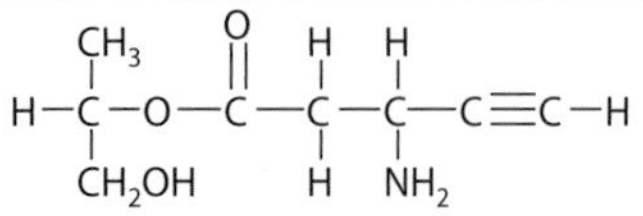	**2** hydroxyl, ester, (primary) amine, alkynyl

10.2

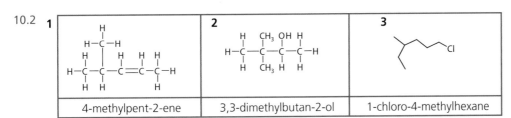

1	**2**	**3**
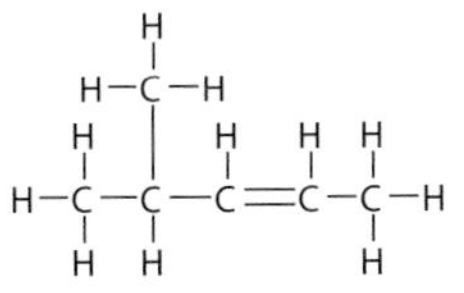	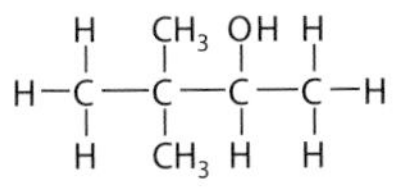	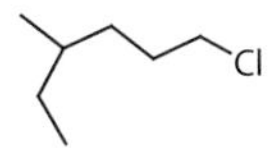
4-methylpent-2-ene	3,3-dimethylbutan-2-ol	1-chloro-4-methylhexane

10.3

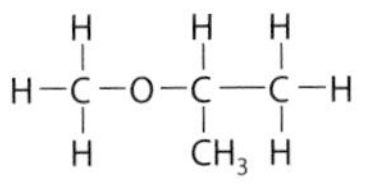

2-methoxypropane

10.4 **1** 3

2 3 (ignoring *cis–trans* isomers)

3 3

4 4

10.5

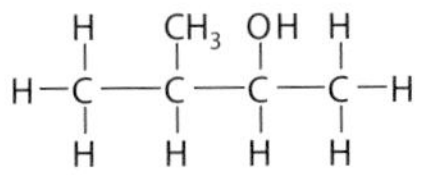	**1** secondary
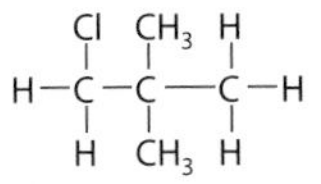	**2** primary
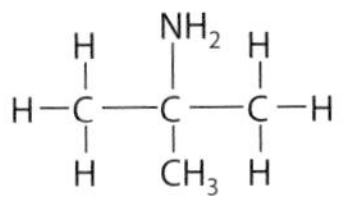	**3** primary

10.6 **1** $2C_2H_6 + 3O_2 \rightarrow 4CO_2 + 6H_2O$

2 $2C_3H_8 + 7O_2 \rightarrow 6CO + 8H_2O$

10.7 **1**

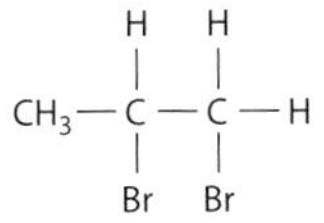

2

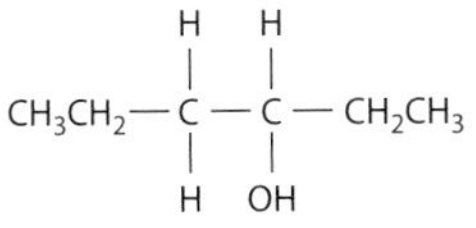

10.8 **1**

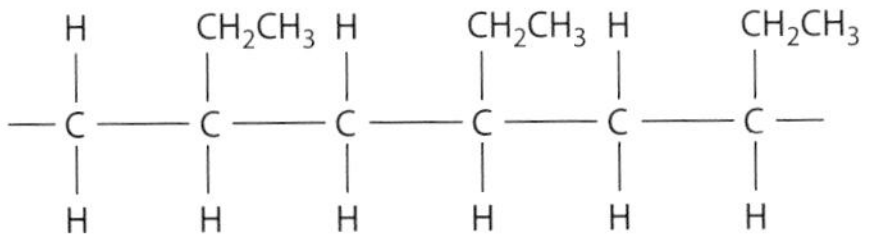

2

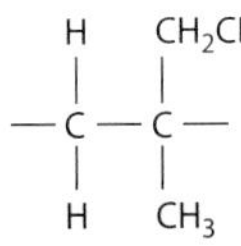

Repeating unit

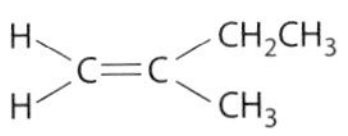

monomer

10.9 **1**

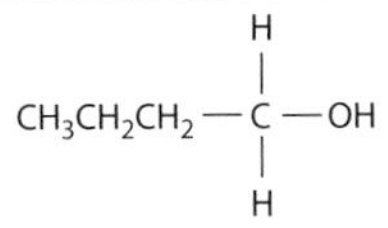	butan-1-ol
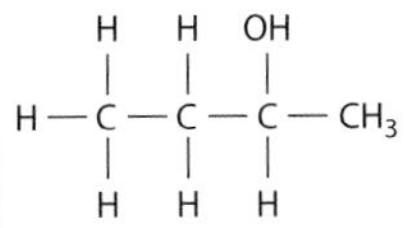	butan-2-ol
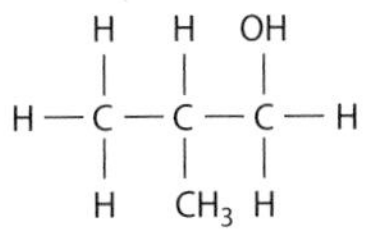	2-methylpropan-1-ol
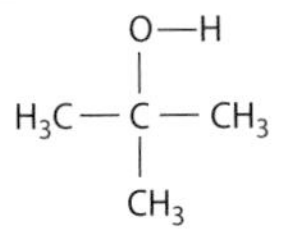	2-methylpropan-2-ol

2

	Partial oxidation	Complete oxidation
propan-1-o1	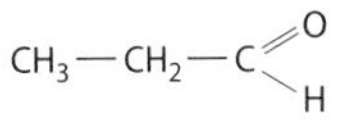	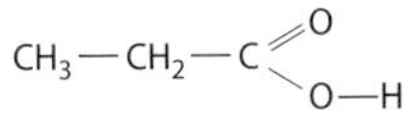
propan-2-o1		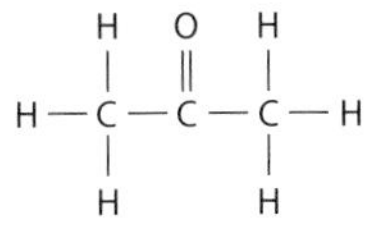
2-methylpropan-1-ol	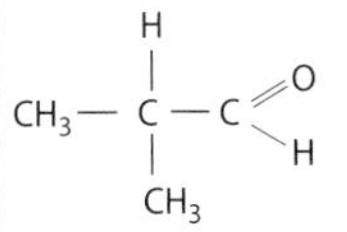	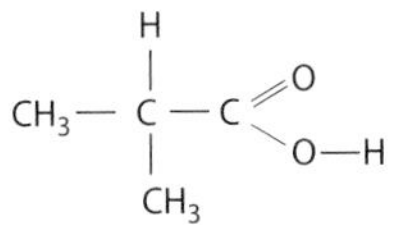
2-methylpropan-2-ol	Not oxidised	

10.10 **1**

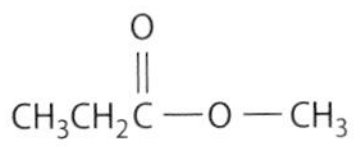

methyl propanoate

2

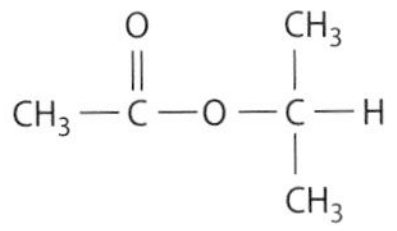

1-methyl ethanoate

10.11

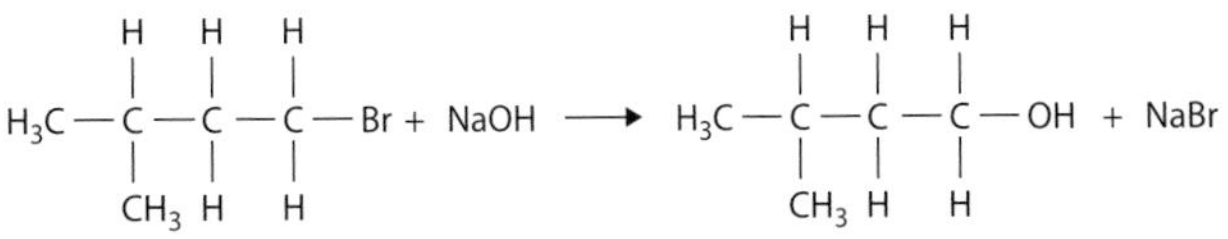

10.12

1

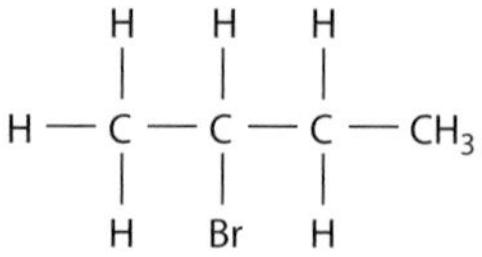

2

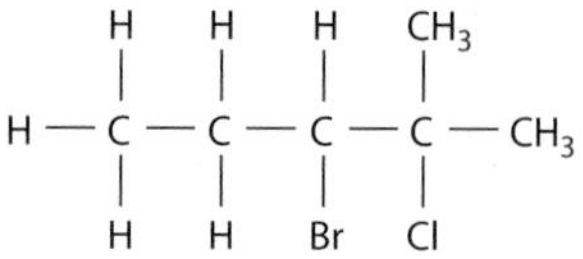

10.13 $CH_3CH_2CH_2CHO + NaBH_4$ butan-1-ol
$(CH_3)_2CHCOCH_3 + LiAlH_4$ 3-methylbutan-2-ol

10.14 $CH_4 + Cl_2 \xrightarrow{UV} CH_3Cl + HCl$

$CH_3Cl + NaOH \xrightarrow[\text{heat}]{NaOH(aq)} CH_3OH + NaCl$

$CH_3CH_2CH_3 + Cl_2 \xrightarrow{UV} CH_3CH_2CH_2Cl + HCl$

$CH_3CH_2CH_2Cl + NaOH \xrightarrow[\text{heat}]{NaOH(aq)} CH_3CH_2CH_2OH + NaCl$

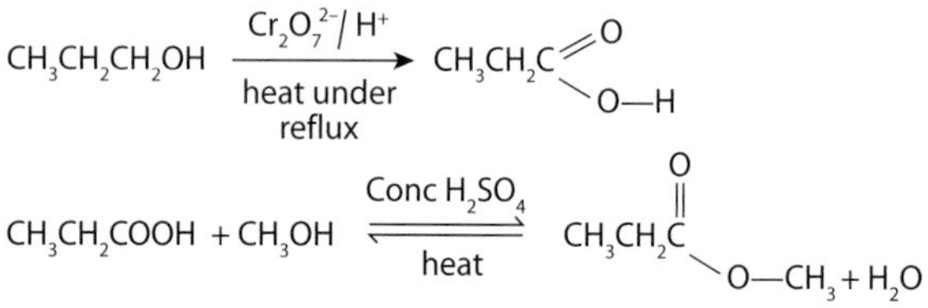

10.15 **1** and **3** do, **2** does not.

1-chlorobut-1-ene	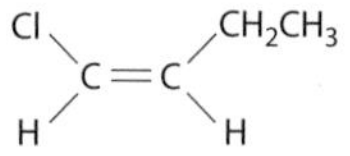	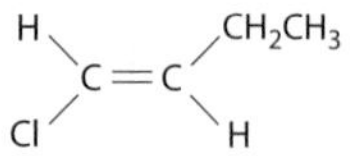
	Z	*E*
2-methylbut-2-ene	No	
3-methylpent-2-ene	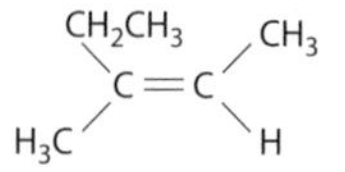	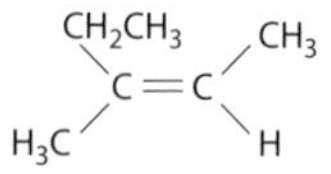
	Z	*E*

10.16 **1** No
2 Yes
3 Yes

11 Measurement, data processing and analysis

11.1 **1** 23.784 ± 0.005
2 680 ± 10
3 $1.34 \times 10^{-3} \pm 2 \times 10^{-5}$

11.2 B -1.50×10^2 kJ mol^{-1}

11.3 2.3 ± 0.1 cm^3 s^{-1}

11.4 **1** 1
2 4
3 3

11.5

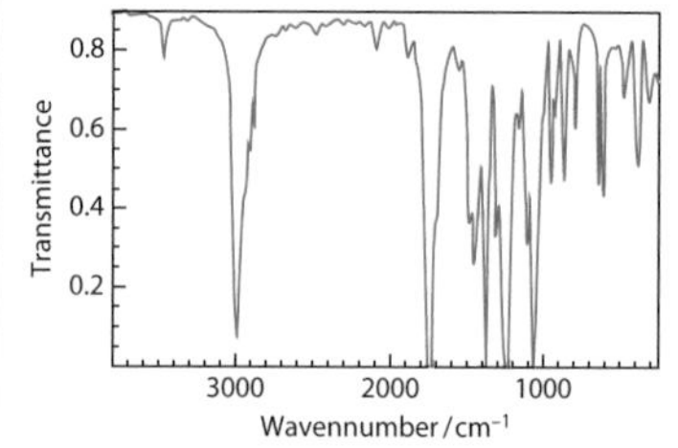	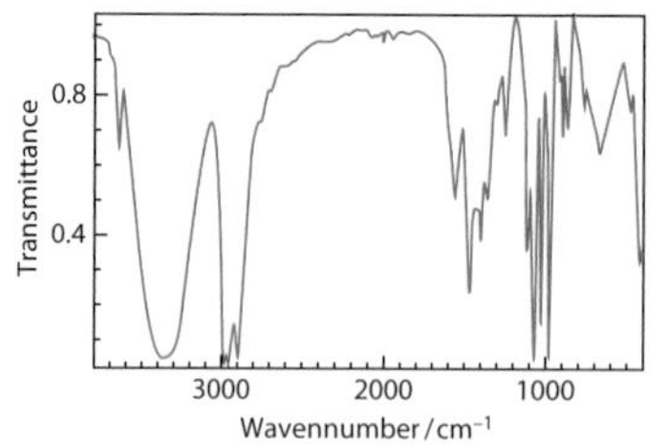	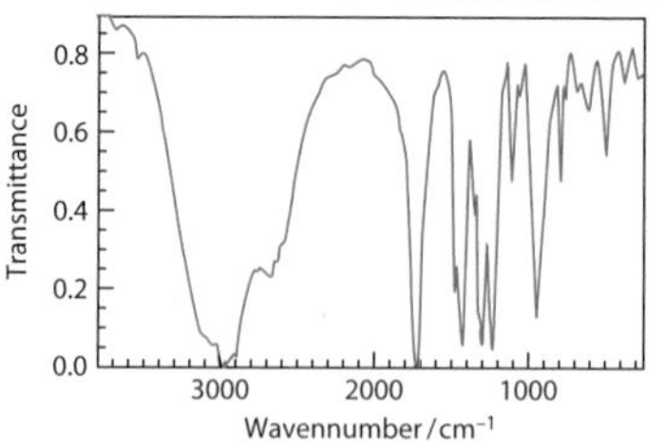
$CH_3COOCH_2CH_3$	$CH_3CH_2CH_2OH$	$CH_3CH_2CH_2COOH$

11.6 **1**

15	29	26	77
CH_3^+	$C_2H_5^+$	$C_2H_2^+$	$C_6H_5^+$

2

29	31	45
$C_2H_5^+$ CHO^+	OCH_3^+	$COOH^+$ $OC_2H_5^+$

11.7

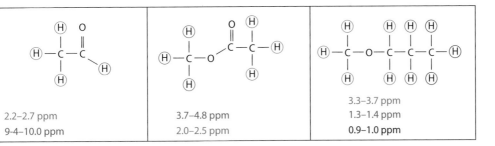

	1	2	3
	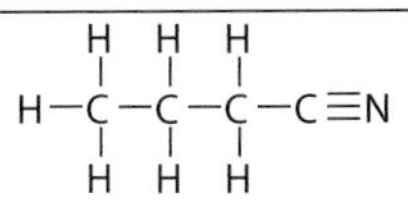	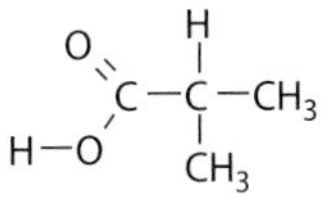	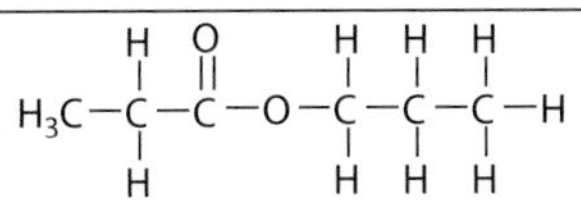
number of signals	3	3	5
ratio of areas	3:2:2	1:1:6	3:2:2:2:3

11.8

	1	2	3
	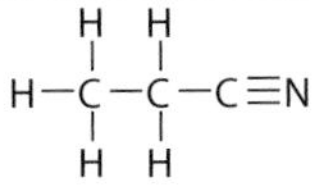	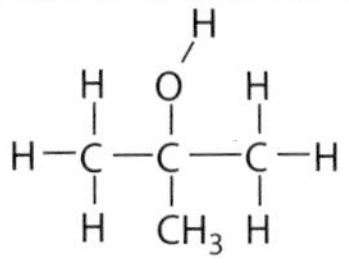	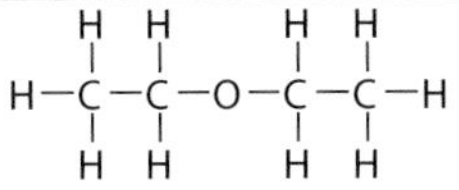
splitting pattern	triplet and quartet	2 singlets	triplet and quartet

11.9

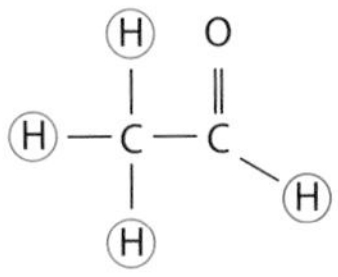 2.2–2.7 ppm 9-4–10.0 ppm	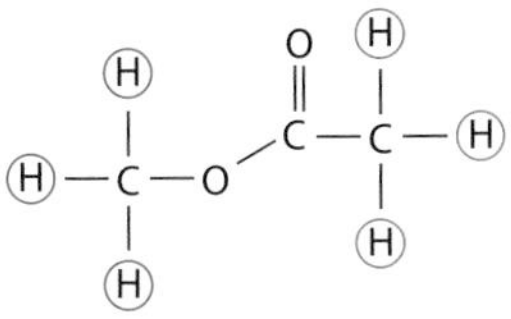 3.7–4.8 ppm 2.0–2.5 ppm	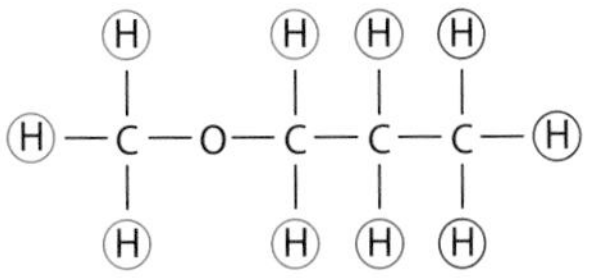 3.3–3.7 ppm 1.3–1.4 ppm 0.9–1.0 ppm

11.10 **1** CH_3COCH_3

2 $CH_3OCH_2CH_3$

A Materials

A.1 **1 a** 67–73% ionic; 27–33% covalent

b 100% covalent

c 48–52% ionic; 48–52% covalent

d 32–38% ionic; 62–68% covalent

2 a Polar covalent

b Covalent

c Ionic

d Polar covalent

A.2 **1** $ZnO + C \rightarrow Zn + CO$ or $2ZnO + C \rightarrow 2Zn + CO_2$

2 $PbO_2 + 2C \rightarrow Pb + 2CO$ or $PbO_2 + C \rightarrow Pb + CO_2$

A.3 **1** 1.3 g **2** 0.16 g **3** 15 g

A.4 **1** Paramagnetic **2** Diamagnetic **3** Paramagnetic

A.5 **1** 100%

2 51.1%

3 75.0%

A.6 **1** 2.73×10^{-10} m

2 34.4°

A.7 **1** 1.54 g cm^{-3}

2 7.53 g cm^{-3}

A.8 **1** $BaSO_4$ 1.0×10^{-10}; $Ni(OH)_2$ 5.6×10^{-16}

2 $PbCO_3$ 2.72×10^{-7} mol dm^{-3}; Ag_2SO_4 1.44×10^{-2} mol dm^{-3}

3 $[Ag^+(aq)]^2[SO_4^{2-}(aq)] = 4.0 \times 10^{-8}$ – this is smaller than K_{sp}, therefore silver sulfate will not precipitate.

Answers to Test yourself questions

B Biochemistry

B.1 Catabolism involves the breakdown of larger molecules into smaller molecules, e.g. during respiration:

$$C_6H_{12}O_6 + 6O_2 \rightarrow 6CO_2 + 6H_2O$$

Anabolism involves the synthesis of larger molecules from smaller molecules, e.g. during photosynthesis:

$$6CO_2 + 6H_2O \rightarrow C_6H_{12}O_6 + 6O_2$$

(any other appropriate metabolic examples would be accepted.)

B.2 1

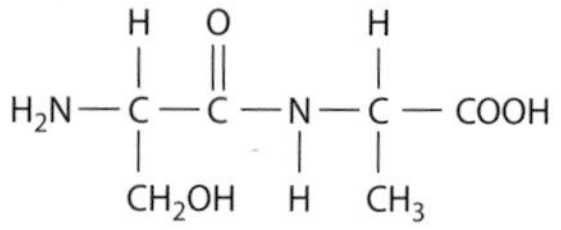

Ser-Ala

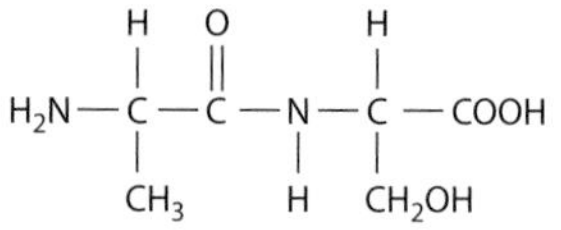

Ala-Ser

B.2 2

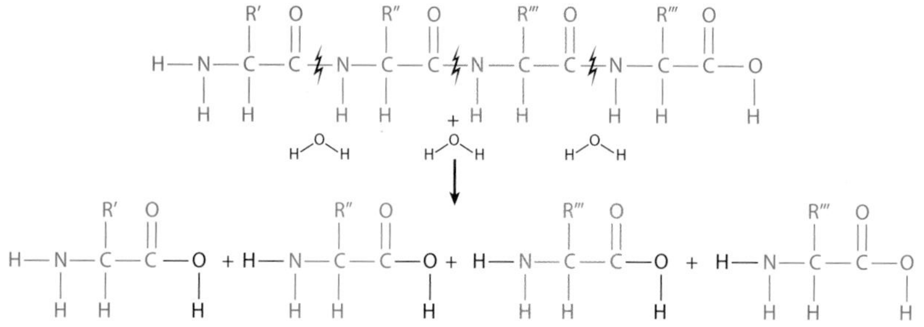

B.3

1

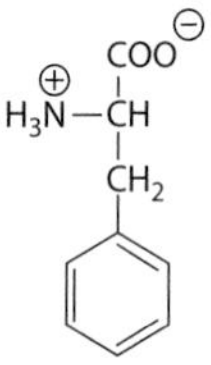

2

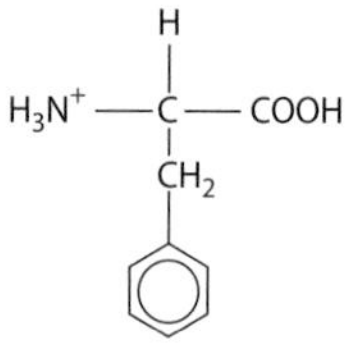

3

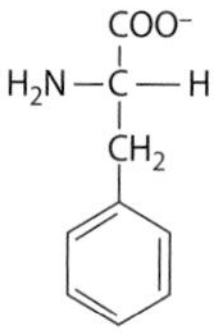

B.4 When substrate concentration is low, a small increase in substrate concentration leads to a large increase in the rate of activity.

This is because there are a large number of unoccupied binding sites.

As substrate concentration continues to increase, the increase in the rate of activity gets smaller.

This is because the number of available binding sites is decreasing.

At very high concentrations of substrate, enzyme activity stops increasing and remains constant.

This is because **all** the binding sites on the enzymes are full.

B.5 Triglyceride: oxidation state of C = −1.56

Sucrose: oxidation state of C = 0

Oxidation state of carbon in triglyceride is **more negative** than in sucrose, therefore carbon/triglyceride is more reduced.

B.6 1 a

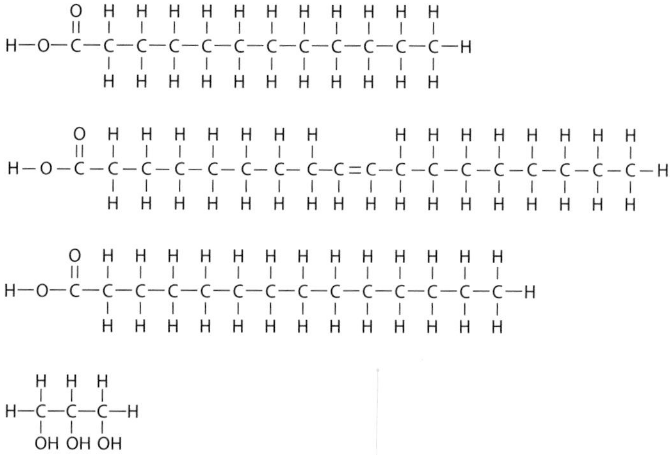

B.6 1 b

$CH_3(CH_2)_{10}COO^-K^+$

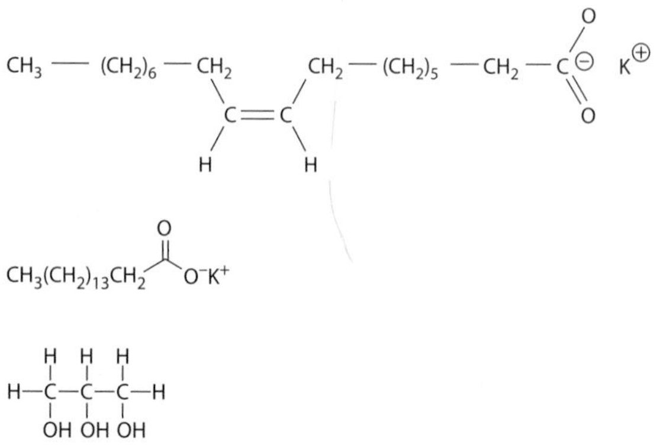

2

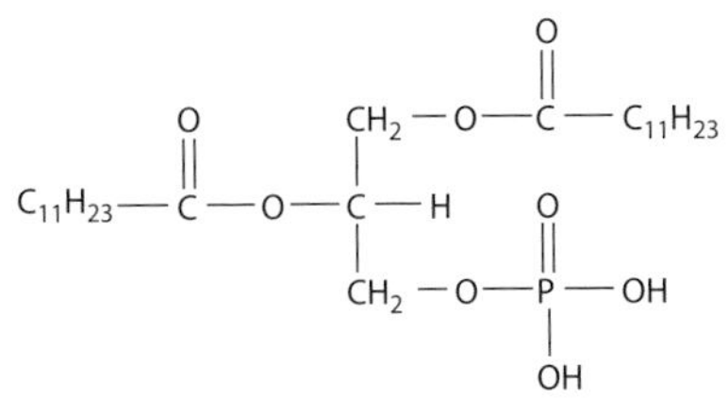

B.7 **1** It contains no C=C double bonds/it is fully saturated

2 120

3 5.0 g

B.8 **1** Fat-soluble

2 Water-soluble

B.9 Causes: Malnutrition/poor diet/famine/poverty/lack of sunlight/poor access to/supply of supplements

Solutions: Additional of vitamins to foodstuffs/genetic modification of organisms/medical programs/education/spend more time outdoors.

Each solution should be accompanied by a brief explanation to fulfil the 'outline' command term.

B.10 **1** **a** 3'-T C A A G T T T C A G C A G G -5'

b 3'- C C T A G C A T C A G A T G A -5'

2 DNA contains deoxyribose, RNA contains ribose

DNA is double-stranded, RNA is single-stranded

DNA uses thymine as a base, RNA uses uracil instead of thymine

B.11 **1** Paper chromatography works on the principle of partition, TLC works based on adsorption

TLC uses silica/alumina stationary phase, paper chromatography uses water on fibres

TLC is faster

TLC gives better resolution

TLC plates can be washed and reused, chromatography paper is single use

2 **a** Chlorophyll turns from green to olive-brown

Mg^{2+} ion in porphyrin ring replaced by two H^+ ions

Creates pheophytin complex

b Anthocyanins turn red in acid.

Acidic conditions shift equilibrium in direction of the flavylium ion.

3 CO is a competitive inhibitor of oxygen binding to hemoglobin.

Hemoglobin has a higher affinity for CO than O_2.

O_2 saturation of hemoglobin significantly decreased in presence of CO.

Cellular respiration cannot occur, leading to toxicity/death.

B.12 **1** D **2** L **3** L

C Energy

C.1 **1** Biomass advantages: renewable/low carbon footprint/reduces landfill

Biomass disadvantages: releases CO_2 when burnt/food shortages/increases food prices/climate-dependent/yield variable

2 Tidal advantages: renewable/very efficient/produces no greenhouse gases

Tidal disadvantages: expensive upfront costs/landlocked countries unable to use/disruption to marine ecosystems/only works during tidal surges

3 Hydroelectric advantages: renewable/no greenhouse gas production/control over rate of energy production/dams last for many decades/dammed lake can be used for other activities

Hydroelectric disadvantages: expensive to build dam/large initial carbon footprint/intentional flooding displaces wildlife and humans/affects downstream users of river water

4 Geothermal advantages: renewable/no greenhouse gases produced/can be used continuously/low running costs

Geothermal disadvantages: only possible in specific geographical locations/high initial cost

C.2 **1** Specific energy = 29.7 kJ g^{-1}

Energy density = 23 400 kJ dm^{-3}

2 Specific energy = 32.8 kJ g^{-1}

Energy density = 74 500 kJ dm^{-3}

3 Specific energy = 141.6 kJ g^{-1}

Energy density = 12.7 kJ dm^{-3}

C.3 **1** **a** hexane **b** 2,2-dimethylpentane

c 1,4-dimethylbenzene

2 **a** C_2H_4 **b** C_3H_6 **c** $C_{13}H_{28}$

C.4 $^{2}_{1}H + ^{3}_{1}H \rightarrow ^{4}_{2}H + ^{1}_{0}n$

C.5 **1** $^{99}_{38}Sr$

2 $^{133}_{48}Cd$

C.6 **1** 4.22×10^9 y

2 12.7 d

C.7 **1 a** $C_4H_9COOC_2H_5 + C_3H_7OH \rightleftharpoons$ $\mathbf{C_4H_9COOC_3H_7 + C_2H_5OH}$

b $CH_3CH(CH_3)COOCH_3 + CH_3CH(OH)CH_3 \rightleftharpoons$ $\mathbf{CH_3CH(CH_3)COOCH(CH_3)_2 + CH_3OH}$

2 a

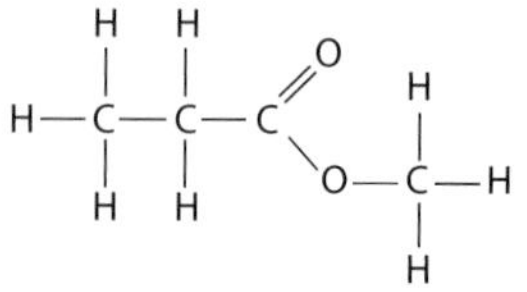

b

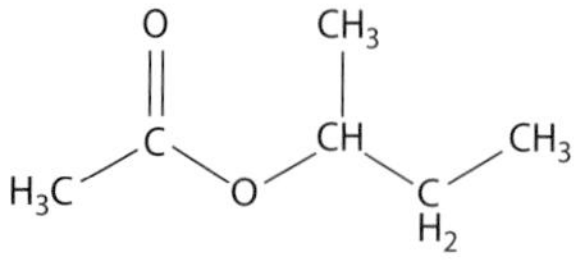

C.8 HF, CH_3Cl, SO_2, H_2O, CO

C.9 0.965 (96.5%)

C.10 **1 a** −0.79 V

b +1.50 V

2 a 4.75 mol dm^{-3}

b 0.210 mol dm^{-3}

C.11 **1** +0.06 V

2 +0.02 V

C.12

		mass defect/u	nuclear binding energy/J	nuclear binding energy per nucleon/J
1	^{12}C	0.098938 u	1.48×10^{-11}	1.23×10^{-12}
2	^{79}Br	0.736784 u	1.10×10^{-10}	1.39×10^{-12}
3	^{235}U	1.915048 u	2.86×10^{-10}	1.22×10^{-12}

C.13 **1 a** 2.37×10^{8} kJ mol^{-1}

b 2.30×10^{8} kJ mol^{-1}

2 5.70×10^{9} kJ mol^{-1}

C.14 **1 a** 0.234

b 0.0114

c 1.13×10^{-7}

2 a 1.00×10^{6} y

b 45.3 y

3 a 4.03 d

b 12.1 min

C.15 **1 a** 0.712

b 1.75

2 a 1.52×10^{-4} mol h^{-1}

b 3.65 cm^3 h^{-1}

3 a 48.96

b 25.63

C.16 **1**

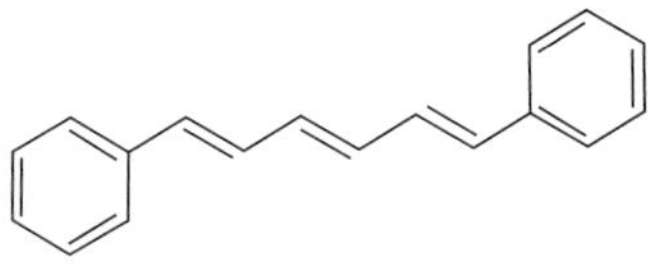

2

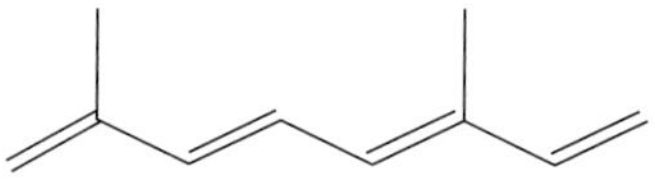

D Medicinal Chemistry

D.1 Any one of:

Determination of LD_{50} value means animals are killed.

All animals are terminated at end of study.

Unethical as animals have no choice in participation.

Examples of drugs with no toxicity in animal models but toxicity in humans.

D.2 **1** intravenous injection

2 topical

3 rectal/topical/inhalation (ignore injection)

D.3 **1** carboxyl, phenyl

2 ester

D.4 **1** Overprescription of antibiotics;

Addition of antibiotics to livestock feed;

Patients failing to complete full course of prescribed drugs;

2 Beta-lactam ring;

Ring breaks open;

Allows molecule to (covalently) bond to transpeptidase enzyme;

Enzyme cannot create cell wall cross-links;

D.5 **1** 28

2 Any three of:

(tertiary) amine

ether

alkenyl

phenyl

3 Treating diarrhoea/coughing

4 Diamorphine contains two ester groups whereas morphine contains two hydroxyl groups;

Diamorphine is more lipophilic/fat-soluble;

Diamorphine crosses blood–brain barrier more easily/rapidly

D.6 1 $NaHCO_3 + HCl \rightarrow NaCl + CO_2 + H_2O$

$Ca(OH)_2 + 2HCl \rightarrow CaCl_2 + 2H_2O$

2 0.2 mol

3 aluminium hydroxide;

it can neutralise 3 moles of acid per mole compared to 2 moles of acid per mole of magnesium hydroxide

D.7 1 0.0343 mol g^{-1}

2 0.400 dm^3

3 4.75

D.8 1 Any two from:

viruses have no cell organelles/cell membrane/cell wall/cytoplasm;

viruses are much smaller than bacterial cells;

viruses need a living cell to multiply;

viruses exhibit no metabolic functions or ability to grow or feed;

viruses contain only simple genetic information (DNA or RNA);

viruses multiply and mutate far more rapidly;

viruses are always parasitic;

2 lack of cellular structures means fewer drug targets;

viruses mutate more rapidly than bacteria;
drugs likely to damage host cell as well as virus;

3 HIV mutates very rapidly;

Drugs also damage/kill host cells;

HIV infection rates high in impoverished populations;

HIV kills helper T-cells which weakens immune system;

4 Any two from:

ether

carboxamide

alkenyl

amine

Oseltamivir only – ester

Zanamivir only – hydroxyl/carboxyl

D.9 1 High-level waste emits a greater quantity of radiation;

High-level waste emits radiation for a longer period;

2 Antibiotics enter the water supply;

Trace amounts of antibiotics ingested by people/livestock increases the chances of bacteria developing antibiotic-resistance

3 Shikimic acid comes from star-anise, which is renewable;

Synthesis from shikimic acid requires fewer steps and so less energy;

Shikimic acid can be made by fermentation of glucose;

This method requires water as a solvent and a low temperature;

D.10 1 Taxol isolated from bark of Pacific Yew tree which is endangered/killed in the process

Synthesis is a multi-step process with low yield/uses large amounts of energy

Taxol has many (11) chiral carbons

2 Chiral auxiliary bonds to molecule

Forces subsequent reactions to be stereospecific

Chiral auxiliary removed from molecule

3 Plane-polarised light shone through both enantiomers

Direction and angle of rotation are recorded

Enantiomers rotate the plane of plane-polarised light by an equal amount but in opposite directions

D.11 1 ${}^{10}_{5}B + {}^{1}_{0}n \rightarrow {}^{7}_{3}Li + {}^{4}_{2}\alpha$

2 ${}^{212}_{83}Bi \rightarrow {}^{208}_{81}Tl + {}^{4}_{2}\alpha$

3 ${}^{90}_{39}Y \rightarrow {}^{90}_{40}Zr + {}^{0}_{-1}e$

D.12 1 148 mg

2 17.7%

D.13 All primarily soluble in hexane except zanamivir, which is water-soluble

D.14 1

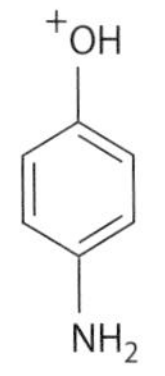

2 Paracetamol contains both N–H and O–H bonds which give absorption band/peaks at similar wavenumbers.

3 5

D.15 1 $2Cr_2O_7^{2-} + 3CH_3CH_2OH + 16H^+ \rightarrow 4Cr^{3+} + 3CH_3COOH + 11H_2O$

2 +6 to +3

3 −2 to 0

D.16 $O_2 + 4H^+ + 4e^- \rightarrow 2H_2O$

INDEX

Index

Index